Scientific Research on the Pictorial Arts of Asia

Proceedings of the Second Forbes Symposium at the Freer Gallery of Art

Scientific Research on the Pictorial Arts of Asia

Proceedings of the Second Forbes Symposium at the Freer Gallery of Art

Edited by Paul Jett, John Winter, and Blythe McCarthy

Archetype Publications

in association with the Freer Gallery of Art, Smithsonian Institution

First published 2005 by Archetype Publications Ltd.

Archetype Publications Ltd.
6 Fitzroy Square
London W1T 5HJ

www.archetype.co.uk

Tel: 44(207) 380 0800
Fax: 44(207) 380 0500

Smithsonian
Freer Gallery of Art

ISBN 1-873132-74-3

British Library Cataloguing in Publication Data
A catalogue record for this book is available from the British Library.

Managing Editor for Scholarly Publications: DeeDee Clendenning
Edited by Ann Hofstra Grogg
Cover design by Kate Lydon
Cover Illustration: A section of a handscroll by Katsushika Hokusai known as *Miscellaneous Subjects: Fishes, Flowers, Figures, Landscapes, Etc*. F1902.42

Typeset by Kate Williams, Swansea
Printed and bound in Italy by Printer Trento srl

Contents

Foreword

In 2001, the Freer Gallery of Art hosted the First Forbes Symposium on scientific research in the field of Asian art. That symposium was of a general nature and did not focus on any specific type of material. With the Second Forbes Symposium held in 2003, we decided to begin concentrating on specific topics in the realm of Asian art; to begin, we chose the subject of the pictorial arts of Asia. One reason for this choice is that the Department of Conservation and Scientific Research of the Freer and Sackler galleries is particularly strong in this area at the present time. We have a very active East Asian Painting Conservation Studio that is now deeply involved in the training of conservators in the traditional methods of Japanese and Chinese painting conservation. We have also dedicated a large part of our scientific research effort to the study of paintings, work that is supported by the Andrew W. Mellon Foundation (and to which we are very grateful). The scientists' main areas of research are in the use and development of organic pigments in Asia, the physical properties of carbon-based black ink, and the fairly neglected area of the materials used in later Chinese scroll paintings. We hope and expect that this research will produce results of important and long-lasting meaning to scholars in the field and to the interested public.

Future Forbes symposia will continue this trend of focusing on the impact of scientific methods of research on various subjects within Asian art. The next symposium is planned for September 2005 on the topic of the sculptural arts of Asia. These symposia are the realization of the wishes of John Thacher, who bequeathed to the Freer Gallery the funds to support these meetings and named them after his friend and colleague, Edward Waldo Forbes. We are happy to present the symposia in honor of Mr. Forbes and Mr. Thacher, and we do so with much gratitude for what both of them did to encourage the incorporation of science and conservation into the study of art.

Julian Raby
Director
Freer Gallery of Art and Arthur M. Sackler Gallery

Acknowledgments

I would like to thank the members of the Department of Conservation and Scientific Research (DCSR) for all of their efforts in organizing and coordinating the Second Forbes Symposium. Conservation scientists Blythe McCarthy and Janet Douglas and conservation technician Jai Alterman, who strove diligently to ensure a successful event, deserve special mention. Audiovisual specialist Andy Finch proficiently managed a variety of presentation media used by the speakers as well as various last-minute modifications. John Gordy, head of Digital Media, was a tremendous help with informing the public of necessary changes to the schedule. Special events coordinator Caroline Bedinger proficiently coordinated the logistics of the symposium. Joseph Swider and Andrew Hare of the DCSR also assisted with the speakers and other guests. Senior conservation scientist John Winter played a critical role in planning the symposium and in selecting and editing the papers. Editor Ann Hofstra Grogg took the drafts of the papers and polished them into their final form. Editorial and conservation research intern Özge Gençay Üstün assisted enormously in preparing the manuscript for print production. The editorial efforts were coordinated by DeeDee Clendenning, managing editor for scholarly publications; her diligence in shepherding the proceedings through to the end has made a tremendous difference and I cannot thank her enough for her hard work. I would also like to thank Dr. Julian Raby, director of the Freer and Sackler galleries, and Dr. James Ulak, deputy director, for their support of the symposium and the publication of its proceedings, and for their continuing support for the efforts of the DCSR; it is greatly appreciated.

The symposium was not without problems. A hurricane—a very rare occurrence in Washington, D.C.—struck at the time of the symposium. Though there was a forced delay of a day and a half, the meeting went ahead, and all the papers were delivered. And while the attendance at the meeting was reduced, it was gratifying to see how many people did turn out. Throughout this ordeal, the speakers could not have been more gracious, good humored, understanding, and cooperative. They worked very hard to make the meeting a success, and for whatever success the meeting enjoyed, the credit mostly goes to them.

Paul Jett
Head, Department of Conservation and Scientific Research
Freer Gallery of Art and Arthur M. Sackler Gallery

List of Contributors

CHEN Yuansheng
Shanghai Museum
Shanghai, China

Sandra A. CONNORS
Carnegie Mellon University
Pittsburgh, Pennsylvania, USA

Elizabeth I. COOMBS
Independent Paper Conservator
Cranston, Rhode Island, USA

Yasushi EGAMI
National Research Institute for Cultural Properties
Tokyo, Japan

Jacki ELGAR
Museum of Fine Arts
Boston, Massachusetts, USA

Kathleen M. GARLAND
Nelson-Atkins Museum of Art
Kansas City, Missouri, USA

Jennifer GIACCAI
The Walters Art Gallery
Baltimore, Maryland, USA

Yasuhiro HAYAKAWA
National Research Institute for Cultural Properties
Tokyo, Japan

Agnieszka HELMAN-WAŹNY
Nicolaus Copernicus University
Torun, Poland

Sandhya S. JAIN
Los Angeles County Museum of Art
Los Angeles, California, USA

Roger S. KEYES
Carnegie Mellon University
Pittsburgh, Pennsylvania, USA

Hans LEISEN
University of Applied Sciences
Cologne, Germany

Marco LEONA
The Metropolitan Museum of Art
New York, New York, USA

Christine MACKAY
National Museums and Galleries of Wales
Cardiff, Wales, UK

Sadatoshi MIURA
National Research Institute for Cultural Properties
Tokyo, Japan

Tetsuo MIYAKOSHI
Meiji University
Tokyo, Japan

Anne Nishimura MORSE
Museum of Fine Arts
Boston, Massachusetts, USA

Jesse MUNN
Library of Congress
Washington, D.C., USA

Alison MURRAY
Queens University
Kingston, Ontario, Canada

Richard NEWMAN
Museum of Fine Arts
Boston, Massachusetts, USA

Yasuhiro OKA
Oka Bokkodo Co., Ltd.
Kyoto, Japan

Ekaterina PASNAK
Queens University
Kingston, Ontario, Canada

Esther von PLEHWE-LEISEN
University of Applied Sciences
Cologne, Germany

Mark RICHTER
Bavarian State Department of Historical Monuments
Munich, Germany

Aditi Nath SARKAR
Independent Scholar
Kolkata, India

Shiho SASAKI
Intermuseum Conservation Association
Cleveland, Ohio, USA

TSAI Yu-lin
National Center for Research and Preservation of Cultural Properties
Taipei, Taiwan

Nicole TSE
University of Melbourne
Melbourne, Australia

Season TSE
Canadian Conservation Institute
Ottawa, Ontario, Canada

TUNG Kuo-king
National Center for Research and Preservation of Cultural Properties
Taipei, Taiwan

John TWILLEY
Independent Conservation Scientist
Hawthorne, New York, USA

Rangachar VASANTHA
Sri Krishnadevaraya University
Anantapur, India

Paul M. WHITMORE
Carnegie Mellon University
Pittsburgh, Pennsylvania, USA

John WINTER
Freer Gallery of Art/Arthur M. Sackler Gallery
Smithsonian Institution
Washington, D.C., USA

XIE Yulin
Shanghai Museum
Shanghai, China

M. Brigitte YEH
University of Southern California
Los Angeles, California, USA

Hyejung YUM
Cornell University
Ithaca, New York, USA

Introduction

Frequently, works of art have lost their context: we often no longer have an understanding of why a piece was made, or where it was made, or by whom it was made. Frequently, works of art have suffered physical losses: they have changed over time, both by the processes of nature and by their being used or adapted for new artistic or cultural purposes. And frequently, in the case of the pictorial arts, works of art have lost their intended pictorial effect. That effect depends primarily on a thin skin of pigment atop a ground or other support, and that thin skin is easily lost or altered. Among the areas where scientific research on works of art can be applied, perhaps it is the pictorial arts that hold the most potential to reward our attempts to better know and understand them.

It was because of this potential that the theme chosen for the Second Forbes Symposium at the Freer Gallery of Art was the pictorial arts of Asia. The presentations combined scientific methods of study with research on historical texts on painting and the materials used. Most of the papers are about materials and techniques, but some address questions of conservation and deterioration. In selecting the papers we hoped to broadly represent the pictorials arts of Asia; not surprisingly, some areas are far better represented than others due to the fact that much more research is being conducted in those areas. Many of the papers deal with the physical nature of paper itself or with works of art on paper. This preponderance of works relating to paper seems appropriate given paper's invention and extensive use in Asia, and East Asia in particular.

We have included two papers that may seem out of place given the topic: one on the wall decorations at Angkor Wat and the other on the use of polychromy in Japanese sculpture. But we interpreted the term "pictorial" in its broadest sense. We felt that what makes a work of art pictorial is not two-dimensionality. It is rather the effect of the work of art, which can be pictorial even though it entails painted reliefs, as in a Cambodian temple, or the figures of gods and guardians, as in a Japanese temple. Further, it is also the case that many of the painting materials are the same for sculpture, architecture, or works of art on paper or textile, and so the study of those materials in one context is relevant to the studies of painting materials in other contexts.

In choosing the papers, we not only wanted to bring together a great diversity of subjects in the area of pictorial arts of Asia; we also wished to bring together colleagues from around the world. We hoped the symposium would, to some degree, document the research being carried out at this point in time. We also felt it was important to have a symposium of this type so that the speakers could meet, discuss their work, and, we hope, build relationships that lead to future collaboration and have a synergistic effect on researchers' work. We further hope that the publication of these proceedings will continue that effort to bring together those interested in the pictorial arts of Asia and to assist them in their studies.

Paul Jett

A Technical Study of the *Hokkedō konpon mandara*

Jacki Elgar, Anne Nishimura Morse, and Richard Newman

ABSTRACT The *Hokkedō konpon mandara* is considered one of the most important early East Asian Buddhist paintings held in any collection outside of Asia. Since the painting was acquired by the Museum of Fine Arts, Boston, in 1911, it has been the subject of continuing debates about its origins, date, physical condition, and iconographic significance. The painting has undergone several campaigns of restoration, one of which is documented by an inscription that was once attached to the reverse. Dated to 1148, the inscription suggests that the painting was used in the Lotus Hall (Hokkedō) of Tōdai-ji in Nara, Japan. Chinkai, a monk and respected artist, was commissioned to restore the painting, particularly the lower half. Recently comprehensive examination was conducted on the *Hokkedō konpon mandara* to clarify the original composition that has become obscured by damage and a severely darkened original support and to distinguish the original from the various campaigns of restoration, most notably that of 1148. Infrared reflectography was used to interpret and clarify indecipherable sections of the painting as well as to reveal original underdrawings. Pigment samples were taken, and these results were compared with known pigments used throughout the history of Japanese pictorial arts.

Introduction

The Historical Buddha Preaching on Vulture Peak, or *Hokkedō konpon mandara* (fig. 1), as it has been known in Japanese since at least the twelfth century, has been universally acclaimed as "one of the most important landmarks in the history of East Asian art to be seen in the West" (Rosenfield 1996). Although it has suffered extensive damage and discoloration, the painting remains the earliest surviving monumental landscape composition on a woven support from anywhere in East Asia. The work measures more than 143 cm in width and 107 cm in height. In the center Sakyamuni, the Historical Buddha, preaches to his assembled followers and attendant bodhisattvas. The setting is Vulture Peak, described in the sixteenth chapter of the *Lotus Sutra* as a pure land in which there are "gardens and groves,

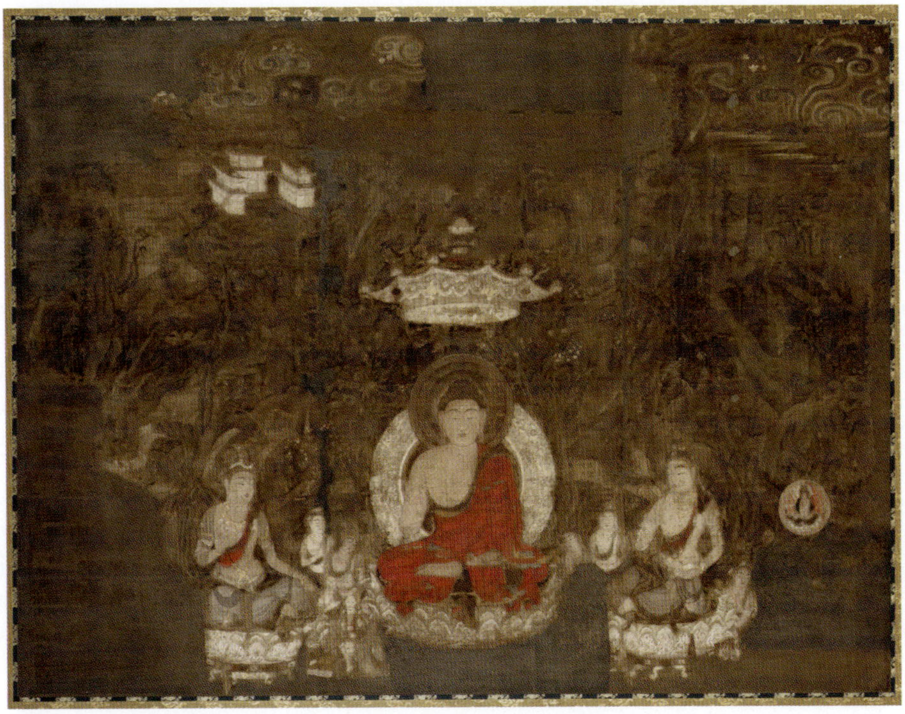

Figure 1. Hokkedō konpon mandara. *Museum of Fine Arts, Boston, 11.6120*

halls and towers, variously adorned with gems, as well as jeweled trees with many blossoms and fruits, wherein the beings play and amuse themselves; where the gods beat their divine drums, making melodies most skillfully played, and rain down māndārava-flowers" (Hurvitz 1976, 243). In 1995, a technical examination was begun on the *Hokkedō konpon mandara*, with the aim of identifying previous campaigns of restoration, and the painting's intricate background was documented using infrared reflectography (IRR). Over the next eight years the authors of this paper continued to gather new information about the painting by means of a wider range of techniques and more advanced analytical equipment. As a result, they were able to form a better understanding of the painting's original appearance and subsequent history.

Previous Assessments

The *Hokkedō konpon mandara* was acquired by the Museum of Fine Arts, Boston, in 1911 as part of an unprecedented gift of more than three thousand Japanese paintings from William Sturgis Bigelow, a prominent Boston physician. Bigelow had purchased this particular work from the well-known art trading company Kiryū kōshō kaisha, in 1886, during his lengthy residence in Japan. When he donated the painting to the museum, it was in a hanging-scroll format, but the work was remounted as a panel in 1936. Ever since Okakura Kakuzō, adviser to the Department of Chinese and Japanese Art, first published the *Hokkedō konpon mandara* in the *Museum of Fine Arts, Bulletin* in 1906, there have been extensive debates about the painting's iconography and provenance. Prior to World War II, scholars theorized that it was painted in China during the Tang dynasty (618–907) because of the handling of the soaring peaks and twisted trees in the background landscape. Since the 1950s, however, most scholars have assigned it a Japanese origin and have dated it to the late Nara period (710–794)—an era when Japanese art was heavily influenced by that of China. In support of this dating, scholars have focused on the treatment of the landscape elements in seventh–ninth century wall paintings at the cave complex of Dunhuang in western China (caves 103, 112, 332, and 369), the Northern Song (960–1127) copy of the eighth-century *Emperor Minghuang's Journey to Shu*, and Japanese eighth-century screens in the Shōsō-in in Nara depicting women beneath trees. The handling of the foreground figures has been cited as evidence for the Japanese provenance of the painting. The only documented evidence for the *Hokkedō konpon mandara*'s early history is an inscription that was once attached to the back of the work when it was still mounted in hanging-scroll format. Written in 1148 by Kanshin, the director of temple affairs (*bettō honmu*) at Tōdai-ji in Nara, this inscription states:

> *Hokkedō konpon mandara*. The previously mentioned mandala is a pictorial vision (*hensō*) of Vulture Peak and is a real Indian painting. The sections below where the Historical Buddha is seated have been entirely destroyed. They have been lost through natural causes or deliberately cut off. Many years have passed and its date is now unknown. Therefore, in the third month of Kyūan 4 (1148), I ordered Chinkai, a monk of this temple and lecturer (*ikō*) and Dharma Master (*hōshi*), to repair it because he is naturally gifted in his family style, and talented at painting. I have written these details in a summary fashion in order to preserve them for future generations.

This inscription has provoked intense speculation about the extent to which the present painting reflects the hand of an earlier, presumably eighth-century, painter and the extent to which it represents the work of Chinkai. In the mid-1930s, Yashiro Yukio stated that Chinkai must have retouched all of the forms currently legible against the darkened support—the figures, daises, canopy, background pavilions, and clouds (Yashiro 1935). Subsequent studies by Kameda Tsutomu and Matsushita Taka'aki in 1950 and 1956 reasserted this view, although Matsushita believed that the bodhisattva in the lower right corner with his hands held in prayer preserved his original appearance (Kameda 1950; Matsushita 1956). In 1978, Akiyama Terukazu and Yanagisawa Taka undertook extensive visual and pioneering scientific analyses, using infrared photographs and high magnification. Aside from obvious repairs in the Buddha's robes, Akiyama found no evidence to support the claims made in the 1930s (Akiyama 1978), while Yanagisawa asserted that Chinkai's involvement was limited to the addition of fine ink lines to make the landscape clearer, texture strokes for the craggy cliffs, and outlines for the twisted trees (Yanagisawa 1978).

Visual Examination and Analysis: Comparison with the *Kusha mandara*

The 1148 inscription not only raises important questions for historians of early Japanese painting but is also crucial to an understanding of the role that the *Hokkedō konpon mandara* played in the development of Buddhist painting practice during the revival of Nara Buddhism in the twelfth and thirteenth centuries. One of the best-known paintings from this later period is the monumental *Kusha mandara* at Tōdai-ji, depicting Sakyamuni and two attending bodhisattvas surrounded by the Ten Patriarchs of the Kusha school, one of the schools of Nara Buddhism. Four Divine Kings (*tennō*) stand in the corners of the composition, while Indra and Brahma occupy the central left and right edges of the painting respectively. Kameda Tsutomu has shown that the composition of the *Kusha mandara* is derived from three different sets of paintings (Kameda 1972). The Ten Patriarchs were based on the door paintings of the *Portable Shrines of the Six Schools* (*Rokushū zushi*) that adorned the Great Buddha Hall of Tōdai-ji until their destruction in 1180. The Indra and Brahma and the Four Divine Kings were derived from paintings, now lost, that adorned the doors of the portable shrine housing Kegon sutras in the Ordination Hall (Kaidan-in) at Tōdai-ji. Most important for this discussion, the central triad was clearly drawn from the seated Sakyamuni and the two attendants in the *Hokkedō konpon mandara*.

Visual examination confirms the close relationship between the central figures of both *Kusha mandara* and *Hokkedō konpon mandara*, which is evident in the similarity of their poses, the arrangement of their robes, and the treatment of the upper sections of the pedestals. Furthermore, the dimensions of selected key features of the deities in the *Kusha mandara*—such as the length of the faces—are almost identical to those in the *Hokkedō konpon mandara*. As with all Buddhist deities, the positioning of the hands of the figures is governed by strict iconographic rules. In the case of the central deity of the *Hokkedō konpon mandara* and *Kusha mandara*, the proper right hand is held with the thumb and right index finger forming a circle and the proper left held in the lap—a gesture associated with the preaching of the Buddhist law. The disposition of the red robes, which fall across the proper left shoulder, down across the chest in a sinuous curve, and then over the legs, further underscores the affinity between the two paintings. Close visual examination of the left attendant bodhisattva in both paintings reveals other similarities in the handling of the figures. Not only are the deities positioned in the same manner, but details such as the crown with its curved plaque, the arrangement of the drapery across the chest, and the highlighting of the folds across the knees are also clearly related. These points of commonality led Kameda to go as far as to conclude that the twelfth-century scroll must also have been executed by Chinkai.

Examination and Analysis

Noninvasive examination techniques applied to the *Hokkedō konpon mandara* included observation under ultraviolet light (UV) to distinguish between repairs and retouched areas, based on differences in fluorescence; infrared reflectography (IRR) to reveal intricate lines of the landscape that had been obscured by the darkened support and underdrawings hidden by paint layers; and x-radiography to reveal long-lost features of the design, based on differences in the density of the pigments.

Support

In his first studies of the *Hokkedō konpon mandara*, Yashiro Yukio described the original support as silk. Later he revised his identification and asserted that the support was hemp, an opinion repeated in publications by later scholars. To identify the fibers of the support (a simple plain weave with 17–18 weft threads and 22–23 warp threads per centimeter), a sample was taken for analysis and examined under a polarizing light microscope; it exhibited transverse dislocation marks at frequent intervals along the fiber. Since these properties are characteristic of bast fibers such as hemp, flax, and ramie, the red plate test on the polarizing light microscope was utilized in a further effort to define more precisely the material from which the support was made. This test revealed a pattern of color shifts that indicated an "s-twist" fiber such as flax or ramie, rather than a "z-twist" fiber such as hemp (fig. 2). Cross-sections of the fibers were then prepared for microscopic examination, which, however, proved inconclusive. Further analysis by a fiber specialist, Debora Mayer, confirmed that the fibers were bast and had an "s-twist," even though a test using a cuprammonium hydroxide solution (Schweizer's Reagent), which can distinguish between hemp and flax, also proved inconclusive. Assessing the morphological features of the fiber, Mayer concluded that its transverse markings were more consistent with ramie, which is native to tropical Asia and has been cultivated in China and Japan since prehistoric times. That *jōdai-gire*, Japanese "ancient fabrics" dating from the seventh and eighth centuries, were constructed exclusively from silk or ramie fibers (Matsumoto 1984) supports Mayer's identification.

Pigments

Noninvasive pigment analysis was carried out by energy-dispersive x-ray fluorescence (XRF), and forty-five samples were taken for more detailed analysis by polarizing light microscopy (PLM), Fourier transform infrared microspectroscopy (FTIR), electron probe microanalysis (EPMA), and scanning electron microscopy–energy-dispersive x-ray spectrometry (SEM-EDS). The thinness of the paint layer

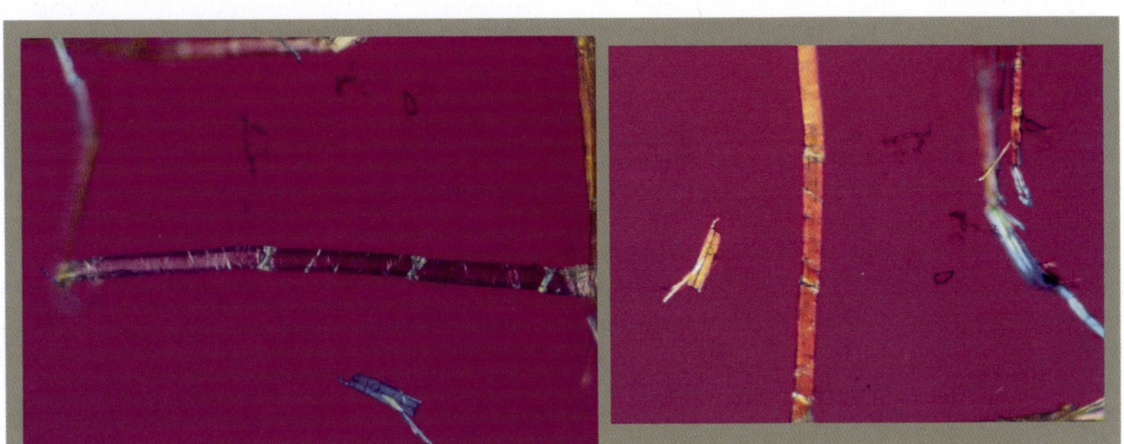

Figure 2. *Results of red plate test. Original fibers exhibit an "s-twist" color shift.*

and the deteriorated nature of the original support made it difficult to obtain pure samples. Thus some samples contained pigments of more than one color. The samples did, however, fall into several categories based on visible color—primarily greens, blues, reds, pinks, and whites.

Green and blue samples were taken from the landscape background, drapery elements of the figures, or the mandorla of the main figure. All the greens tested contained malachite; all blues contained azurite. Under a stereobinocular microscope a number of small islands of green and/or blue pigment were visible scattered throughout the darkened background. This observation lends support to the hypothesis that the landscape was originally much more colorful than it is now, although this analysis cannot compute the original extent of the color or help determine whether the painting was done in the "blue-and-green" style of the Tang dynasty. Areas in the drapery hem of the attending bodhisattvas that appear in visible light as blue in tone and the centers of the flowers in the mandorla behind the central Buddha were identified as containing azurite. Similar traces of azurite that are not visible without the aid of high magnification were also found in the hairline of the right bodhisattva.

Red and pink samples were taken from the three-petaled flower in the upper right corner; the mandorla, drapery, proper right elbow, and face of the Buddha; and the drapery elements and dais of the proper right bodhisattva. The predominant red pigment was vermilion, but an organic red (not specifically identified) was documented in the outline of the inner circles of the lower section of the mandorla, and red lead was found in the pink bands of the drapery of the attendant bodhisattvas.

Eighteen white samples were taken, in the form of minute chips from which cross-sections could not be prepared, from the buildings, clouds, and waterfalls in the background; skin of five of the attendants; the dais of the proper right bodhisattva; and the mandorla, canopy, and skin of Sakyamuni. These were documented and analyzed by SEM-EDS (table 1). One of the white pigments discovered was lead white (basic lead carbonate), which has been utilized in East Asian paintings from at least the seventh century. The analyses also identified a second lead-containing white—lead chloride—which appears in very fine-grained aggregates (figs. 3 and 4). Individual pigment particles were a micrometer or less in size, and individual crystals often seemed to be prismatic or needlelike. Lead to chlorine atomic ratios ranged from about 1 up to 4. Lead white was also present in a number of the samples, intermixed with lead chloride, so the lead to chlorine atom ratios did not always accurately reflect the composition of the lead chloride pigment. Our results point to the possible presence of laurionite (a basic lead chloride), but another (or other) specific compound may also be present. Winter (1981) discovered the use of laurionite on Japanese paintings dating from the twelfth century onward but did not have the opportunity to analyze any Japanese paintings dated before the twelfth century. Finally, many white-pigmented areas were found to contain aragonite (see fig. 3), derived from shells, which was commonly used in Japanese paintings only from about the fifteenth century. The presence of this material in most of the samples taken from thickly painted areas, such as figures and architectural elements, is a strong indication that all of these areas were extensively repainted at a considerably later date. Lead-containing whites were also found in many of these same samples, possible remnants of earlier (if not original) paint. As Akiyama (1978) and Yanagisawa (1978) previously noted, cut gold (*kirikane*) can be found in the lines radiating outward from behind the central figure's head and on his robe. No conclusions can be made about the date of the *kirikane*, but logic would dictate that the gold cannot be original. The cut gold behind the Buddha's head is in an area of abrasion and discoloration to the original support, and the cut gold in the Buddha's red robe is in an area that most agree has had some repainting.

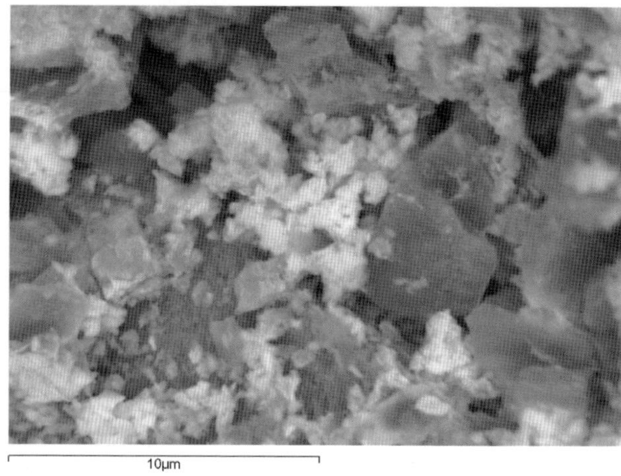

Figure 3. *Photomicrograph of sample 1, backscattered electron image taken in a scanning electron microscope. Large angular particles are aragonite, while the lighter gray clusters of smaller particles are lead chloride white.*

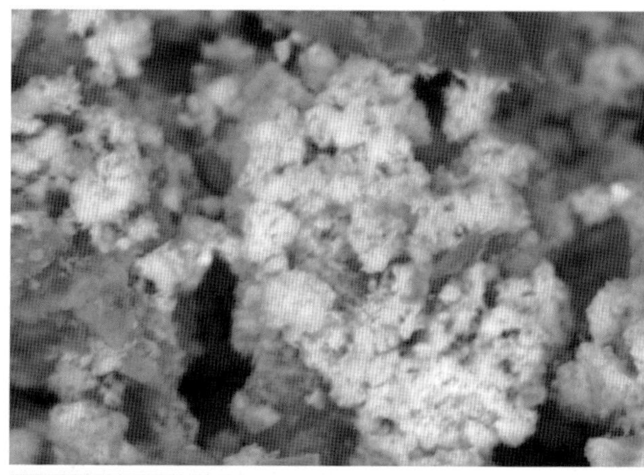

Figure 4. *Photomicrograph of sample 6, backscattered electron image taken in a scanning electron microscope. The sample consists primarily of small elongated grains of lead chloride white.*

The painting was photographed under UV light. UV photographs revealed the extent of the damage to the original support, as the various materials including several fabrics and a grayish blue paper used to fill losses fluoresced differently. Four distinct fabrics were identified by thread count and weave.[1] Under magnification and cross-polarized light they were identified as silk. Also discernible under UV light

Table 1. *Summary of analyses of white pigments in white or pink paint areas*

Sample	Description	FTIR microspectroscopy[a]	SEM-EDS analysis[b]
1	Face, small figure on left 1	Aragonite, some hydrocerussite and gypsum	Ca + Pb-Cl
2	Face, small figure on left 2	Aragonite, some oxalate and gypsum	Ca, a little Pb-Cl
3	Face, main figure	Hydrocerussite	Pb-Cl
4	Face, main figure	Hydrocerussite	Pb-Cl
5	Neck, small figure on left 3	Hydrocerussite and aragonite	Pb-Cl, a little Ca
6	Face, small figure on left 4	Oxalate, some aragonite and hydrocerussite	Pb-Cl, a little Ca
7	Face, small figure on left 5	Oxalate, some cerussite	Pb-Cl, a little Ca
8	White waterfall on left	Oxalate and aragonite	Ca + Pb-Cl
9	Face, small figure on right 1	Aragonite and oxalate	Ca + Pb-Cl
10	Face, small figure on right 2	Aragonite and oxalate	Ca + Pb-Cl
11	Face, large figure on right	No inorganics identified	Pb-Cl
12	Neck, small figure on right 3	Aragonite and oxalate	Ca, a little Pb-Cl
13	Hand, large figure on right	Aragonite and oxalate	Pb-Cl
14	Pale green on right	Oxalate	Ca + Pb-Cl
15	Mandorla, small figure on right 4	Aragonite and oxalate	Ca + Pb-Cl
16	White water on right	Oxalate	Pb-Cl
17	White clouds on right	Oxalate	Ca + Pb-Cl
18	White form on left	Oxalate, some unidentified carbonate	Pb-Cl

NOTES

a. Specific inorganic compounds identified are indicated by their mineral names (gypsum, aragonite, hydrocerussite). In some cases, only the class of compound could be determined (such as oxalate).

b. Only the elements of significance for specific identification of the white pigment or pigments in the samples are indicated. Pb-Cl indicates the presence of a compound or compounds that contain(s) both elements.

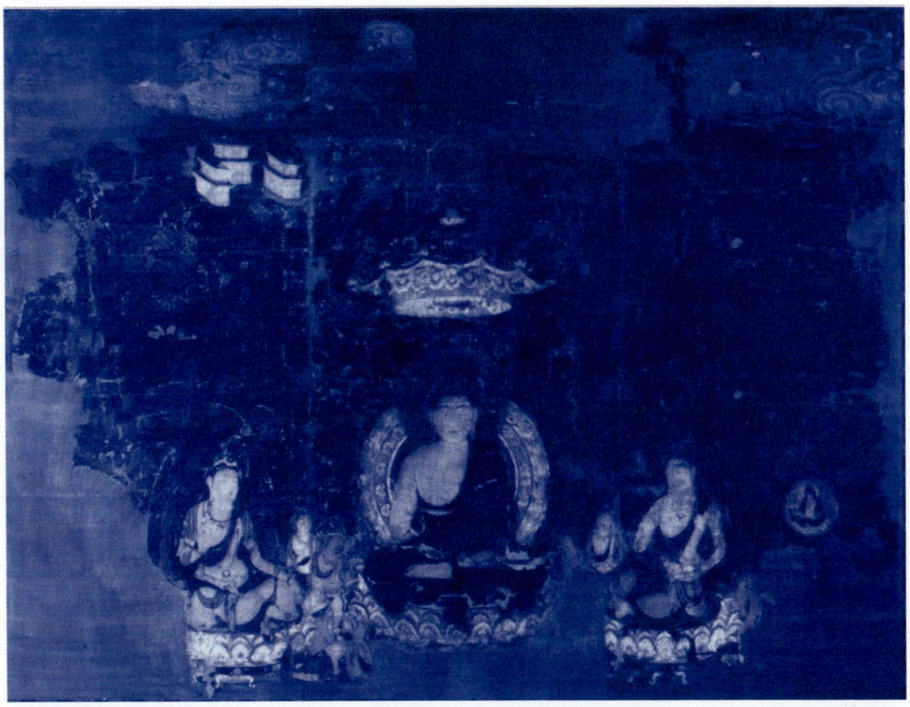

Figure 5. Hokkedō konpon mandara *photographed under ultraviolet light*

was the mottled surface of some of the painted figures in the foreground of the painting. What is perceived in visible light as a continuous layer of paint was revealed under UV light to contain irregularities, which can be explained as either losses or repainted areas (fig. 5).

The *Hokkedō konpon mandara* was further examined using IRR, a noninvasive technique that permits visualization and photographic documentation of underdrawings.[2] The infrared reflectogram for the *Hokkedō* is comprised of 182 images that were assembled on a computer (fig. 6) to produce a very well-defined image that helps scholars to understand better the intricate landscape and precise iconography of the figures. In the upper right corner, for example, rock promontories and sheltered inlets with twisted trees are discernible, as well as a seascape that is given further definition by lines of undulating waves (fig. 7), while in the upper left corner a waterway can be seen coursing around Vulture Peak and behind the buildings (fig. 8). IRR also

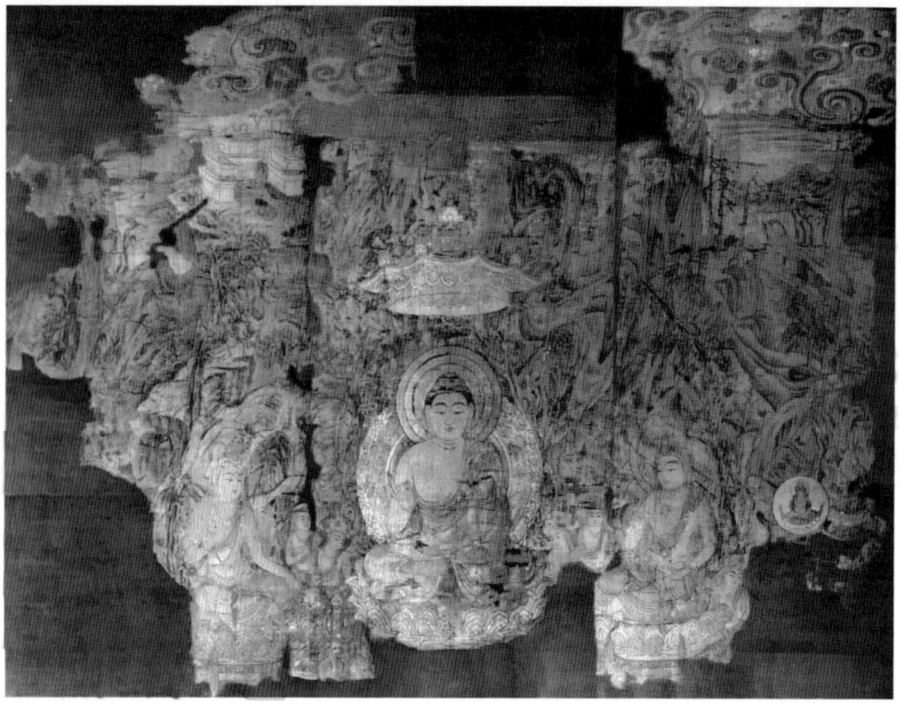

Figure 6. *Infrared reflectogram of* Hokkedō konpon mandara

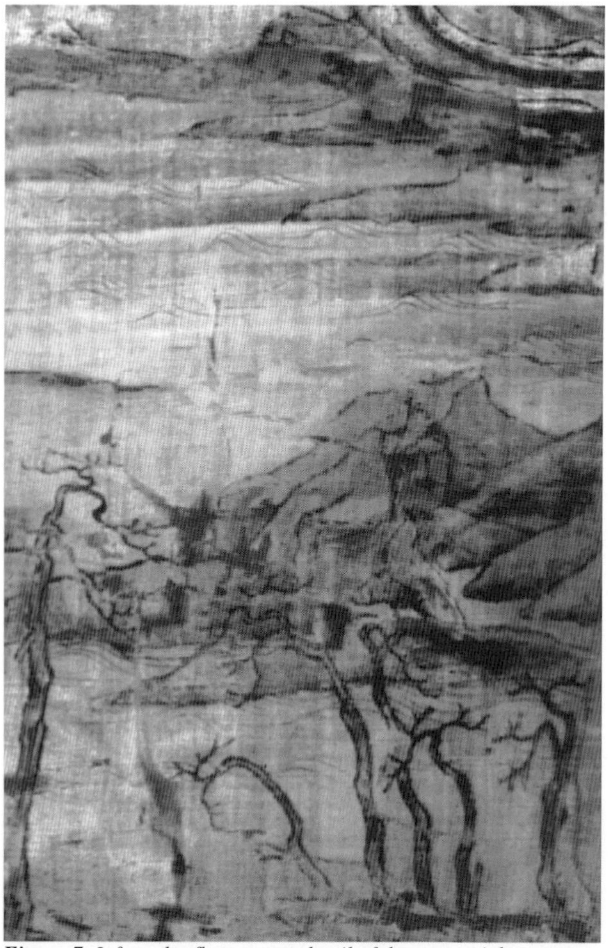

Figure 7. *Infrared reflectogram, detail of the upper right corner*

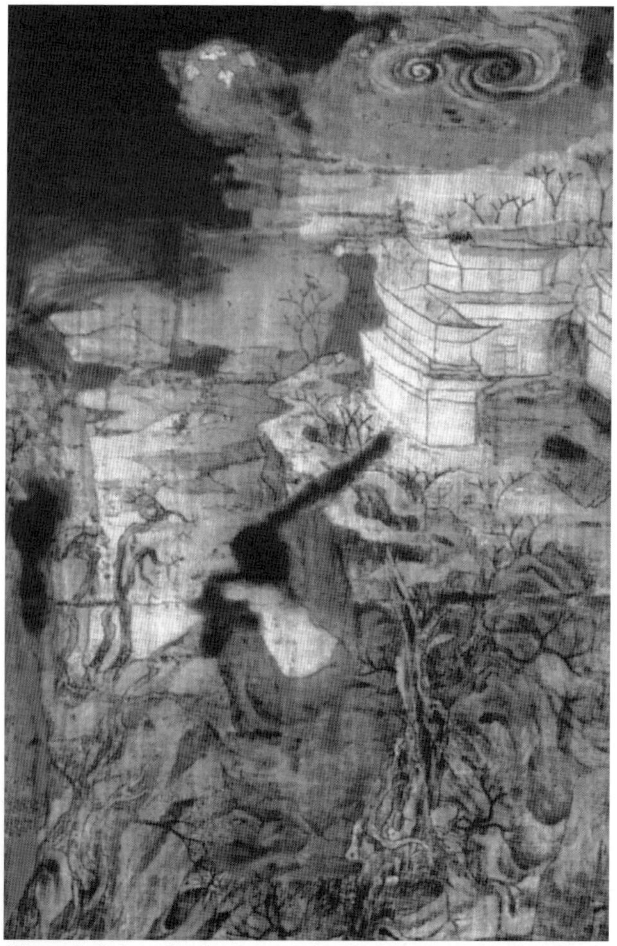

Figure 8. *Infrared reflectogram, detail of the upper left corner*

clarified details of the background at the center-right of the Buddha, revealing an attenuated pine tree with vines entangled through its gnarled branches (fig. 9). This tree bears a striking resemblance to the vine-entangled trees in a set of eighth-century paintings depicting Tang-style women from the Shōsō-in (Nara National Museum 1999).

Extensive analysis was also conducted on each individual deity in an effort to confirm to what extent the figures have been repainted. During the examination of the bodhisattva on the left, details of the crown that are difficult to decipher in visible light were documented. As previously noted, in visible light the crown appears to have a simple tapering plaque at its center. However, IRR (fig. 10) revealed this headdress to be more complex and similar to a crescent moon motif found in the well-known eighth-century image of a bodhisattva on hemp from the Shōsō-in (Nara National Museum 1995). IRR examination of another area from the same bodhisattva showed that there are many delicate lines in the underdrawing of the sash that crosses the body (fig. 11). These lines greatly contribute to the suggestion of volume in the drapery, a feature that is absent from the existing painted image. The infrared reflectogram of the central figure revealed damage to the original support and numerous drawn lines below the painted surface of Buddha's robes (fig. 12). All of these discoveries contribute to the hypothesis that the robes have undergone several campaigns of restoration.

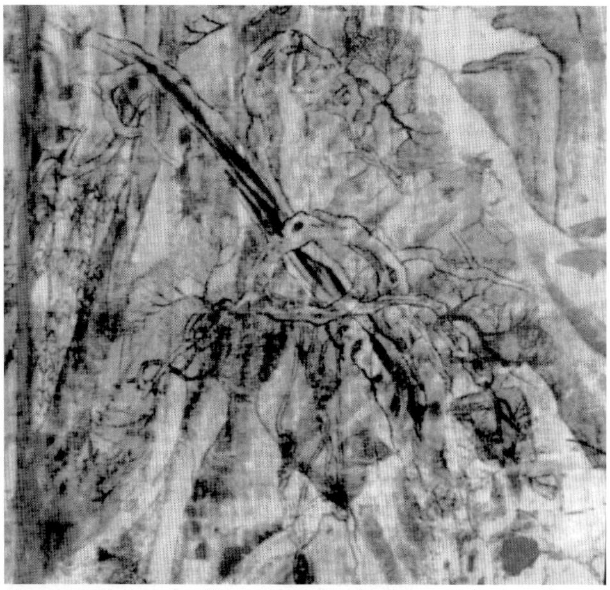

Figure 9. *Infrared reflectogram, detail of tree with vines (in the background to the center-right of the Buddha)*

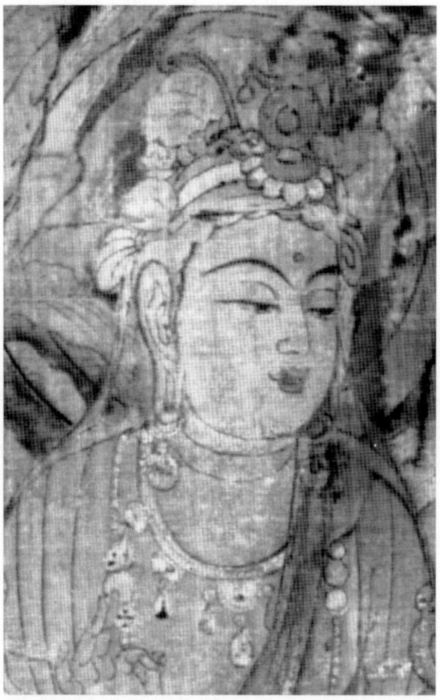

Figure 10. *Infrared reflectogram, detail of large bodhisattva on the left*

Mandorla

Previous scholars have commented on the appearance of the Buddha's mandorla, speculating that during the twelfth century it might have received touches of restoration, predominantly in the ink lines, that did not affect its overall outline. In visible light, there are traces of white and red pigments outside the concentric circles of the mandorla. IRR was not helpful in this case since it only clarified the black outlines of the flowers within the concentric circles of the mandorla, but x-radiography detected small traces of lead-containing pigments just outside a mandorla very different from the one visible today. It appears that an earlier, if not the original, mandorla had a scalloped perimeter, possibly created by large flowers, unlike the currently visible mandorla, which has a smooth circular outline and is comprised of concentric circles containing flowers.

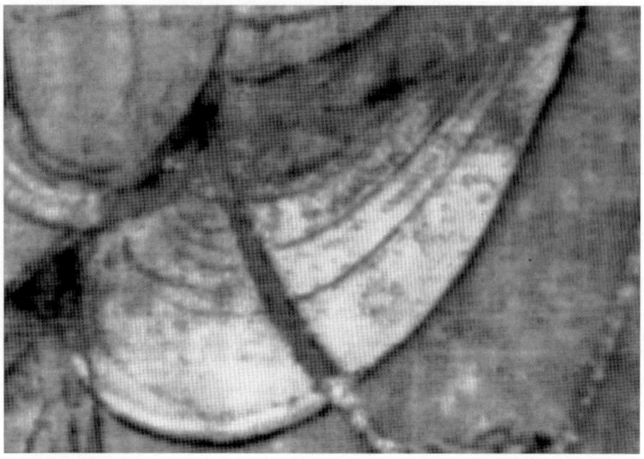

Figure 11. *Infrared reflectogram, detail of large bodhisattva's body sash on the left*

Subsidiary Figures

Since the 1950s there has been a consensus among scholars that the small attendant figure in the lower right corner retains his original eighth-century appearance. Modern technology, however, presents evidence to the contrary and suggests that this figure has been subject to at least two attempts at restoration. Under IRR the crown appears to be drawn with two different perspectives: some elements are depicted straight on, while in others the head appears to be slightly tilted upward. A comparison of the infrared reflectogram with the UV image helps to explain this anomaly. The

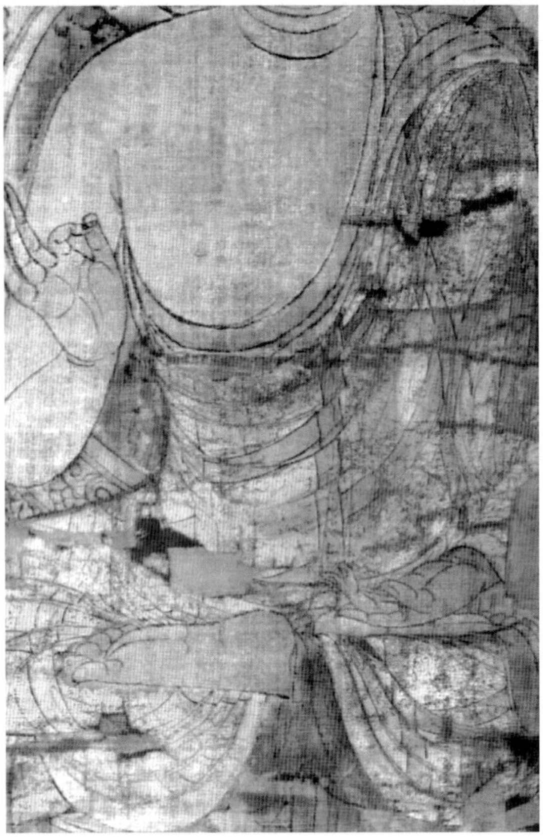

Figure 12. *Infrared reflectogram, detail of the Buddha's robes*

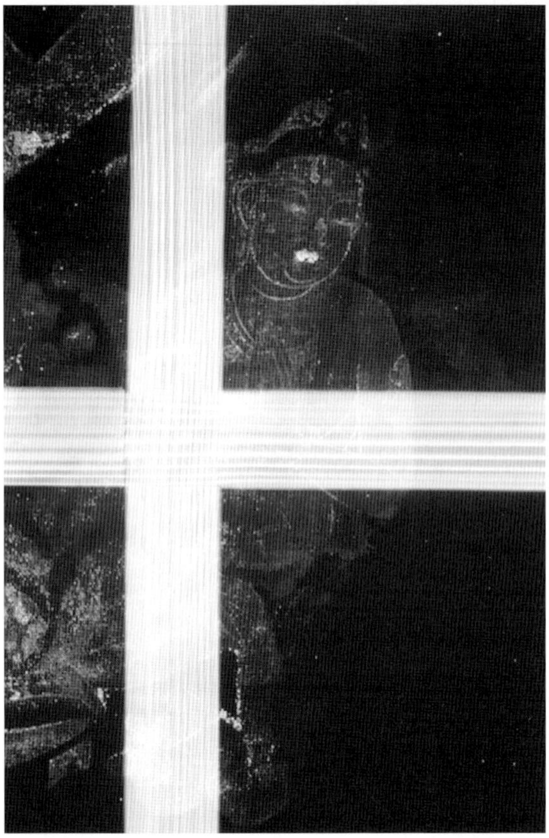

Figure 13. *X-radiograph, detail of small attendant figure in the lower right corner*

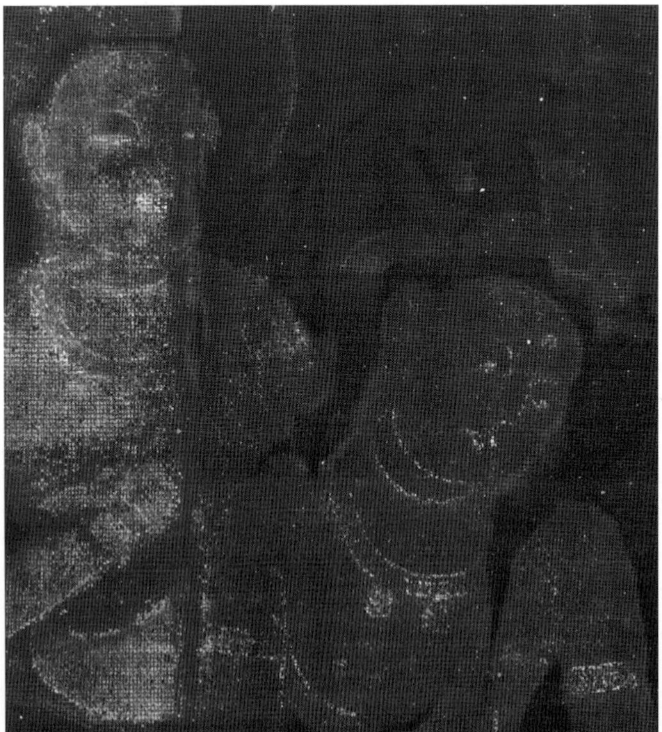

Figure 14. *X-radiograph, detail of two small attendant figures to the left of the Buddha*

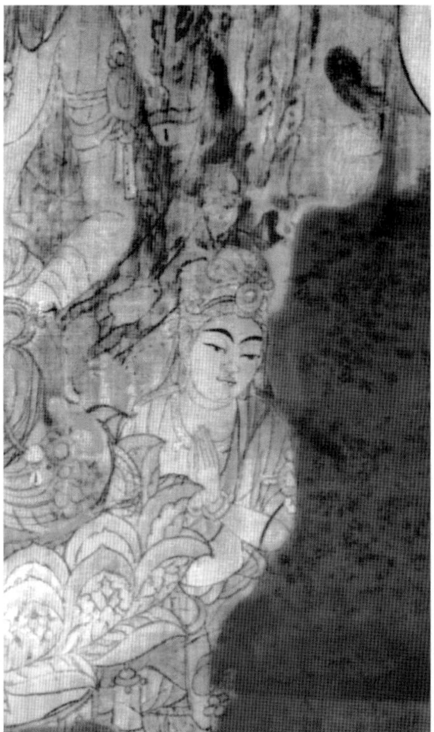

Figure 15. *Infrared reflectogram, detail of lower right corner. Note that the missing guardian figure's halberd and flywhisk are visible in the upper right of this image.*

face of the figure appears blotchy, and there is an area that fluoresces brightly, suggesting that the original head and crown were aligned differently. An x-radiograph showed limited amounts of lead-containing whites, confirming that this figure is not original but a composite of subsequent restorations (fig. 13).

When the figures just to the left of the Buddha are examined in visible light, it seems as though a white pigment has been universally applied to describe their bodies. In UV light, however, only a section of one of the figures fluoresces brightly, suggesting that one or possibly both figures have been repainted. An x-radiograph was taken to detect where, if any, lead-containing whites were present in these two small figures. The heavier lead white pigment is bright white, whereas the less absorbing aragonite white, commonly found in Japanese paintings from the fifteenth century onward, appears gray. The x-radiograph of these two attendant figures revealed limited amounts of lead white. The body of the figure at the right fluoresces gray, with only its red lead outlines fluorescing brightly (fig. 14).

The inscription cited earlier states that "the sections below where the Historical Buddha is seated have been entirely destroyed," indicating that as early as the twelfth century certain areas of the original composition no longer existed. Reference to iconographically similar works, such as the wall painting in cave 321 at Dunhuang, indicates that the assembly of followers was once much larger, a conclusion supported by the discovery with IRR of remnants of a guardian figure at the lower right corner (fig. 15).

Conclusions

The technical studies conducted at the Museum of Fine Arts, Boston, uncovered irrefutable evidence of several campaigns of restoration to the *Hokkedō konpon mandara*, of which only the 1148 repair had previously been noted. Unfortunately, even modern scientific analyses cannot specifically date the work of the various restorers, but the detection of the presence of aragonite indicates that the painting was retouched at least once since the fourteenth century. Comparisons of the icon with the twelfth-century *Kusha mandara* performed by the authors clarified the damaged state of the *Hokkedō konpon mandara* around, if not at the time of, Chinkai's restoration. As stated earlier, under visual examination the bodhisattvas at the left of each painting seemed remarkably similar, but IRR revealed that the headdress of the attendant in the *Hokkedō konpon mandara* originally included a distinctive crescent moon. Despite the faithfulness with which he executed other parts of the headdress, the artist of the *Kusha mandara* was obviously unable to see this crescent moon because by the twelfth-century many of the original details of the figures in the *Hokkedō konpon mandara* had already been obscured. In addition, the detection with IRR of fine-lined underdrawings executed in a Nara period style that renders the figures with more volume leads to the conclusion that the appearance of the figures visible today is greatly indebted to later restorations.

Some questions posed by the painting remain unanswerable. It is impossible to ascertain which areas of the figures Chinkai retouched and which a later extremely skilled artist repainted. Nor can we determine when the ink lines of the landscape elements and the mandorlas were reinforced. The authors urge art historians, equipped with the new information provided here, to use the time-honored tradition of disciplined visual analysis to formulate more hypotheses concerning the history of this important painting. Furthermore, we hope that this study will provide critical new documentation on the restoration techniques used on Buddhist paintings in Japan at different periods.

Notes

1. Two types of silk were used for the large areas of loss along the perimeter of the painting: one is a tightly woven plain weave with 34–35 (weft) × 37–38 (warp) threads/cm^2; the other is also a plain weave with 30 (weft) × 25 (warp) threads/cm^2. Another type of silk used for fills can be seen in the Buddha's *kesa*. This silk is also a plain weave but has thicker warp threads than weft.
2. An infrared television camera equipped with an AF Micro Nikkor 1:2.8/60 mm lens on a Hamamatsu C2741 vidicon was used at the Museum of Fine Arts, Boston.

References

Akiyama Terukazu. 1978. *Shaka Ryōjusan seppō zu (Hokkedō konpon mandara)* [The Historical Buddha Preaching on Vulture Peak]. Tokyo: Gakken kenkyūsha.

Hurvitz, L., trans. 1976. *Scripture of the Lotus Blossom of the Fine Dharma (The Lotus Sutra)*. New York: Columbia University Press.

Kameda Tsutomu. 1950. Hokkedō konpon mandara shinron. *Bukkyō geijutsu*, no. 6 (February).

Kameda Tsutomu. 1972. Nara jidai no sōshizō to Kusha mandara zu. *Nihon bukkyō bijutsushi josetsu.* Tokyo: Gakugei shorin.

Matsumoto Kaneo. 1984. *Jōdai-gire: 7th and 8th century textiles in Japan from the Shōsō-in and Hōryū-ji.* Kyoto: Shikosha.

Matsushita Taka'aki. 1956. Hokkedō konpon mandara ni tsuite. *Bijutsu kenkyū*, no. 186 (May).

Nara National Museum. 1995. *Shōsō-in ten.* Nara: Nara National Museum.

Nara National Museum. 1999. *Shōsō-in ten.* Nara: Nara National Museum.

Rosenfield, J. M. 1996. Session statement for *The Historical Buddha Preaching on Vulture Peak*: A closer look. Paper presented at the annual meeting of the College Art Association, Boston, February.

Winter, John. 1981. Lead white in Japanese paintings. *Studies in Conservation* 26:89–101.

Yanagisawa Taka. 1978. Shaka Ryōjusan seppō zu (*Hokkedō konpon mandara*). *Butsuga.* In: *Nihon no shihō*, vol. 1. Tokyo: Mainichi shinbunsha.

Yashiro Yukio. 1935. Bosuton bijutsukan shozō Hokkedō konpon mandara ni suite. [About the *Hokkedō konpon mandara* in the Boston Museum]. *Bijutsu kenkyū*, no. 37 (January). (in Japanese)

Materials for "Gold" and "Silver" Tints in Pictorial Ornamentation of Twelfth-Century Japanese Manuscripts

Yasushi Egami

ABSTRACT The use of gold and silver in manuscript ornamentation has a long history in East Asia, known from eighth-century China, Korea, and Japan. Gold and silver first adorned the inside and outside of purple or indigo covers of sutra scrolls, signifying their great value and nobility. In tenth-century Japan combined gold and silver pigments were applied to the decoration of secular manuscripts as well as sacred works. By the twelfth century a variety of alternative materials was also employed to achieve nuances in "gold" and "silver" tints. Ocher as a substitute, mixtures of silver pigment and yellow dye, or silver pigment coated with yellow or orange translucent colorants were introduced to create "gold" tints, while silver leaf or a mixture of mica dust and pale ink were employed for "silver." These developments, previously unrecorded, have recently been detected using modern techniques such as x-radiography and close-up photography. This paper discusses instances of these innovations, particularly those belonging to the first half of the twelfth century, which were analyzed in the context of art-historical evolution. The author suggests that these innovations correspond to the beginning of the so-called Age of the Retired Emperors, which occurred around 1100, when parallel attitudes toward diversity and new aesthetics in lacquerware and in the literary world were also introduced. Retired Emperor Shirakawa, who occupied this position for more than forty years after reigning as emperor from 1072 to 1086, was an influential figure in this climate of political and cultural change.

General History of Gold and Silver Ornamentation in Japanese Manuscripts

Gold and silver tints have long been common in the pictorial ornamentation in Japanese manuscripts. The idea of using gold and silver tints was introduced from China and Korea, perhaps in the eighth century, in the form of gold-and-silver decorative patterns and frontispieces outside and inside the indigo or purple covers of sutra scrolls. This sort of sutra ornamentation has been seen through the ages in Japan. "Gold" and "silver" pictorial designs in various other forms have also been applied, however, to religious and secular manuscripts.

The earliest surviving Japanese gold-and-silver manuscript ornamentation is that on both sides of the purple cover of a *Bonmōkyō* handscroll, a Buddhist scripture, kept in the Shōsō-in repository, Nara. The outside of the cover depicts a landscape, while the inside shows flower patterns in both gold and silver. The text itself, transcribed in the Nara period (710–794), is written in ink on white paper.

A well-known work from the Heian period (794–1185), sometimes referred to as the classical era of Japan, is a copy of *Man'yōshū*, the oldest anthology of Japanese poetry compiled in the previous period. This copy, later named the *Katsura Version*, was made in the middle of the eleventh century. The content of the poetic text and the ornamental motifs beneath it are largely unrelated. As precious and contrasting colors, gold and silver pigments were used for the natural motifs beneath the ink-written text on a scroll formed by joining paper sheets of varied hues.

Still later examples include many calligraphy works by Hon'ami Kōetsu (1558–1637), a talented artist of the Momoyama period (1573–1615), which were done on white paper handscrolls embellished by the prominent painter Tawaraya Sōtatsu (?–1643?) or his studio in bold gold-and-silver patterns.

Gold and silver thus became a traditional means of decorating manuscripts. Most of this decoration, as in the works mentioned above, was done with ordinary metal pigments of gold and silver, although in the final case, the silver seems to be partly and unevenly blended with pale ink and the gold is partly blended with brown colorant in order to break the monotony and subdue the glitter that may spoil the charm of the calligraphy.

Early Use of Alternative Materials in the *Jōyuishikiron* and the *Sanjūrokunin-shū*

Modern research technologies have enabled identification of instances in which diverse materials were used to give more delicate nuances to "gold" and "silver" tints in the pictorial ornamentation of manuscripts, a trend that gained momentum from the outset of the twelfth century, the last century of the Heian period.

One early example is the *Jōyuishikiron* owned by Kōfukuji Temple, a copy of a Buddhist scripture set that most probably dates to the first or second decade of the twelfth century (Egami 1971). The outside of the blue cloth covers of the scrolls is decorated in scattered sprays of plants seemingly in "gold," while the inside shows a landscape depiction also in "gold." But minute examination with the naked eye and close-up photographs reveal that the "gold" on the outside and inside can, in fact, be divided into two kinds: (i) ordinary gold pigment, (ii) a mixture of silver and yellow dye. Many segments painted with the latter have been spoiled by a phenomenon known as *gin'yake* (literally, "silver burn"), and remnants of the yellow dye in this mixture can be detected around these broken portions.

Another example is a bound-format version of *Sanjūrokunin-shū*—more formally *Sanjūrokunin-kashū* (Anthology of the Thirty-Six Poets)—which I believe was produced in 1112 (Egami 1991), a date most scholars accept today. This set, originally in thirty-eight volumes, is perhaps the most elaborate manuscript ever made in Japan. In addition to a variety of papers both dyed and undyed, as well as a number of adornment techniques including paper patching, pattern printing, and the sprinkling of gold and silver leaf, all of the more than two thousand pages are embellished with painted patterns or pictorial compositions or both. The set was transferred from the Imperial Court to Nishi-Honganji Temple in 1549.

Based on an indirect account in a nobleman's diary, the literary historian Hitaku Kyūsojin has suggested that this set of decorative volumes was made to commemorate the sixtieth birthday of the Retired Emperor Shirakawa in 1112 (Kyūsojin 1953, 1960). A number of facts further indicate that the work was presented on that occasion by the Retired Emperor to then 9-year-old Emperor Toba. About fifty painters were involved in its production (Egami 1991), and it features a complex use of "gold" and "silver" tints. Careful observation reveals that the intricate tints are created both by gold and silver pigments and by many other materials.

In an important pictorial composition with a design of a wheel symbolizing the sixty-year life cycle of the Retired Emperor (fig. 1), silver leaf was used to describe the petals of plum blossoms. Close-up photography and x-ray radiography confirm its presence. As the silver leaf is much thinner than the layer of silver pigment, it shows up only as a slight shadow in the print, or reverse density image, of an x-ray radiograph (fig. 2). Silver pigment rendering plum trees on the reverse, on the other hand, leaves a remarkably darker image in the same printed x-ray radiograph. The silver leaf depicting the lower blossoms and buds features darker hues, which were perhaps intentionally devised by the artist, but in the upper reaches the effect is not present and this silver leaf is therefore believed to be a silver alloy containing some gold.

Replaced or altered gold and silver pigments occurs often throughout the work. Scattered patterns of birds and plant sprays on some pages, for instance, substitute ocher pigment for the gold usually paired with silver. A yellowish or orange colorant is seen on other pages superimposed on certain areas painted in silver to create the effect of gold wash. The best example is a double-page design in which a snowscape is arranged by patching together pieces of white-coated paper to render snow-covered ground and a piece of deep blue paper to represent water (fig. 3). Silver dust sprinkled over the blue paper suggests falling snow. Pines, flying geese, and water ripples are painted in gold and silver. Silver pigment is applied along the edges of the water, but here the silver is covered with yellowish dye. A comparison of the same shoreline segment in the color photograph and a printed x-ray radiograph (fig. 4) shows that the stronger whitish reflection in the former corresponds to the darker shadows in the latter. On the other hand, a kind of pigment—almost certainly silver—is coated with a translucent colorant in a beach scene with pines and a boat at rest (figs. 5 and 6). Here the design is basically made up of silver pigment, supplemented by a few tones of translucent orange colorant along the water's edge and over some other motifs. The parts

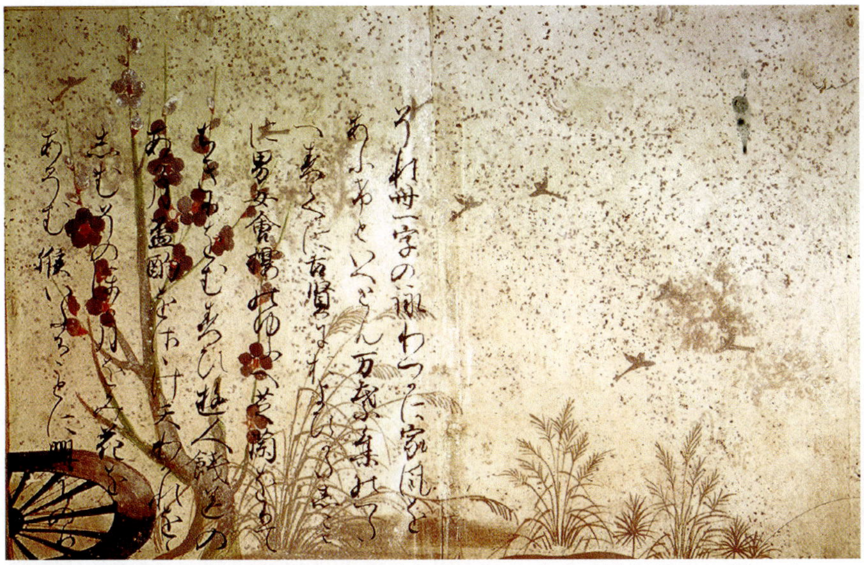

***Figure 1.** Pages from the Yoshinobu volume 1,* Sanjūrokunin-kashū. *Nishi-Honganji Temple*

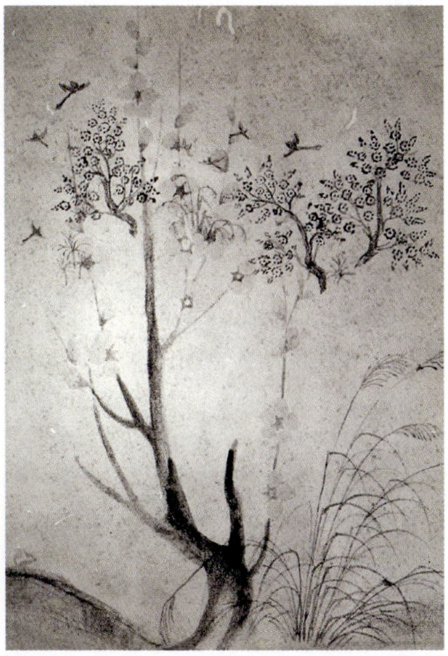

***Figure 2.** Reverse density image of the x-ray radiograph of the left page of figure 1*

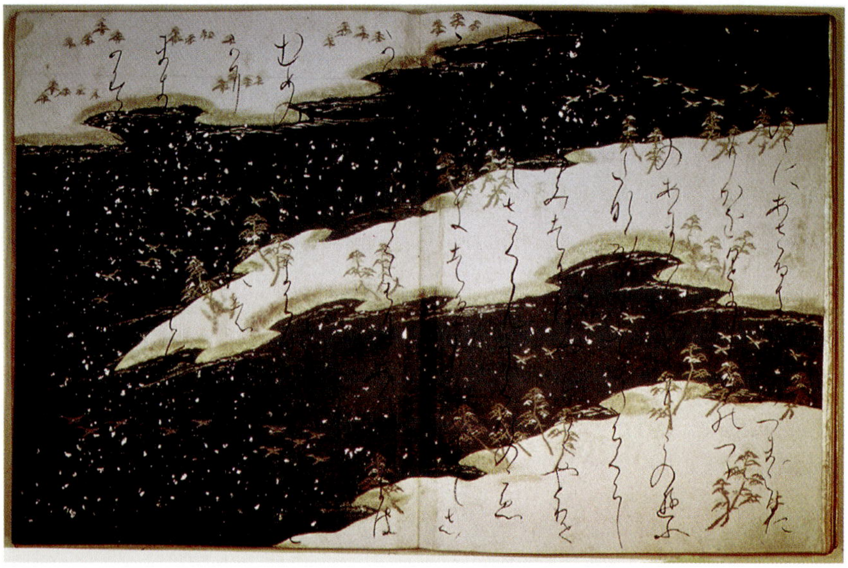

Figure 3. *Pages from the Akahito volume,* Sanjūrokunin-kashū. *Nishi-Honganji Temple*

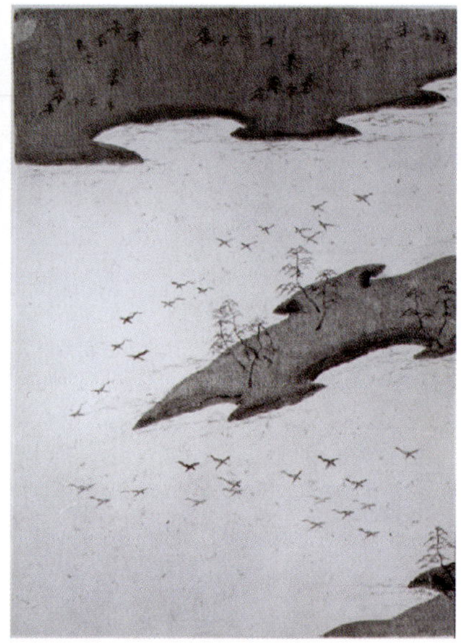

Figure 4. *Reverse density image of the x-ray radiograph of the left page of figure 3*

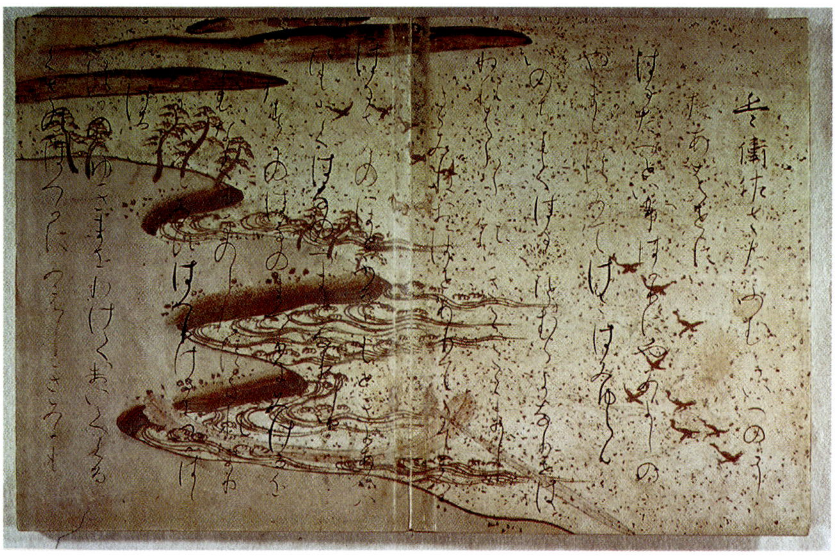

Figure 5. *Pages from the Tadamine volume,* Sanjūrokunin-kashū. *Nishi-Honganji Temple*

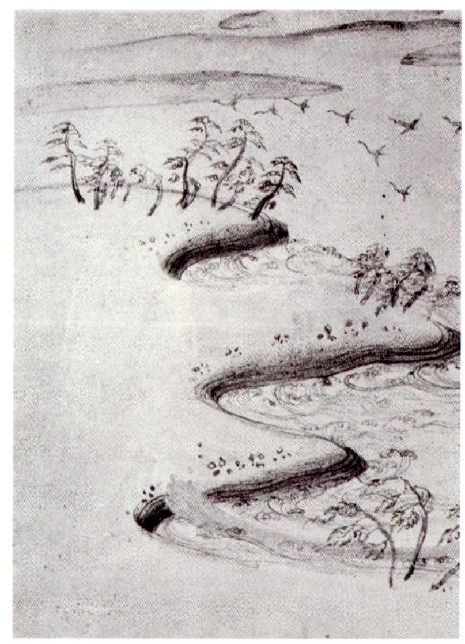

Figure 6. *Reverse density image of the x-ray radiograph of the left page of figure 5*

of the shore where silver pigment is coated with an orange colorant must have originally taken on the appearance of gold.

The similar combination of silver and yellowish dye in the paintings on three small cypress fans handed down by the Itsukushima Shrine (fig. 7) are of a more calculated nature. These works are attributed to the second quarter of the twelfth century based on a comparison with datable works (Egami 1992). Silver dust is sprinkled over the entire white-sized surface of the three fans, and then two kinds of colorants seem to have been applied in wash along the shores in the landscapes—yellow on the land portions of the shores and blue on the water portions (fig. 8). This method must have been convenient for arranging "gold" and "silver" tints side by side in small compositions of about 30 cm in width. Other more advanced scientific research methods may offer additional information.

An Amateur Device in the *Ashide Rōeishū*

Lastly, the decorative characteristics of the *Ashide Rōeishū* are considered. It is a version of the *Wakan-rōeishū* (fig. 9), an anthology of selected poems in Japanese and Chinese transcribed by the celebrated calligrapher Fujiwara no Koreyuki (fl. ca. 1150–ca. 1175) in 1160. The version takes its name from the term *ashide-e*, a colored pictorial design of sketchy natural motifs mingled with patterned shapes that were derived from letters. Judging from the brushwork and the fact that the silvery pigment in this design is a mixture of mica dust and pale ink, I believe that the painting was almost certainly done by the calligrapher himself (Egami 1982). Fine, transparent flat flakes peculiar to the mineral mica (Japanese, *unmo*) are visible in close-up photographs of this pigment (fig. 10). When applied to a paper surface, seaweed extract (Japanese, *funori*), which is easily affected

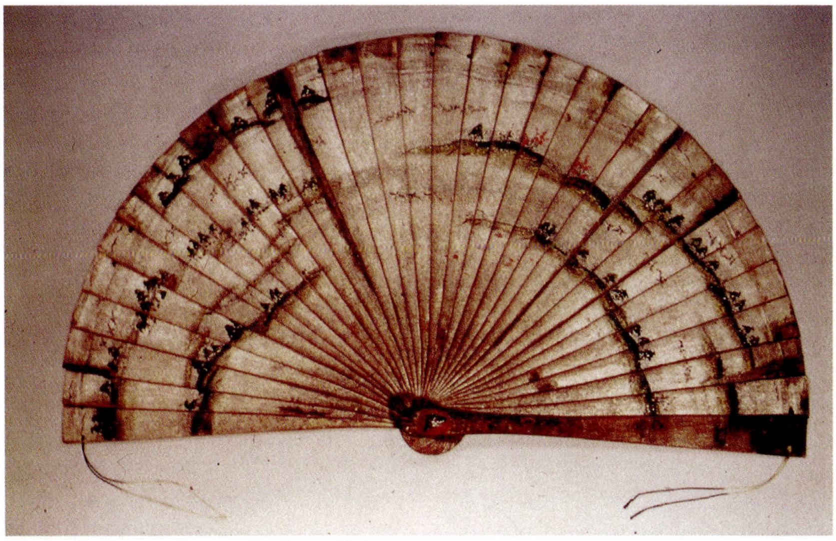

Figure 7. *Cypress fan. Itsukushima Shrine*

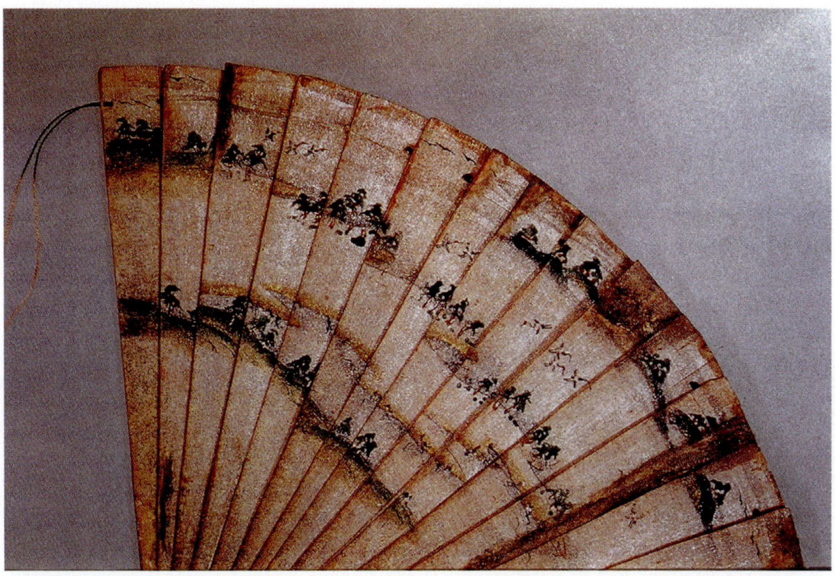

Figure 8. *Detail of figure 7*

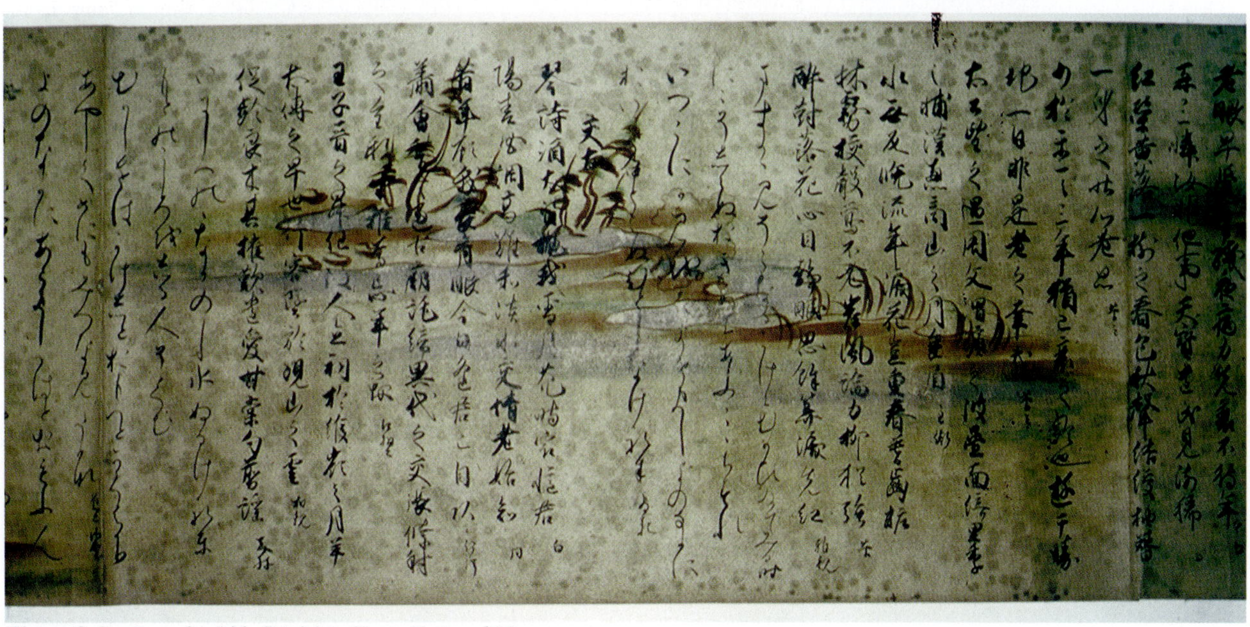

Figure 9. *Portion of* Ashide Rōeishū. *Kyoto National Museum*

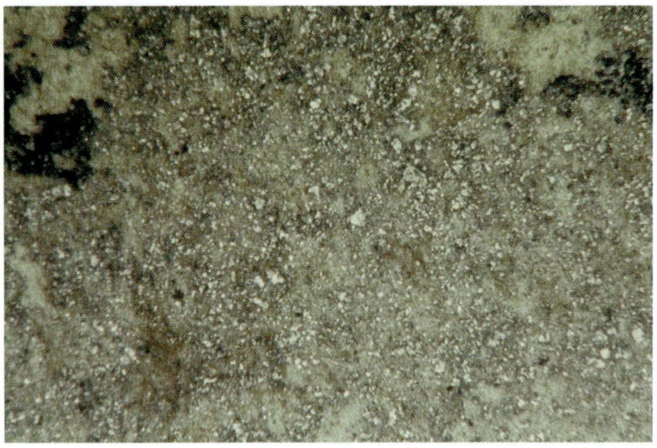

Figure 10. *Photomacrograph of mica flakes on* Ashide Rōeishū

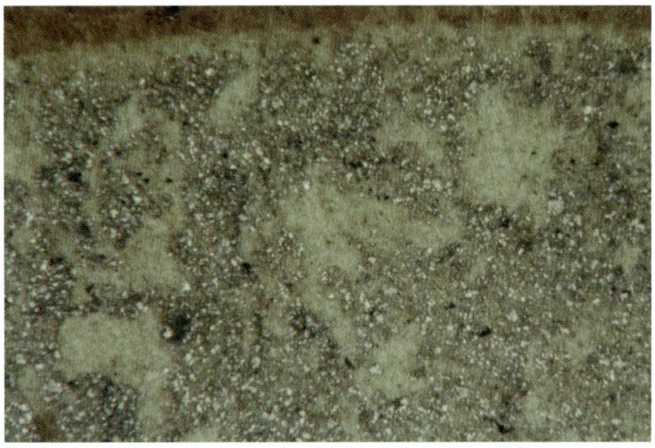

Figure 11. *Photomacrograph of mica flakes on* Ashide Rōeishū

by water, is added to the mica dust or mica powder. A distinctive deterioration of the silvery pigment in this work is characteristic of this substance (fig. 11). Furthermore, the places where the pigment has been removed do not in any way manifest the phenomenon of *gin'yake* or "silver burn"—a darkening of the paper. It is tempting to think that the calligrapher deliberately selected the mixture in order to avoid the oxidation of the pure silver that would conflict with the beauty of the ink calligraphy. It is possible that Fujiwara no Koreyuki was the first person to use a mica dust–pale ink mixture as a surrogate for silver, though there is no way to be certain.

The Relationship between Stylistic Evolution and Political Change

The sudden emergence of sophisticated methods to induce a wide range of "gold" and "silver" tints in the pictorial ornamentation of manuscripts at the beginning of the twelfth century deserves attention. It is closely related, I believe, to the rapid change that occurred in the political and cultural climate of the time. Generations of Fujiwara family regents at court from the end of the ninth to the end of the eleventh century—particularly from the late tenth to the mid-eleventh century—held a virtual monopoly on political and cultural initiatives. Subsequently, what historians call the Age of the Retired Emperors, when retired emperors were at the center of politics and culture, continued for more than a century. One of the most important figures at the very opening of this age was Emperor Shirakawa, who reigned from 1072 to 1086 and who acted as Retired Emperor from 1086 to 1129. I think that the nobles who lived under the leadership of people like Shirakawa and who were sensitive to this transition contributed to the creation of an aesthetic standard free of rigid conventions and promoted the prevalence of a markedly new trend of art that emphasized diversity and harmony.

The stylistic evolution that occurred in manuscript ornamentation is seen almost simultaneously in lacquerware in a similar form. For example, a legged lacquer chest owned by Kongōbuji Temple and widely considered to be from the end of the eleventh century (or more probably from the beginning of the twelfth century) has a *maki-e* design of water and irises accomplished with ordinary gold and a paler gold alloy called *aokin* combined with mother-of-pearl inlay (Nakasato et al. 1968). The color scheme is quite close to the pictorial ornamentation that uses two distinct kinds of gold tints in the *Jōyuishikiron*.

A lacquered horizontal harp owned by Kasuga Shrine and considered to be from about the same time as the *Sanjūrokunin-shū* has a *maki-e* design of natural motifs done with dust of gold, silver, *aokin*, and copper (Emoto 1980). The unusual use of copper dust for *maki-e* is again indicative of the innovative mood of the period and reminds us of the type of color experimentation seen in the *Sanjūrokunin-shū*.

Evidence of the cultural shift that took place around the beginning of the twelfth century was not confined to

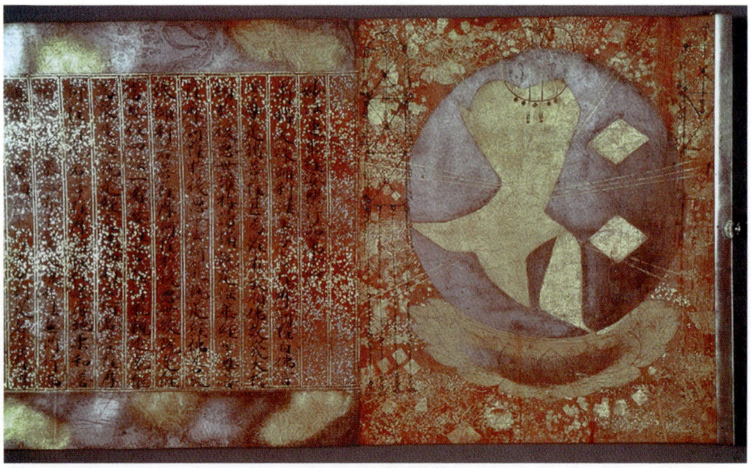

Figure 12. *Portion of Anrakugyō-hon scroll,* Heike-nōkyō. *Itsukushima Shrine*

ornamentation. Voluminous anthologies of past and contemporaneous poems, compiled from time to time by order of emperors and retired emperors, exhibit the same inclination to seek change. Retired Emperor Shirakawa commissioned leading poet Minamoto no Toshiyori (1055?–1129) to compile *Kin'yō-wakashū*, the fifth of seven such anthologies of the Heian period, in 1124. The poet submitted the first draft within the year, but the emperor refused to accept it mainly on the grounds that many of the poems therein, including works by poets of earlier generations, were old-fashioned. The second draft finished in the following year largely with contemporaneous poems was again rejected for the reason that it was too radical. The third edition completed with considerable revision was finally accepted in 1126 or 1127. Takeo Matsuda, a leading authority on the anthology, has indicated that this literary product in the formative phase of the so-called age of the Retired Emperors—when the power of the Fujiwara regency was in decline—opened a new epoch in the history of Japanese verse (Matsuda 1984).

The diversification in "gold" and "silver" tints in pictorial works during this period was characterized not by diversity for diversity's sake or even diversity for mere decorativeness. The tints were adopted to create a more subtle depiction of natural motifs, representing a conscious attempt to move away from the fixed concept of "precious colors." This trend appears to have been short-lived, however. In the ornamentation of *Heike-nōkyō* (fig. 12), a sutra set that Taira no Kiyomori, head of the Taira family, dedicated to the Itsukushima Shrine in the 1160s, silver and particularly gold resumed their original place as symbols of the greatest value and prestige. It is my hope that this study will encourage further studies and contribute to a comprehensive examination of Japanese art.

References

Egami, Yasushi. 1971. Kōfukuji-zō konshikinji Jōyuishikiron no shōgonga [Ornamental painting of the Buddhist scripture *Jōyuishikiron* owned by the Kōfukuji]. *Bijutsu Kenkyū*, no. 277 (September):1–17.

Egami, Yasushi. 1982. Ashide Rōeishū no shitae ni tsuite [Ornamentation painting of the *Wakan-rōeishū* manuscript transcribed by Fujiwara no Koreyuki]. *Bijutsu Kenkyū*, no. 320 (June):1–11.

Egami, Yasushi. 1991. Nishi-Honganji-bon Sanjūrokunin-shū ni okeru Yoshinobu-shū [Ornamentation of the Yoshinobu volumes of the *Sanjūrokunin-kashū* handed down by the Honganji]. In *Akiyama Terukazu Hakase Koki Kinen Bijutsushi Ronbun-shū*, ed. Akiyama Terukazu Hakase Koki-Kinen Ronbushū Kankōkai, 143–67. Kyoto: Benridō.

Egami, Yasushi. 1992. *Senmenga (Kodai Hen)*. Nihon no Bijutsu series 319. Tokyo: Shibundō.

Emoto, Yoshimichi. 1980. Kobunkazai no ekkususen-bunsekihō ni yoru zaishitsu sokutei shiryō III [X-ray analysis data on materials of ancient cultural properties III]. *Hozon Kagaku* 19:35–41.

Kyūsojin, Hitaku. 1953. Nishi-Honganji-bon Sanjūrokunin-shū. In *Sanjūrokunin-shū to Kunōji-kyō*, ed. Kyoto Kokuritsu Hakubutsukan, 1–7. Kyoto: Benridō.

Kyūsojin, Hitaku. 1960. *Sanjūrokunin-shū*. Tokyo: Hanawa Shobō.

Matsuda, Takeo. 1984. Kinyō-wakashū. In *Kokushi Daijiten*, 4:703. Tokyo: Yoshikawa Kōbunkan.

Nakasato, Toshikatsu, Rikuo Ishikawa, and Saburō Tatsuta. 1968. Heian jidai shitsugei gihō shiryō II [Technical data of lacquer work in the Heian period II]. *Hozon Kagaku* 4:61–89.

Scientific Analysis Used in the Treatment of the *Map of the World and Famous Cities* Screens and Resulting New Perspectives on the Paintings

Yasuhiro Oka, Sadatoshi Miura, Yasuhiro Hayakawa, and Tetsuo Miyakoshi

ABSTRACT The *Map of the World and Famous Cities*, a pair of eight-fold screens from the Imperial Household Agency, Museum of Imperial Collections, Tokyo, Japan, was treated in the Oka Bokkodo Asian Art Conservation Studio, Kyoto, in 1996. Initial examination of the paintings before treatment revealed a palette and surface gloss that differ from traditional Japanese paintings. X-radiography, emissiography, and infrared photography were used to carry out a detailed analysis of the paintings when the screens were dismantled. In addition, x-ray fluorescence spectrometry and x-ray diffraction were used for analysis of the pigments. Analysis of the binder, using gas chromatography–mass spectrometry, indicated the presence of a drying oil and lacquer in addition to animal skin glue. The results of these analyses influenced the treatment plan of the painting and illuminated its history.

Introduction

The aim of this paper is to describe, from a conservator's viewpoint, the results of scientific analyses conducted in the process of conserving the *Map of the World and Famous Cities*, paintings on paper mounted on a pair of eight-fold screens with colored paintings on a paper surface (figs. 1 and 2), in the Imperial Household Agency, Sannomaru Shozokan (Museum of Imperial Collections), Tokyo, Japan. These results had an influence on the conservation treatment plan for the screens and on the research of their art history. We would also like to refer to the current situation surrounding the study of paintings of exotic subjects like the *Map of the World and Famous Cities*.

Map of the World and Famous Cities Screens

The *Map of the World and Famous Cities* form together a large decorative painting mounted on a pair of eight-fold screens. On the left screen are a painting of a world map and forty-two paintings of people depicting cultures of different countries; on the right screen are paintings of eight rulers, a

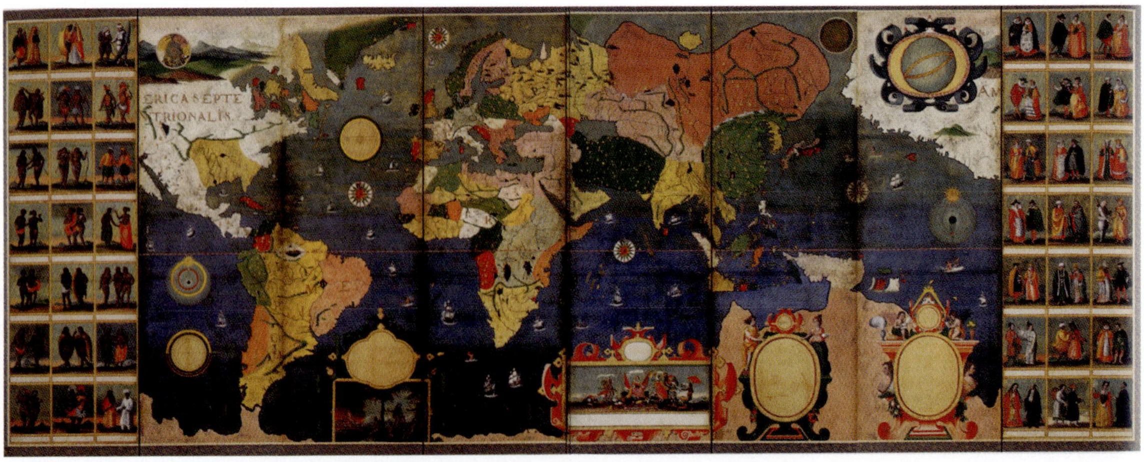

Figure 1. Map of the World and Famous Cities, *left. Imperial Household Agency, Museum of Imperial Collections, Tokyo, Japan*

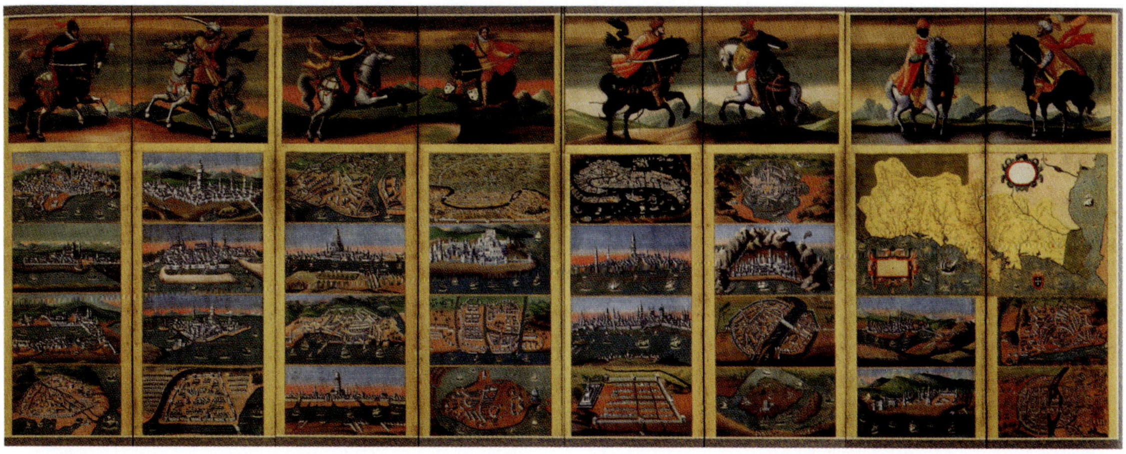

Figure 2. Map of the World and Famous Cities, *right. Imperial Household Agency, Museum of Imperial Collections*

map of Portugal, and twenty-eight cities. Few historical facts are known about this work, other than that it was produced after the early seventeenth century. The original designs for the maps painted on these two screens are presumed to be from the early seventeenth century and produced by either W. J. Blaeu in 1607 or Petrus Plancius in 1609 (Miyoshi 2003).

The screens required substantial repair. Not only were there cracks in the paper hinges (a common form of damage in folding screens), but there was also deterioration in the form of the partial separation of the paper painting support from the lining paper, in addition to some flaking and peeling of the paint layer. The conservation treatment of these screens took one year and was performed by The Oka Bokkodo Asian Art Conservation Studio, Kyoto, in 1996 (Oka Bokkodo Company 1999).

Visual Examination

Initial visual examination, prior to any treatment, revealed to the conservators that the surface finish of the paint layer on the screens was quite different from that of other *nihonga*, or Japanese-style paintings. In terms of art history, several researchers had already classified the *Map of the World* under what is known as "early Western-style paintings" (Sakamoto et al. 1980). This classification was based on two factors: first, it is very clear that what is portrayed was strongly influenced by Western culture; and, second, the surface finish of the paint layer is different from that of most traditional Japanese-style paintings. It should be noted, however, that this classification was based not on scientific analysis but on related literature and the style of expression.

The surface finish of the paint layer on these screens looks smooth and glossy, which, as mentioned above, is very different from that of *nihonga* executed in water-soluble pigments and/or mineral pigments with an animal skin glue binder. With this point in mind, it was considered necessary to examine the artwork as thoroughly as possible before it was actually treated, in order to identify the kinds of pigments and binding media that had been used for coloring the screens.

First, the intensity of fluorescence under ultraviolet light was observed. When ultraviolet light was shone on the painting, low-intensity fluorescence was emitted from areas that appeared to consist of coarse particles of azurite (*gunjō*) and malachite (*rokushō*), while areas with pigments of fine particles emitted comparatively high-intensity fluorescence, regardless of color.

Samples of various pigments bound with animal skin glue were used for comparison. The degree of fluorescence from these samples revealed a wide range of disparity depending on color, and the amount of fluorescence emitted on the whole was much lower than for the pigments used on the screens. This comparison with samples of pigments using animal skin glue led the conservators to presume that the kind of pigment and/or binding media used for the screen was different from that used for Japanese paintings produced with traditional techniques.

Scientific Analysis of Pigments and Binding Media

Prior to any conservation treatment, Sadatoshi Miura and Yasuhiro Hayakawa of the Tokyo National Research Institute for Cultural Properties, and Tetsuo Miyakoshi of the Science and Engineering Department at Meiji University, were commissioned to determine whether the pigments and binding media, observed to have different properties under visual examination and ultraviolet light exposure, are actually different from those that are commonly used for *nihonga*.

With the authorization of the Sannomaru Shozokan Museum, the samples necessary for examination were taken from under the silk borders of the screen. Twelve 2 × 2 mm samples were taken from the screens as follows: nine from the sixth panel from the right on the left screen, one each from the second and third panels from the right on the left screen, and one from the eighth panel from the right on the right screen. All samples had pigments on one side of a paper support.

X-ray fluorescence spectrometry (XRF) and x-ray diffraction (XRD) were employed to analyze the pigments, and an x-ray radiograph and x-ray emissiogram of the map of

Europe on the fifth panel from the right on the left screen were examined (Miura 1999). In addition, the binding media were analyzed by pyrolysis gas chromatography–mass spectrometry (PGC-MS).

Miura and Hayakawa conducted the pigment analysis and, using XRD, confirmed that basic copper carbonate had been used for greens and mercuric sulfide for reds, while XRF analysis proved that the blue pigments contain copper compounds. For XRD analysis, the samples were mounted on sample holders and placed in a Mac Science MXP18VA x-ray diffractometer. The XRF analysis was conducted in air, with the samples placed directly on polypropylene film. The instrument used for this XRF analysis was Seiko Instruments' SEA5230E x-ray fluorescence spectrometer (Hayakawa 1999).

Tetsuo Miyakoshi, of the Science and Engineering Department at Meiji University, selected PGC-MS to analyze the binding media (fig. 3). The instruments used were Hewlett Packard's HP6890 gas chromatograph and HP5972A mass spectrometer in conjunction with Frontier Laboratories' RY-2010 pyrolyzer. Animal skin glue was found in all the samples, and oil was detected in all but one sample. The peak for palmitic acid was higher than that for stearic acid and, because these analysis results are similar to those for flaxseed oil (fig. 4), paulownia wood oil (fig. 5), and soybean oil analyzed by the same method, it was presumed that a vegetable drying oil had been used as a binding medium in many sections of the paintings, in addition to animal skin glue. Furthermore, two of the samples revealed the mass chromatogram and mass spectrum unique to *urushiol*, which is present in lacquer, although the amount was minute. Since the amount of lacquer seems small, it is presumed that it was added not to increase gloss but for other purposes (Miyakoshi 1999).

Effects of the Scientific Analysis on the Treatment Plan and on Art-Historical Research

The Effects on the Treatment Plan

Any basic repair of a painting that is mounted on a screen usually requires the removal of the painting support from the

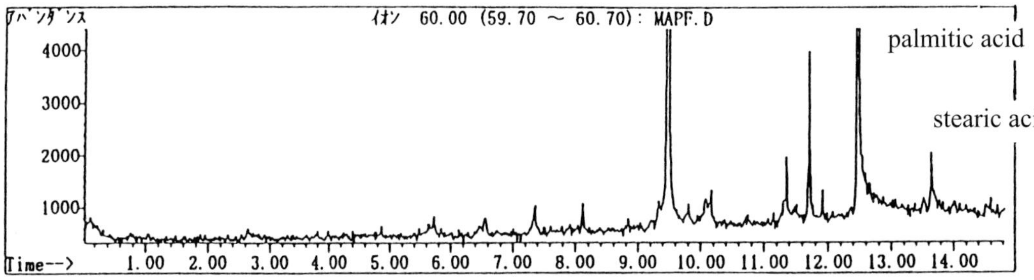

Figure 3. *PGC-MS scan of a charcoal brown (*koge-cha*) sample taken from the lower edge of the sixth panel from the right on the left screen*

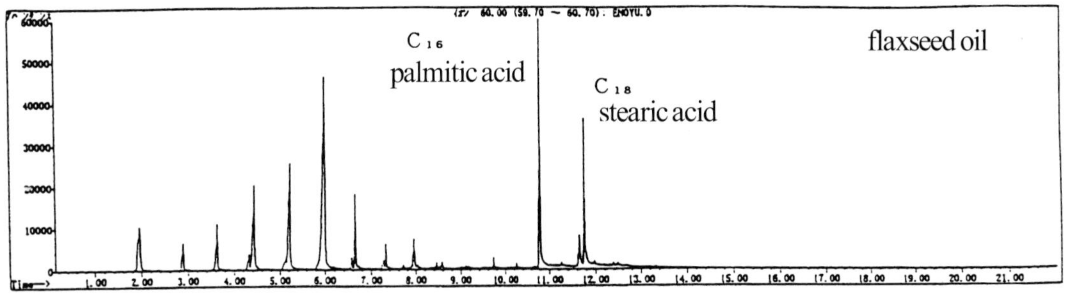

Figure 4. *PGC-MS scan of flaxseed oil*

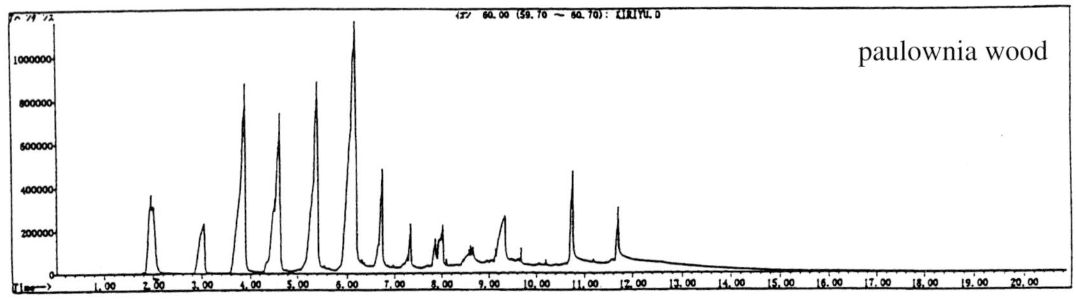

Figure 5. *PGC-MS scan of paulownia wood oil*

wooden core and the complete removal of the lining paper. The most effective way to remove the lining paper is to soften the paste that binds the lining paper to the painting support by infiltrating a large amount of water from the back. However, with this method, water will inevitably permeate to the surface of the painting. As the results of the analyses of *Map of the World* indicated that unusual materials were mixed with the pigment, it was thought that a large amount of water might result in some visual change on the surface of the painting. It was also a known fact from our past experience that, if an artwork has some oil content in its surface layer, the unique luster due to the oil can be decreased or lost when water infiltrates the surface during conservation treatment. Therefore, the conservators decided to treat the painting so as not to allow any water to seep to the surface of the painting support while the lining paper was being removed.

The Effects on Art-Historical Research

The *Map of the World* has been classified as an example of early Western-style paintings on the basis of its subject matter. The scientific analysis conducted this time indicated that a vegetable drying oil had been used in the binding medium, a different practice from that of traditional Japanese paintings. The presence of this oil suggests that the screens were created in a setting influenced by Western artistic practices. Such a setting might be the Society of Jesus seminary in Japan, where music and painting were taught as a proselytizing activity, in addition to Latin, Portuguese, Japanese literature, and mathematics. This seminary, founded in Azuchi in 1581, moved to Kyoto, Takatsuki, and then Osaka. Eventually it was located in Kyushu, where it moved from Hirado to Kitsuki and then to Nagasaki. In December 1581 it was amalgamated with another seminary that had been established in 1580 in Arima on the Shimabara Peninsula in Kyushu. The seminary moved from place to place in that area, and, when it was located in Hachirao, from 1591 to 1595, it seems to have had a printing machine brought in by the Tensho mission, which had returned to Japan in 1590 and taught not only painting but also copperplate engraving.

In addition, *Lettera Annua Giapone*, annual reports on Japan written by the missionary Pedro Gomez during 1593 and 1594 (Takeda 2002), describe how, in the studio of the seminary on Shimabara Peninsula, eight people studied *nikawa-e* (literally, "animal skin glue painting"), eight studied oil painting, and five studied copperplate engraving. They appear to have studied coloring techniques and patterns by copying paintings, woodblock prints, and illustrated books. The description (still extant) of the studio refers to the high level of skill they acquired in animal skin glue painting and oil painting. Historical literature, including the above reports, reveals that the instruction of painting techniques was widespread from the late sixteenth century until an edict banning Christianity was issued at the beginning of the seventeenth century.

Antonio Prenestino, who was in Bungo Kyushu (on the east side of the island of Kyushu), wrote to Possevino, on 11 November 1578, that the "Japanese preferred 'a portrait of an armament knight' or 'a scene of a naval battle.'" A letter from Luis Frois to Acviva, 15 January 1584, mentions "the dominants, Daimyos, preferred the paintings of 'the Roman cities and a parade of the Pope' and those paintings were very effective to show the prosperity of the Roman Catholic Group" (Sakamoto 1973). These records demonstrate that the seminary provided various kinds of paintings like the *Map of the World* to the Japanese population.

A Review of the *Fifteen Mysteries of Saint Mary*

The painting entitled the *Fifteen Mysteries of Saint Mary* is another important cultural asset that has been classified as an early Western-style painting, like the *Map of the World*. It is color on paper, mounted as a hanging scroll. Recently discovered, it is remarkably well preserved, having survived the suppression of Christianity. This painting is attracting attention as an indispensable source for information on early Western-style Japanese painting and the history of Christianity in Japan.

Through visual examination and noninvasive analysis, Nobuyuki Kamba ascertained that the pigments used in this painting were lead white for white, yellow ocher for yellow, and malachite for green. The surface finish and yellow colored areas have craterlike holes, resembling burst air bubbles. Such holes would not be found in animal skin glue painting. They are presumed to be the consequence of combining pigment with a mixture of water-soluble and oil-based binding media (Kamba 1998). The result is a surface finish that resembles the finish of a glossy oil painting.

Following Kamba's report, Eri Takeda identified the particular binding medium and analyzed several samples from the *Fifteen Mysteries of Saint Mary* (Takeda 2002). She found that many craterlike holes were present on the surface of the samples that had yellow pigment using a binding medium made by mixing flaxseed oil and animal skin glue. Similar craterlike holes have also been identified in the yellow colored areas of the *Map of the World*. (Unfortunately, there were no yellow-pigmented areas under the silk borders available for sampling.) This finding is of great interest because it supports not only Miyakoshi's analysis, which verified the use of animal skin glue as a binding medium in all the samples taken from the screens, but also the existence of some vegetable drying oil.

Conclusions

In the *Map of the World* conservation project, conservators had the opportunity to collaborate with scientists and art historians. Scientific analysis provided evidence that allowed the conservators to develop a safe approach in treating the painting. In this case, conservators recognized that the paint mixture contained not only animal skin glue but also a vegetable drying oil. As a consequence, the method for removing the lining papers was modified to use only a limited amount of moisture.

When it became clear that an unusual binding media had been used, the case for classifying the *Map of the World* as an early Western-style painting was solidified from the point of view of science, thus corroborating the opinion of art historians who have classified the *Map of the World* as an early Western-style painting. Records such as those from *Lettera Annua Giapone*, give us a historical perspective for the scientific findings and allow us to imagine a setting and time frame in which the *Map of the World* was created.

The surface of the *Fifteen Mysteries of Saint Mary* is similar to the surface of the *Map of the World*. Consequently, it is assumed that a mixture of flaxseed oil and animal skin glue was used frequently as a binding media when early Western-style paintings were painted.

Acknowledgments

The Oka Bokkodo Studio would like to thank authors Dr. Sadatoshi Miura, Dr. Yasuhiro Hayakawa, and Dr. Tetsuo Miyakoshi for permission to describe the scientific research portion of the conservation report of the *Map of the World and Famous Cities* screens. Also, thanks go to Ms. Aya Ohta, curator of the Imperial Household Agency, Museum of Imperial Collections, Tokyo, Japan, who gave permission to publish the results of research on the screens and also permission to use the images, and to Ms. Jennifer Perry for her assistance in translating and editing.

References

Hayakawa, Y. 1999. Analysis of the pigment of the *World Map and Twenty-Eight Cities. Shūfuku* 5:32–37.

Kamba, N. 1998. Research of the *Fifteen Mysteries of Saint Mary* in the collection of Kyoto University. *National Museum of Japanese History Annual Report* 76:175–247.

Miura, S. 1999. Scientific examination of the *World Map and Twenty-Eight Cities. Shūfuku* 5:28–31.

Miyakoshi, T. 1999. Analysis of the binding media of the *World Map and Twenty-Eight Cities. Shūfuku* 5:38–42.

Miyoshi, T. 2003. The original *World Map and Twenty-Eight Cities* screens. *Kobe City Museum Research Bulletin* 19:55–66.

Oka Bokkodo Company 1999. Conservation report, analysis and treatment of the *World Map and Twenty-Eight Cities* screens. *Shūfuku* 5:6–43.

Sakamoto, M. 1973. Early Western-style paintings. *Nihon No Bijutsu* 80:54–55.

Sakamoto, M., T. Sugase, and F. Naruse. 1980. *Nanban-art and Western-style paintings*. 2nd ed. Tokyo: Shogakukan Publications.

Takeda, E. 2002. A copy of the *Fifteen Mysteries of Saint Mary* in the collection of Kyoto University and the study of its painting technique. PhD diss., Department of Art, Tokyo Art University.

Three Polychrome Japanese Buddhist Sculptures from the Kamakura Period: The Scientific Examination of Layer Structures, Ground Materials, Pigments, Metal Leafs, and Powders

Mark Richter

ABSTRACT This study focuses on the scientific examination of ground materials, pigments, and metals identified in ten samples taken from three polychrome thirteenth-century Japanese wooden Buddhist sculptures from the Kamakura period. The aim of this investigation was to gain information about artists' materials and techniques used for polychrome sculpture at this time. A variety of familiar and less familiar ground materials and colorants was identified, including various green copper compounds and white, lead-based pigments. Some difficulties were encountered with the identification of an unusual greenish blue compound present only in paint layers mainly composed of copper green and copper blue pigments. Interestingly, the analysis with x-ray diffraction revealed the presence of moolooite. The pigments, grounds, and metals used for gildings on polychrome sculptures were analyzed with a range of techniques, including microscopic techniques, elemental analysis using energy-dispersive x-ray microanalysis, and wavelength-dispersive x-ray microanalysis in combination with the scanning electron microscope. The crystallographic identification of crystalline materials was executed with x-ray diffraction.

Introduction

In 1999, the Bavarian State Department of Historical Monuments, Munich, in collaboration with the National Research Institute for Cultural Properties in Tokyo, undertook the examination of a small number of thirteenth-century Japanese sculptures from the Kamakura period (1185–1333) in Japan. The unique opportunity to examine these selected sculptures *in situ* in the temples and museums and take samples was made possible by the collaborative German-Japanese Research Project entitled Historical Polychromy.[1] It is very rare to be given the opportunity to sample sculptures in Japan for research purposes.

The objectives of this investigation were to analyze the pigments, metals, and other inorganic materials that were used and to study the polychroming techniques that were employed. The author further undertook a review of literature pertaining to artists' materials used for East Asian sculpture polychromy and painting, with special emphasis on Japan.

Examined Sculptures

The three standing sculptures that were studied in more detail are the *Miroku-bosatsu* from the Shomyo-ji located beside the Kanazawa Bunko Museum in Yokohama City, the *Jizō-bosatsu* of the Ensho-ji near Tokyo, and the *Bikara Taisho* originally from the Johruri-ji in Kyoto.

The *Miroku-bosatsu*[2] (Sanskrit, *Maitreya*) is the main devotional statue of the Shomyo-ji (fig. 1). The temple's history goes back to approximately 1262, when Hojo Sanetoki, a member of the Hojo family that ruled Japan during the Kamakura period and the founder of the Kanazawa Bunko Library, changed the then existing Buddhist hall at the time of his conversion to the Shingonritsu sect. Sanetoki's conversion marked the beginning of prosperous times for the temple, which included the special protection of the Hojo family for three generations. Of great importance was the discovery of an inscription written on the inside of the sculpture's hollowed head indicating that the statue was made during the final year of Sanetoki's life in March 1276. The *Miroku-bosatsu* exhibits a delicate and striking polychromy that is still well preserved despite its age (figs. 2 and 3). It is representative of the justly famous wooden sculptures of the late Kamakura period (Tsuda 1999, 2004).

The *Jizō-bosatsu*[3] (Sanskrit, *Ksitigarbha*), one of the main devotional statues of the Ensho-ji, has an exquisite and well-preserved polychromy of impressive quality (figs. 4 and 5). The sculpture is from approximately 1250 and is believed to have a connection with the school of Kaikei, one of the most famous sculptors of the early Kamakura period.

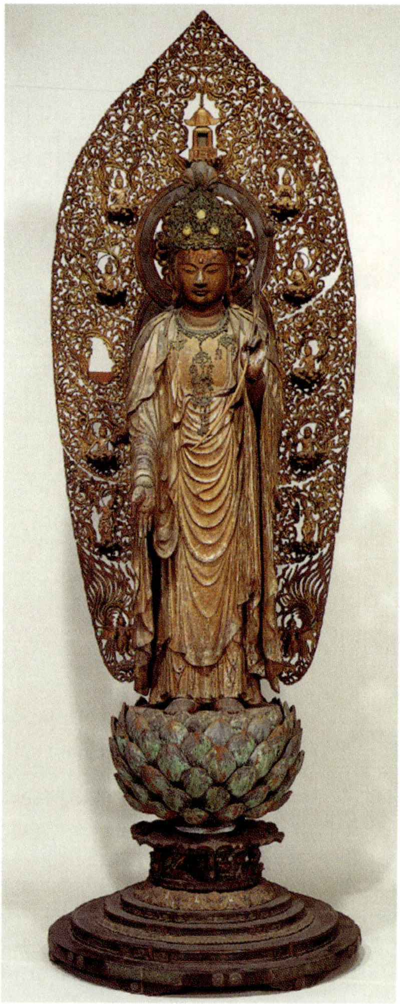

Figure 1. Miroku-bosatsu with Mandorla on Lotus Pedestal. *Shomyo-ji, Yokohama. Important Cultural Property*

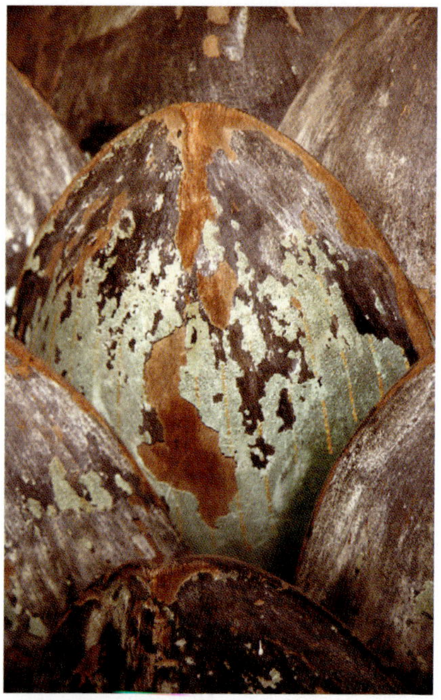

Figure 2. Miroku-bosatsu *(fig. 1), detail showing lotus leaf of pedestal with green polychromy and gold veins executed in* kirikane

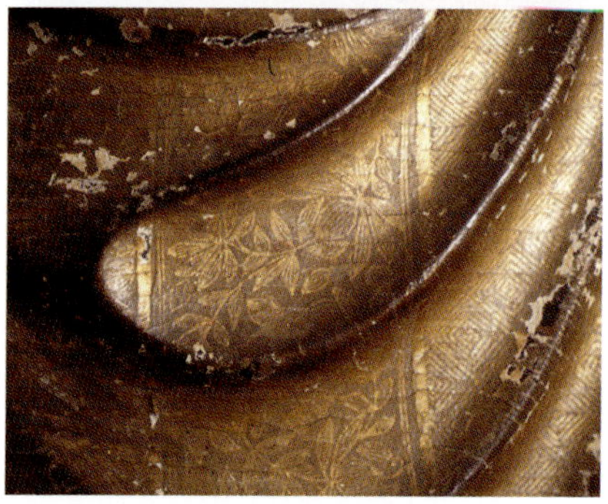

Figure 3. Miroku-bosatsu *(fig. 1), detail showing front of garment with various delicate* kirikane *designs applied on* kindei-nuri

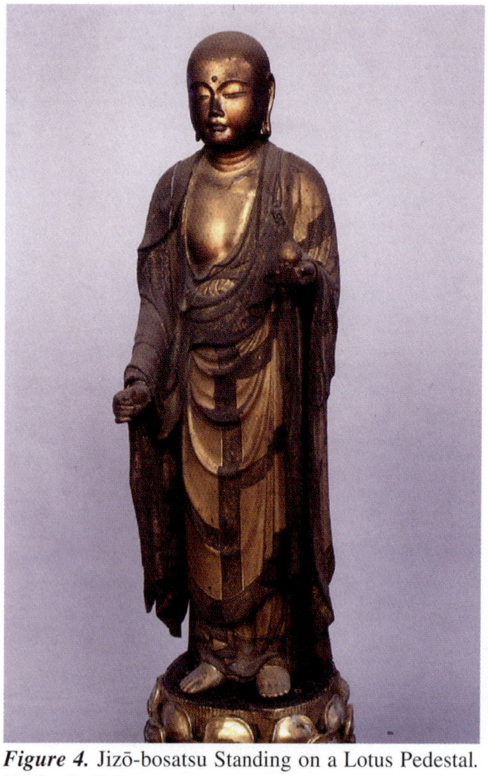

Figure 4. Jizō-bosatsu Standing on a Lotus Pedestal. *Ensho-ji, Tokyo*

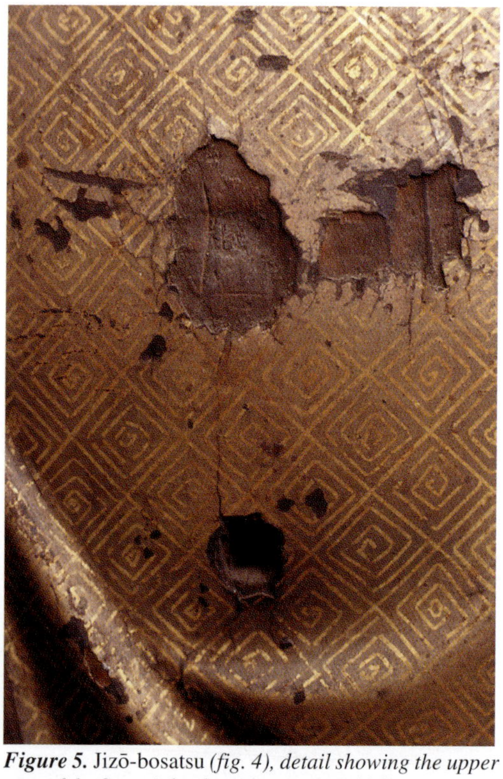

Figure 5. Jizō-bosatsu *(fig. 4), detail showing the upper part of the figure's back, with geometric* kirikane *pattern applied on* kindei-nuri *gilding*

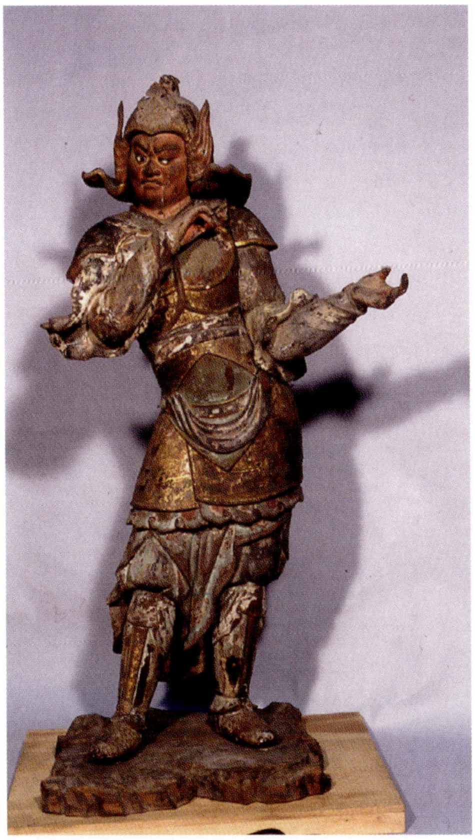

Figure 6. Bikara Taisho, *one of the Juni Shinsho (Twelve Divine Generals). Seikado Bunko Art Museum, Tokyo. Important Cultural Property*

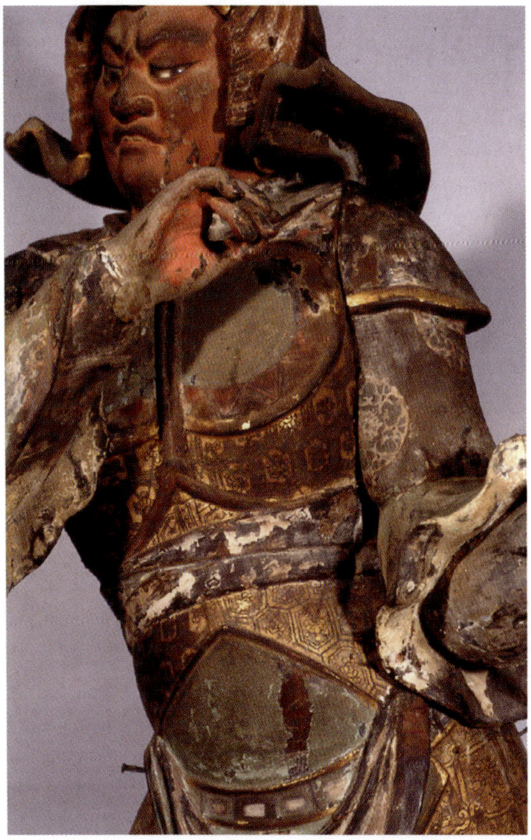

Figure 7. Bikara Taisho *(fig. 6), detail of areas where samples were taken (green areas on waist and breast)*

The standing sculpture of *Bikara Taisho* (Sanskrit, *Vikarala*) from the first quarter of the thirteenth century is attributed to the Kei school (fig. 6). It is said to have been formerly owned by the Johruri-ji in Kyoto but is now part of the collection of the Seikado Bunko Art Museum in Tokyo. *Bikara Taisho* is one of the Juni Shinsho (Twelve Divine Generals) who guard the true worshipers of Yakushi Nyorai (Sanskrit, *Bhaisajya*, the Buddha of healing).[4] He is depicted with red skin, a three-pronged *vajra* (thunderbolt), and a boar on his headpiece (de Visser 1935).[5] The sculpture of *Bikara Taisho* exhibits an exquisite and varied polychromy that in some respects is quite different from that of the other two sculptures (fig. 7).

Pigments Used in Japanese Art

A number of studies of pigments and other artists' materials in Japan relevant to these analyses have been reported in the work by Kazuo Yamasaki, Kyotaro Nishikawa, Yoshimichi Emoto, Masakazu Naruse, John Winter, and Sadatoshi Miura. Yamasaki was responsible for a number of detailed studies in the 1950s and 1960s on the pigments used on a number of Japanese paintings and other works of art including those of the Shōsō-in Treasure House[6] (Yamasaki and Nishikawa 1970; Yamasaki and Emoto 1979). Naruse and Miura provide a good overview of the pigments utilized in polychromy and painting from the sixth to the seventeenth century (Naruse 1996; Miura 2004). Winter was the first to systematically investigate white lead-based compounds used for Japanese, Chinese, and Korean paintings of the twelfth to sixteenth centuries (Winter 1981).

These studies on artists' materials proved to be invaluable for the scientific examinations carried out on three sculptures from the Shomyo-ji, Ensho-ji, and Johruri-ji. The sculptures from the Shomyo-ji and Ensho-ji were part of a previous investigation in 2000 (Richter 2004; Tsuda 2004).

Analytical Methods

The examination of the polychrome surfaces in the temples and the museum was carried out using a portable Zeiss stereomicroscope (magnification up to 40×). Most of the paint samples were mounted in embedding resin and examined as polished cross-sections using microscopic methods that included UV-fluorescence microscopy and polarized light microscopy (PLM).[7] The cross-sections were then examined with x-ray microanalysis using a scanning electron microscope (SEM) equipped with secondary and backscattered electron detectors, an energy-dispersive x-ray spectrometer (EDX), and a wavelength-dispersive x-ray (WDX) spectrometer.[8] In some cases x-ray fluorescence spectroscopy (XRF) was also employed. The crystallographic identification of crystalline materials in selected loose samples was carried out with x-ray diffraction (XRD).[9]

The Identification of Ground Materials, Pigments, and Metals

A total of ten samples—seven from the *Miroku-bosatsu*, two from the *Bikara Taisho*, and one from the *Jizō-bosatsu*—were studied. The scientific examination resulted in the identification of a number of pigments and yielded interesting results concerning the gilding.

Layer Structures

Examination revealed a reinforcement layer of cloth (*asa-nuno*)[10] glued to the wooden substrate of the *Miroku-bosatsu*'s and *Jizō-bosatsu*'s body and garments, followed by a coarse, dark brown layered application of *sabi-urushi*. *Sabi-urushi* is a foundation material often consisting of raw lacquer (*ki urushi*), mixed with *tonoko*, a fine-textured powder of pulverized whetstone or clay that has been kneaded with water (Yamasaki and Nishikawa 1970; Sano 2004). This type of ground has been used to make the final foundation for the polychromy of all three sculptures in this study. The coarse brown ground layer of *sabi* was covered with a thin blackish brown layer of lacquer, which was detected on both the *Miroku-bosatsu* and *Jizō-bosatsu*. The brownish orange fluorescence of this thin layer observed in cross-sections using ultraviolet illumination suggests the use of *urushi* (fig. 8). EDX analysis confirmed the elements iron, silicon, chlorine, potassium, and aluminum in this particular layer of the *Jizō-bosatsu*.

All three sculptures revealed a white layer on the dark brown ground as a final priming on which the polychromy was applied (figs. 9–12). This layer has normally been observed on sculptures of the Heian (794–1185) and Kamakura periods. In the case of the golden garments of the *Miroku-bosatsu* and the *Jizō-bosatsu* exhibiting *kirikane* (cut gold leaf) designs on *kindei-nuri* (gold powder in binding medium), a white preparation consisting of two layers with differing compositions was applied first (fig. 13). The first, very thin layer utilized a lead chloride pigment followed by a thicker layer composed of kaolinite (see section on white pigments, below). Subsequently, a light red preparatory layer was applied on the white priming of both sculptures (see figs. 8–10). This technique was employed in these areas in order to enhance the color of the *kindei-nuri* and the *kirikane* patterns applied on top. The technique is quite similar to gildings executed on European polychrome sculptures and panel paintings, where red preparatory layers have been used since the beginning of the fourteenth century.

The main part of the *Miroku-bosatsu*'s nimbus, foot, and some parts of the lion (lotus pedestal) are gilded with *shippaku* (gold leaf), while black *urushi* is coated on the back. *Shippaku* was also used for the body and the garments of the angels who are part of the nimbus. The cross-section of the sample from the nimbus revealed that the *urushi* adhesive under the gold leaf consists of two layers, which are applied directly on the *sabi-urushi* ground.

The green lotus leaf pedestal of the *Miroku-bosatsu* and the green armor of the *Bikara Taisho* also revealed a multi-layered structure from the dark brown ground upward (fig. 14; see also fig. 2). As in the case of the *Miroku-bosatsu*, the following layer sequence was identified in the *Bikaro Taisho*: 1. very thin white layer on dark brown ground layer; 2. fine pale green underpainting; 3. coarse bright green final layer. The waist armor of the *Bikara Taisho* also exhibits a white layer followed by a pale green layer, a slightly darker green layer, and a final coarse bright green

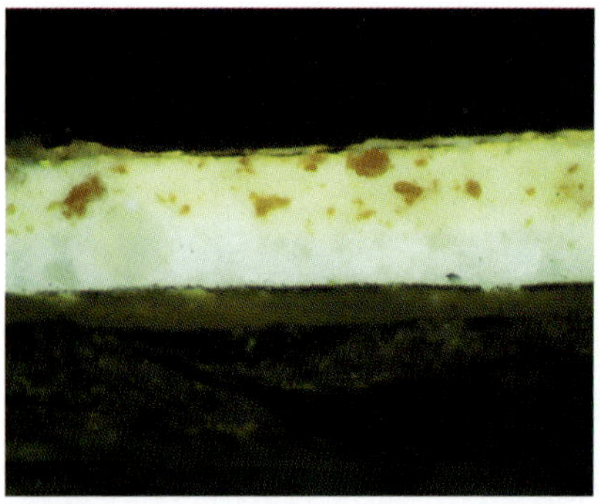

Figure 8. Jizō-bosatsu *(fig. 4), gilded garment below sleeve of right arm. Paint cross-section showing the layer structure: 1. dark brown* sabi-urushi *with brownish red fluorescence; 2. thin black layer of binding medium with distinct brownish red fluorescence (*urushi*); 3. white layer with bluish white fluorescence; 4. light red layer with orange and red particles, bright yellowish white fluorescence; 5. gold powder (*kindei-nuri*); 6. thin layer of binding medium (adhesive for* kirikane*), yellowish white fluorescence; 7. gold leaf (*kirikane*). Viewed in UV light, photographed at a magnification of 500×*

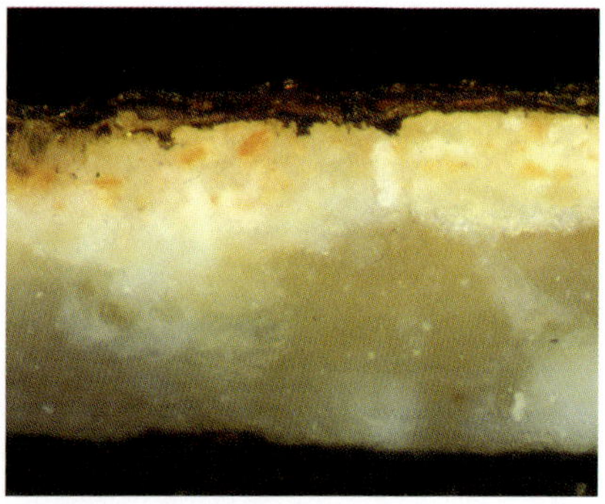

Figure 9. Miroku-bosatsu *(fig. 1), gilded garment, left side below hip. Paint cross-section showing the layer structure: 1. dark brown ground; 2. thin black layer of binding medium; 3. white layer; 4. light red layer with orange and red particles; 5. gold powder (*kindei-nuri*); 6. thin layer of binding medium (adhesive for* kirikane*); 7. gold leaf (*kirikane*). Viewed in normal light, photographed at a magnification of 500×*

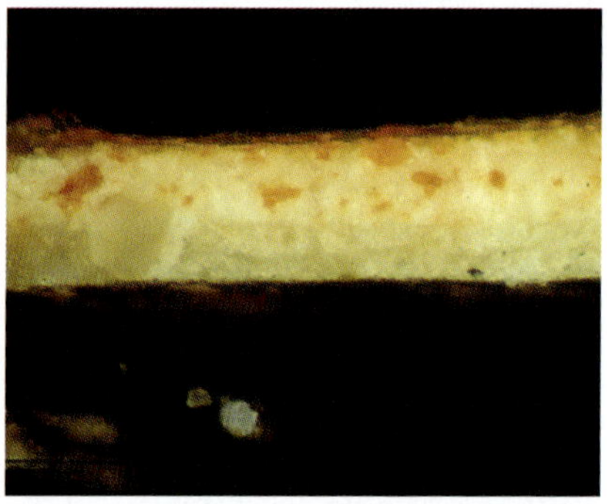

Figure 10. Jizō-bosatsu *(fig. 4), gilded garment below sleeve of right arm (fig. 8). Paint cross-section showing the layer structure: 1. dark brown* sabi-urushi *with brownish red fluorescence; 2. thin black layer of binding medium (*urushi*); 3. white layer; 4. light red layer with orange and red particles; 5. gold powder (*kindei-nuri*); 6. thin layer of binding medium (adhesive for* kirikane*), 7. gold leaf (*kirikane*). Viewed in normal light, photographed at a magnification of 500×*

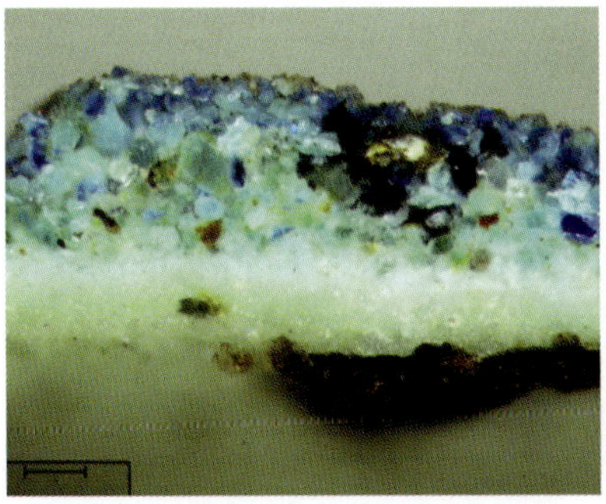

Figure 11. Bikara Taisho *(fig. 6), bluish green area of breast armor. Paint cross-section showing the layer structure: 1. dark brown* sabi-urushi*; 2. homogeneous greenish white layer; 3. coarse green layer consisting of quartz, malachite, brochantite, atacamite, and fluorite; 4. final bluish green layer consisting of azurite and copper green pigments. Viewed in normal light, photographed at a magnification of 100×*

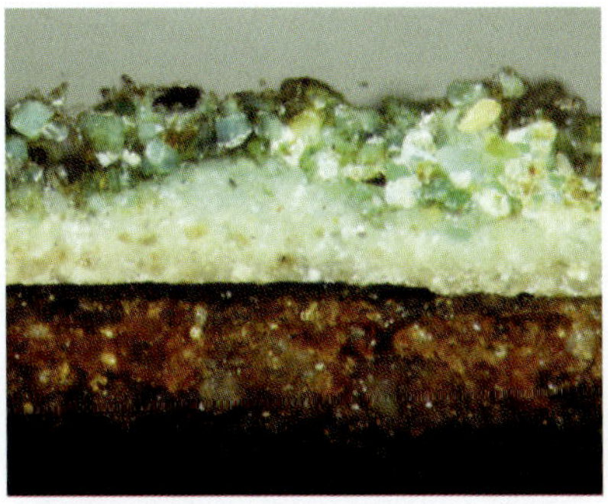

Figure 12. Bikara Taisho *(fig. 6), green area of waist armor. Paint cross-section showing the layer structure: 1. dark brown* sabi-urushi*; 2. thin black layer of binding medium; 3. homogeneous yellowish white layer; 4. light green layer; 5. coarse green layer consisting of quartz, malachite, brochantite, atacamite, and fluorite. Viewed in normal light, photographed at a magnification of 200×*

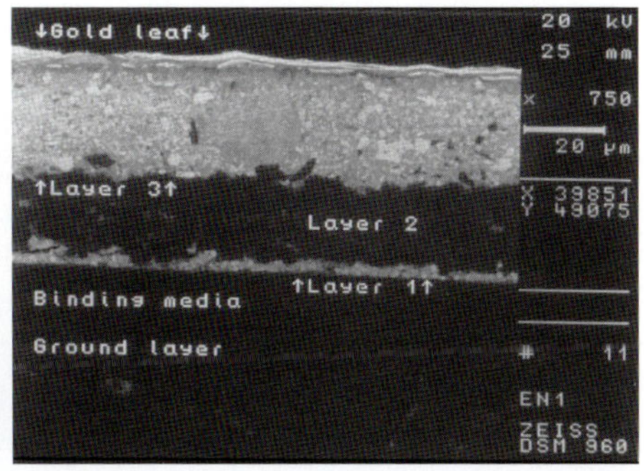

Figure 13. Jizō-bosatsu *(fig. 4), gilded garment below sleeve of right arm (figs. 8 and 10). Scanning electron micrograph of paint cross-section showing the layer structure: 1. ground; 2. layer of binding medium; 3. very thin white layer; 4. white layer; 5. light red layer; 6. gold powder (*kindei-nuri*); 7. thin layer of binding medium (adhesive for* kirikane*); 8. gold leaf (*kirikane*). Photographed at a magnification of 750×*

layer, whereas the bluish green breast armor reveals a light green first layer followed by a green layer and a final bluish green layer (see figs. 11–12).

Ground Layers

The analysis of the dark brown ground (*sabi-urushi*) of the *Miroku-bosatsu* samples produced some unexpected results.

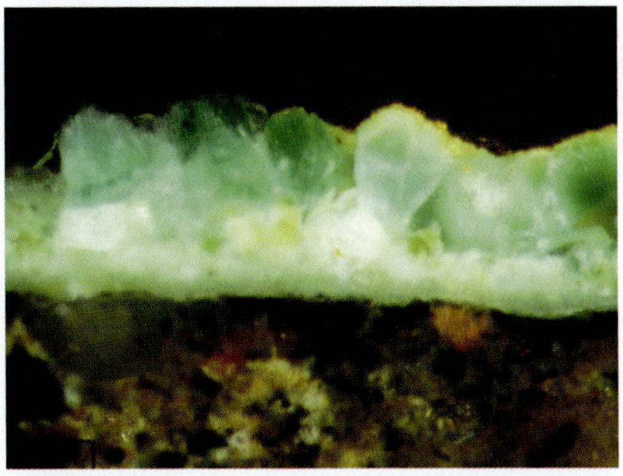

Figure 14. Miroku-bosatsu *(fig. 1), green lotus leaf with* kirikane *stripes. Paint cross-section showing the layer structure: 1. dark brown ground; 2. thin black layer of binding medium; 3. light green layer; 4. coarse bright green layer; 5. gold leaf (*kirikane*). Viewed with normal light, photographed at a magnification of 200×*

XRD and PLM identified a mixture of materials consisting mainly of quartz, anorthite, and nontronite as well as small amounts of jarosite, anglesite, calcite, muscovite, and brochantite. Some particles of wood powder were also observed. A reddish brown bole (ferruginous aluminum silicate), calcite, and quartz were identified in the ground of the *Jizō-bosatsu* and *Bikara Taisho* with EDX.

These results confirmed that the final foundation layer for the polychromy of all three sculptures consists of *sabi-urushi* mixed with *tonoko*, which was also used for lacquerware in Japan. The analysis of the yellow and reddish yellow particles present in the *sabi* ground of the *Miroku-bosatsu* produced unexpected results. Analysis with EDX and XRD confirmed the use of nontronite and jarosite (fig. 15). The main component was identified as nontronite ($Na_{0.3}Fe^{3+}{}_2Si_3\ AlO_{10}(OH)_2 \cdot 4(H_2O)$), a sodium ferric-iron silicate hydroxide hydrate mineral of the clay group, along with a small amount of jarosite ($KFe_3(SO_4)_2(OH)_6$), which is an anhydrous hydroxyl iron sulfate of the alunite group. The small amount of jarosite found in the ground of *sabi-urushi* showed a varied morphology suggesting it is a ground natural mineral. It has a refractive index of $n_D = 1.72$ to 1.81. When viewed in plane polarized light, the jarosite particles were transparent and pleochroic, changing from pale yellow to a warm amber yellow as the stage was rotated. Cleavage surfaces were also visible. The particles were either lath-shaped or rounded. Extinction of the lath-

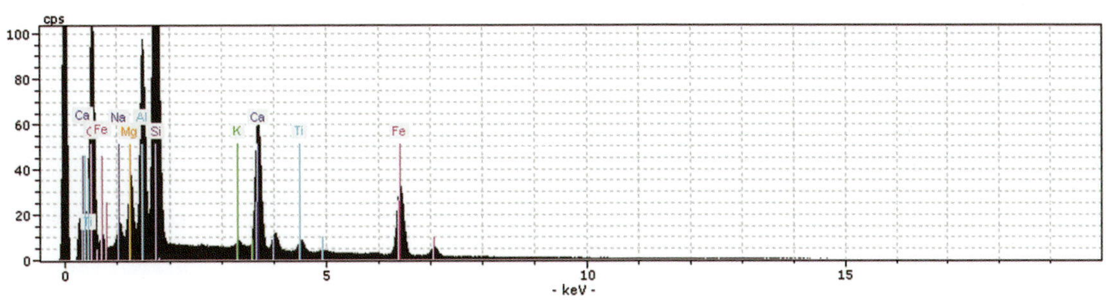

Figure 15. Miroku-bosatsu *(fig. 1), EDX spectrum of a particle of nontronite in the* sabi-urushi *ground*

shaped particles was parallel, and between crossed polars they exhibited strong interference colors.

As far as the author is aware, nontronite and jarosite have until now not been identified in Japanese polychromy and painting. Both compounds are often associated with other iron minerals such as limonite, hematite, and goethite. Jarosite has been identified in sedimentary iron ores from the Gunma iron mine (Gunma prefecture, Honshu Island) in Japan (Akai et al. 1999). Limonite has been identified in Japanese paintings of the seventh-century (Miura 2004). Jarosite has also been confirmed on the Great Buddha of Dafosi from the Tang dynasty (618–907), where it was identified as a finely pulverized pigment in the oldest polychromy consisting of a yellow layer on white ground (Emmerling et al. 1996). Jarosites have also been identified in ancient Egyptian and Greek painting and ceramics (Colinart and Pàgas-Camagna 2001; Wallert 1995; Noll 1991).

White Pigments

The white pigments found on all three sculptures proved to be, for the most part, quite unusual, and included calcite, kaolinite, cerussite, anglesite, blixite, and cotunnite.

Small amounts of calcite ($CaCO_3$) were confirmed by XRD in the ground layers and the following white layer of the *Miroku-bosatu* and *Jizō-bosatsu*. The use of shell white can be ruled out due to the particle form, indicating that chalk or limestone was probably used here. An official document written in 734 on the construction of the West Golden Hall of Kofuku-ji describes the use of limestone as a white ground (Naruse 1996). Calcite was also identified on a number of *gigaku* masks from the Shōsō-in in Nara (Naruse 1996).

Kaolinite ($Al_2Si_2O_5(OH)_4$), a type of white clay (Japanese, *hakudo*), was confirmed in a number of pigmented layers, including, in some cases, the ground. It was identified in the white preparatory layer of the *Jizō-bosatsu* sample and was also confirmed in the pale greenish white underpaint (on the ground) of the bright bluish green breast armor of the *Bikara Taisho*. The use of kaolinite as a white pigment in Japanese painting and polychromy is well known.

Muscovite (Japanese, *kirara*) was also identified with XRD in the ground and in some white layers. It is a potassium aluminum silicate hydroxide mineral ($KAl_2(Si_3Al)O_{10}(OH)_2$) of the mica group and occurs widely distributed. Of special interest was the muscovite used in the ground of the *Miroku-bosatsu*, which clearly revealed barium with XRF. Such a mineral, originally described by Bauer and Berman in 1933, was given the name *barium-muscovite* (Bauer and Berman 1933). This barium-muscovite is different from common micas in its physical appearance: it is fine grained, cleavage is not evident, the color is a deep violet-pink, and it has a dull grainy luster. Muscovite has also been confirmed in Japanese painting as early as the Heian period including two *gigaku* masks of the Shōsō-in, where it was identified with XRD.

EDX showed one white pigment to contain mainly lead in the ground layer (*sabi-urushi*) of the *Miroku-bosatsu*. XRD made it possible to identify this pigment as neutral lead carbonate (cerussite, $PbCO_3$). Gettens et al. (1994) point out that cerussite has rarely been used as a white pigment except for one case, where it was identified in the ground layer of a Chinese polychromed terracotta tomb figure of the Tang dynasty. The white pigment was found to be mainly neutral lead carbonate containing a small amount of basic lead carbonate. Basic lead carbonate (hydrocerussite, $2PbCO_3 \cdot Pb(OH)_2$) has been identified much more frequently than cerussite in seventh–fifteenth-century Japanese painting and polychromy.

The highly interesting composition of the *Miroku-bosatsu*'s ground layer also revealed a small amount of anglesite (lead sulfate, $PbSO_4$). Anglesite has also been identified in Japanese painting from the eighth to eleventh century[11] (Mogi et al. 1975) and in Chinese painting from the Tang dynasty and the Yuan dynasty (1279–1368) or Ming dynasty (1368–1644) (Guoxin et al. 1993; Winter 1981).

Other white pigments identified in the white and light reddish orange preparatory layers of the *Miroku-bosatsu* and *Jizō-bosatsu* also proved to be of interest (see figs. 9 and 10). The analysis of a number of particles with EDX and WDX revealed a distinct peak for chlorine. XRD confirmed the use of cotunnite (lead chloride, $PbCl_2$). Another lead chloride compound called blixite (lead chloride oxide, $Pb_2Cl(O,OH)_{2-x}$) was identified in the greenish white underpainting of the bright green lotus leaf pedestal of the *Miroku-bosatsu*. The EDX analysis of the white layers of the *Jizō-bosatsu* also confirmed the use of a lead chloride pigment and probably kaolinite (see fig. 13). EDX and WDX revealed strong peaks of lead, chlorine, aluminum, and silicon. Lead and chlorine were also identified in the white particles of the overlying red preparatory layer (fig. 16), whereas the red pigment was identified as red lead.

White, lead-based pigments containing chlorine have been found more frequently in Japanese painting and sculpture polychromy than lead white (basic lead carbonate).

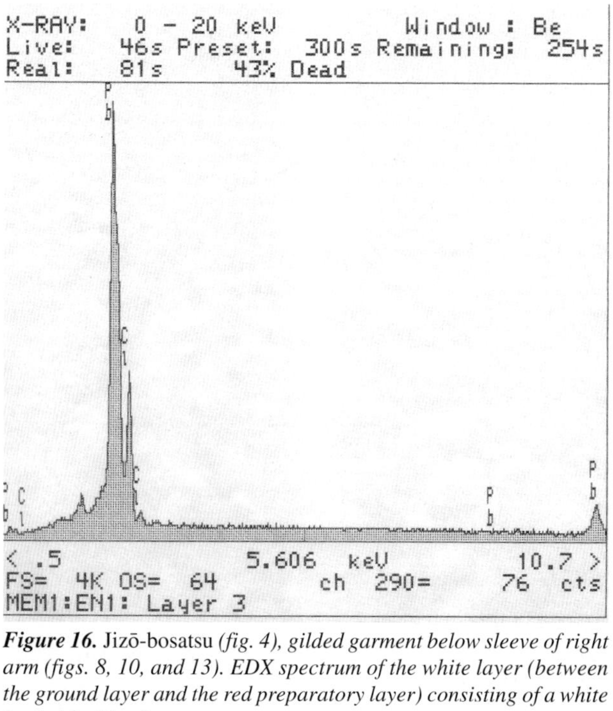

Figure 16. Jizō-bosatsu *(fig. 4), gilded garment below sleeve of right arm (figs. 8, 10, and 13). EDX spectrum of the white layer (between the ground layer and the red preparatory layer) consisting of a white lead chloride pigment*

These compounds include cotunnite, laurionite (lead hydroxide chloride, PbOHCl), and blixite. Winter was the first to systematically investigate lead-containing white pigments used in Japanese, Chinese, and Korean paintings of the twelfth and sixteenth centuries belonging to the collection of the Freer Gallery of Art, Washington, D.C. He identified both laurionite and blixite with XRD in a number of thirteenth–fifteenth-century Japanese paintings (Winter 1981). Naruse (1996) investigated pigments used on objects of the Shosō-in from the eighth century and identified cotunnite, blixite, and laurionite with XRD. In China the polychrome sculptures of the Tiantishan grottoes from the Tang dynasty also revealed the use of laurionite (Guoxin et al. 1993).

Black Pigments

Black pigments were not identified in the samples except in the top bluish green layer of the armor of the *Bikara Taisho*. The morphology of the black particles observed was examined with PLM. Many exhibited an elongated form and a splintery outline indicating the use of soot or charcoal black, which was also utilized for carbon-based inks. Soot, made by burning pine wood in closed chambers, is the earliest recorded type of carbon. Carbon-based inks such as Chinese ink were introduced into Japan as early as the seventh century C.E. (Winter 1996).

Red Pigments

The pigments red lead (Pb_3O_4) and hematite (Fe_2O_3) were confirmed with EDX and XRD in the light reddish orange preparatory layers of the *Miroku-bosatsu* and *Jizō-bosatsu*, which were applied onto the white layer superimposed on the brown ground layers. Both colorants belong to a group of main red pigments (including natural cinnabar) used in the early historical periods. Red bole, a natural ferruginous aluminum silicate, was also identified in the ground layers.

Green Pigments

Green copper compounds were identified in the ground, the light green underpainting, and the top green layer of the *Miroku-bosatsu*'s lotus leaf pedestal (see figs. 2, 14) as well as the bluish green armor of the *Bikara Taisho* (see figs. 7, 11, and 12). The EDX analysis of a number of green particles from both sculptures revealed high peaks for chlorine and sulfur (figs. 17 and 18). XRD confirmed the use of atacamite (basic copper chloride, $Cu_2(OH)_3Cl$), brochantite (basic copper sulfate, $Cu_4SO_4(OH)_6$), and malachite (basic copper carbonate, $2CuCO_3 \cdot Cu(OH)_2$) (fig. 19). The fact that brochantite and atacamite were identified together with malachite on both sculptures could suggest that they are not original pigments but an alteration product of malachite. On the other hand, it should be kept in mind that green copper chloride pigments have also been identified on various Japanese works of art from different periods and that atacamite and brochantite occur together as minerals with malachite (Japanese, *rokusho*) in nature. A puzzling feature is the presence of zinc detected in a number of the brochantite and atacamite particles, traceable with the aid of EDX using the spot analysis facility (see figs. 17 and 18). Zinc was not detected in the malachite particles. Recent examinations of the twelfth-century *Scroll Paintings of the Tale of Genji* using XRF detected a small amount of zinc in the green copper pigment used in the light green areas (Hayakawa et al. 2002). Its presence may suggest the use of artificial pigments, but the presence of quartz in the green layers of both sculptures almost certainly indicates that natural green and blue copper minerals were utilized.

Brochantite has rarely been identified as a pigment in East Asian painting, although it was used together with malachite on a set of seventeenth-century Tibetan *thangka*s (Duffy and Elgar 1995). Using XRD, Naruse (1996) confirmed the use of atacamite on two bronze mirrors from the Shōsō-in, which are decorated with mother-of-pearl inlays on the back. A basic copper chloride, either paratacamite or atacamite, was also identified by Elisabeth FitzHugh in a Japanese ukiyo-e painting attributed to the artists of the Utagowa school who were active from the late sixteenth to the mid-nineteenth century (Scott 2002).

Some difficulties were encountered with a prominent and characteristic peak identified with XRD during the analysis of copper green and copper blue pigments of all three sculptures. The XRD pattern corresponds quite well to that of the greenish blue copper compound moolooite (copper oxalate hydrate, $Cu(C_2O_4) \cdot n(H_2O)$ [fig. 20]). The analysis with XRD also revealed weddellite (calcium oxalate dihydrate, $Ca(C_2O_4) \cdot 2(H_2O)$) in one sample of the *Miroku-bosatsu*.

The presence of copper oxalates in the green and blue pigments is unusual, but not entirely surprising, considering that they have been observed in Chinese wall paintings from the Yuan dynasty. Copper oxalates were identified together with calcium oxalates in Chinese wall paintings from the

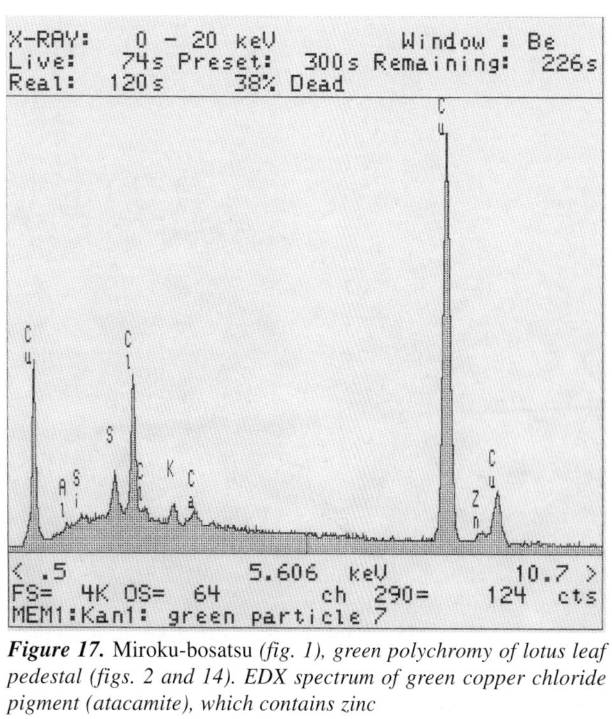

Figure 17. Miroku-bosatsu *(fig. 1), green polychromy of lotus leaf pedestal (figs. 2 and 14). EDX spectrum of green copper chloride pigment (atacamite), which contains zinc*

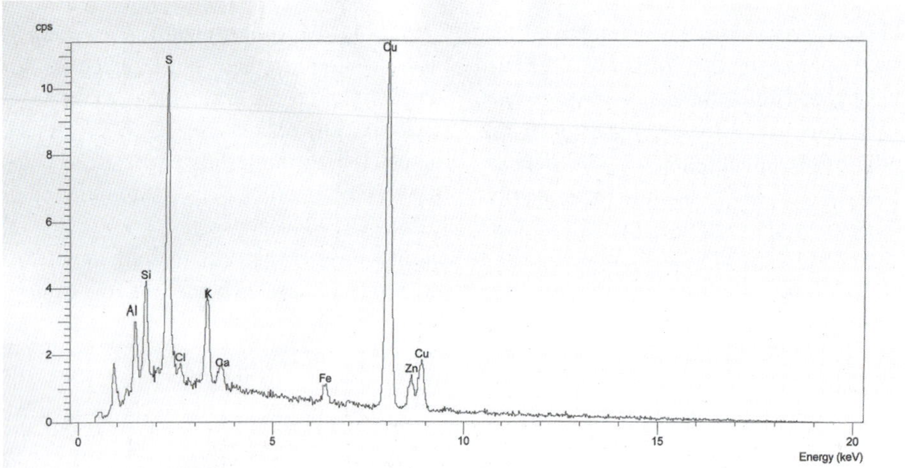

Figure 18. Miroku-bosatsu *(fig. 1), green polychromy of lotus leaf pedestal (figs. 2, 14, and 17). EDX spectrum of green copper sulfate pigment (brochantite), which contains zinc*

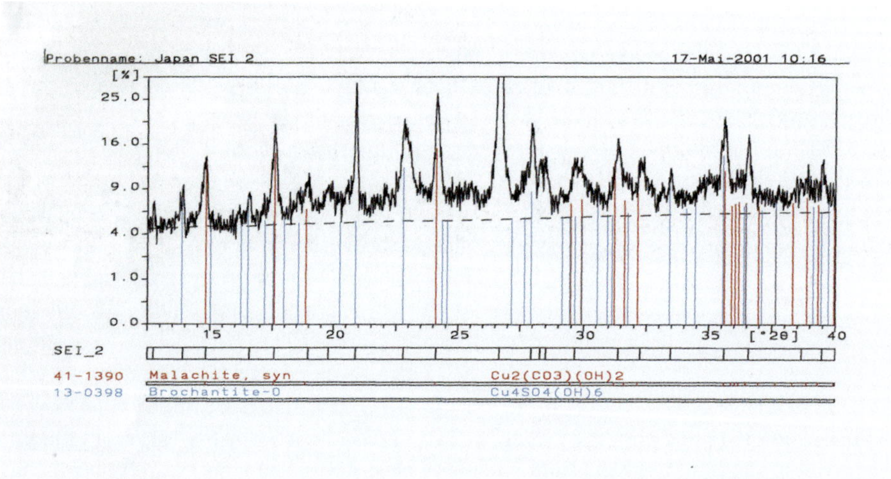

Figure 19. Bikara Taisho *(fig. 6), x-ray diffraction pattern of green layer on waist armor (fig. 12)*

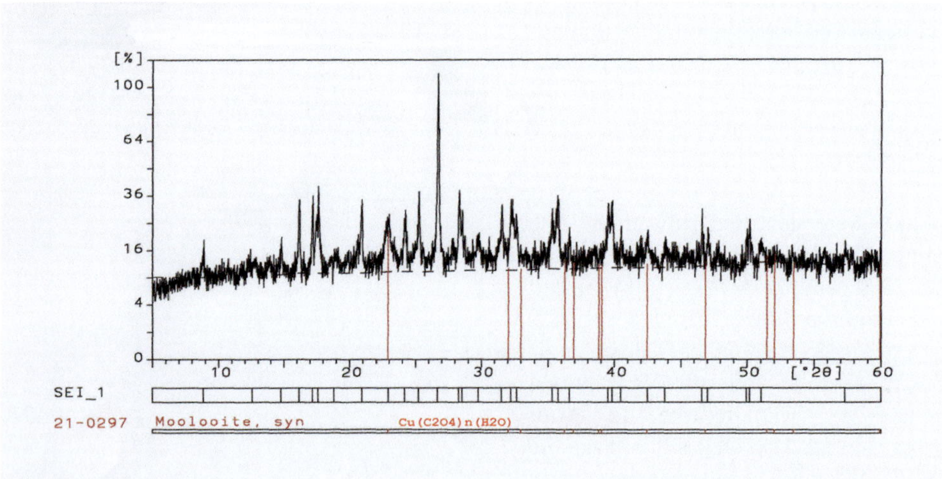

Figure 20. Bikara Taisho *(fig. 6), x-ray diffraction pattern of bluish green area of breast armor (fig. 11) with moolooite*

thirteenth century, and copper oxalate hydrate was identified as a major component of many of the blue and green pigment samples. The mechanism of incorporation into the paint layer may be the reaction product of a copper-containing pigment and oxalic acid. The oxalates could also be of microbiological origin (Moffat et al. 1985). Copper oxalates were also identified in wall painting fragments of the fifth century from the Mogao Temple grottoes at Dunhuang and from the Bingling Temple grottoes, near Lanzhou, Gansu province (Wainwright et al. 1993).

Blue Pigments

Azurite (basic copper carbonate, $2CuCO_3 \cdot Cu(OH)_2$) was identified in the bluish green layer of the breast armor of the *Bikara Taisho* together with brochantite, atacamite, malachite, and fluorite (see fig. 11). It was also used for the blue hair of the *Miroku-bosatsu*. Azurite (Japanese, *gunjō*) has been identified in a great number of Japanese polychromed sculptures dating from the seventh century onward and was the most important blue pigment of East Asian art.

The identification of fluorite is indeed a surprising result because as far as the author is aware it is the first time fluorite (Japanese, *hotaruishi*) has been identified in Japanese art. Why fluorite was used in this specific layer is not completely clear. A likely explanation for its presence is that it is an impurity, since it can be associated with the natural green or blue copper minerals used in this layer. Still, its use as a colorant in a pigment mixture in order to obtain a specific color or hue cannot be ruled out completely.

Other Pigments

Besides the identification of quartz, azurite, brochantite, atacamite, and malachite with EDX and XRD, the bluish green paint layer of the waist armor of the *Bikara Taisho* also revealed the use of what seems to be the purple variety of fluorite (calcium fluoride, CaF_2) (figs. 21–23; see also figs. 11 and 12). In samples of the bluish green underpainting, both colorless and pale purple particles of fluorite are visible; some of them exhibit colored banding (striations), which is quite typical of this variety of fluorite. Colorless particles of fluorite were identified in the dispersed sample used for polarized light microscopy (fig. 24). Some of the particles appear as broken splintery crystals, often with flat tops and conchoidal fractures. Fluorite is isotropic and has quite a low refractive index ranging between n_D= 1.433 and 1.435.

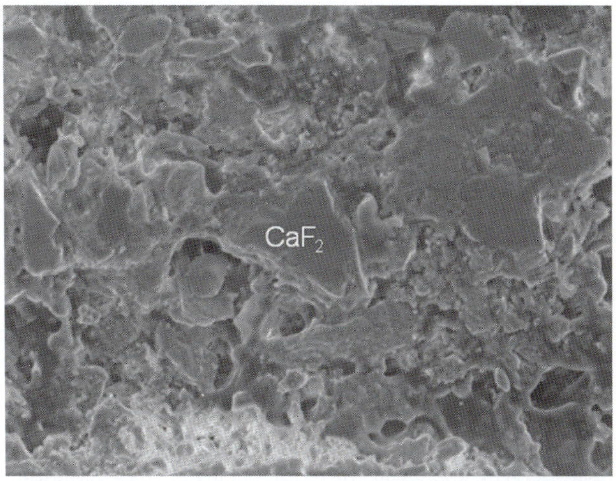

Figure 21. Bikara Taisho *(fig. 6), scanning electron micrograph of cross-section with particle of fluorite in the bluish green layer of the breast armor (figs. 11 and 20)*

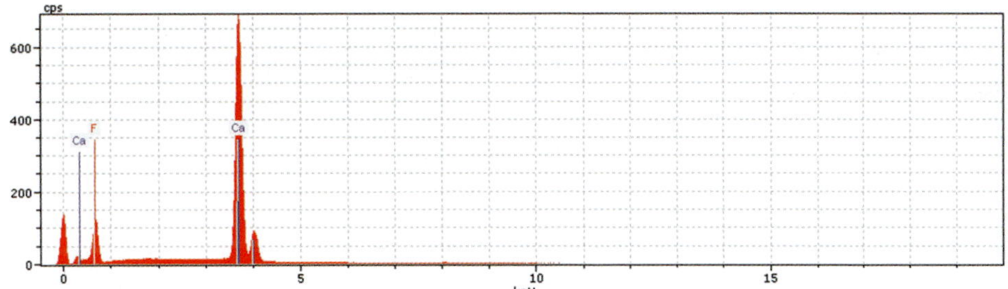

Figure 22. Bikara Taisho *(fig. 6), EDX spectrum of a fluorite particle in the bluish green area of the breast armor (figs. 11, 20, and 21)*

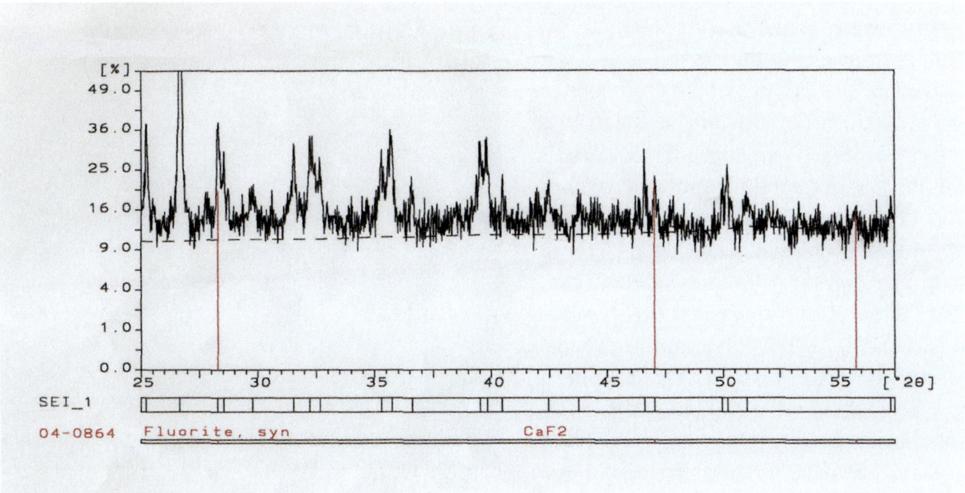

Figure 23. Bikara Taisho *(fig. 6), x-ray diffraction pattern of bluish green area of breast armor (figs. 11, 20–22) with fluorite*

Fluorite is one of the very few naturally occurring purple inorganic pigments that have been identified in art. A dark blackish purple type of fluorite called antozonite was probably used for this pigment. It has been discovered primarily in painting and polychrome works of art from Germany, Tyrol (Austria), Switzerland, Silesia (Poland), Hungary, the Netherlands, and Belgium from approximately 1450 to 1550 (Richter et al. 2001; Spring 2000). In the examined paint layers of these works of art the observed fluorite particles varied in color and appearance. Their color ranged from colorless to blackish purple, and they often showed a colored banding. Fluorite was used either by itself to create a mauve color or as an additive to pigment mixtures used for colored grounds, underpaintings, or final paint layers (e.g., shading) in order to obtain a specific color hue. In wall paintings, fluorite was only used with lime to create a purple color. This is not the case for the panel paintings and polychrome sculptures, where it has mostly been used in a more or less complex pigment mixture (often including lead white).

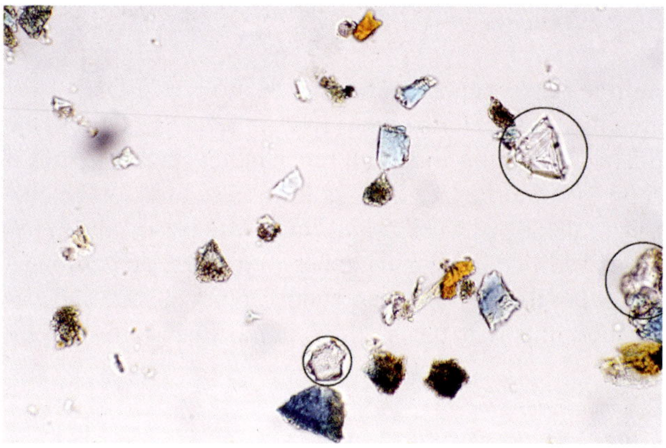

Figure 24. Bikara Taisho *(fig. 6), dispersed pigments of the bluish green layer of the breast armor (figs. 11, 20–23) showing particles of azurite, malachite, and fluorite in transmitted light. Photographed at a magnification of 630×*

Metals

Gold and silver have been used in several ways on Japanese polychrome sculpture and painting, including the techniques *shippaku*, *kindei-nuri*, and *kirikane*. These gilding and decoration techniques were combined in different variations on the three sculptures, especially the *Miroku-bosatsu*.

Shippaku, a technique in which gold foil is applied over the *urushi*-coated surface of a statue, was employed on the golden nimbus, the lotus leaf pedestal, and on some of the attributes of the *Miroku-bosatsu*. The technique of *kindei-nuri* utilizes gold powder mixed with a binding medium. The first reference in a historic source on *kindei-nuri* dates to 1184, where the author mentions a statue of *Amitabha-tathagata* whose body was decorated with *kindei-nuri* rather than with gold foils. Interestingly, the oldest existing example of *kindei-nuri* is the *Miroku-bosatsu zazo* (the seated statue of *Maitreya-bodhisattva*) at Daigoji Sanpo-in made by the Buddhist sculptor Kaikei in 1192 (Yamamoto 2004). The decorative *kirikane* technique was used to draw patterns with thin strips of gold leaf on polychromed surfaces. In this technique, gold foil or silver foil is cut into very fine strips in order to make delicate geometrical and floral motifs, which are applied to the surface (Winter 1985; Rösch 2001; Tsuda 2004). The technique of *kirikane*, which originated in China, was introduced to Japan at an early date and was perfected there in the first half of the eleventh century. It reached its zenith in the twelfth and thirteenth centuries.

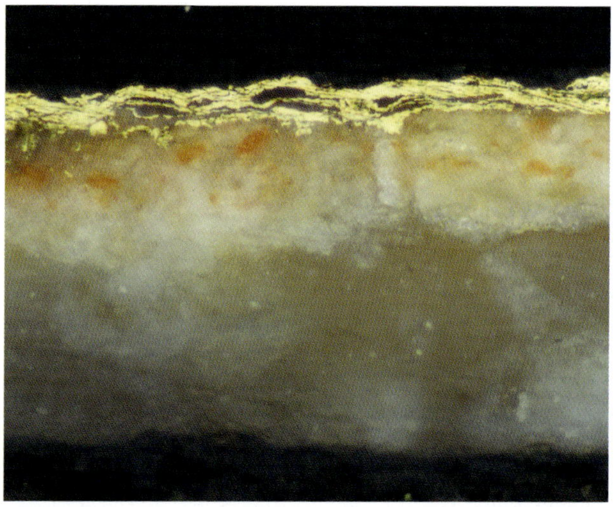

Figure 25. Miroku-bosatsu *(fig. 1), gilded garment, left side below hip. Paint cross-section showing the layer structure: 1. dark brown ground; 2. white layer; 3. light red layer with orange and red particles; 4. gold powder (*kindei-nuri*); 5. gold leaf (*kirikane*). Viewed in unpolarized normal light, photographed at a magnification of 500×*

The examination of the three sculptures led to some interesting observations of the combined use of the *kirikane* and *kindei-nuri* techniques on the various garments (*kun*, *fukken'e*, and *dai'e*) of the *Miroku-bosatsu* and *Jizō-bosatsu* (fig. 25; see also figs. 1 and 8). While polychromy is generally applied on a white priming, in the case of *kindei-nuri* a light red preparatory layer was first applied over the entire white surface to enhance the color of the gold powder and to make the delicate and exquisite *kirikane* designs on the surface of *kindei-nuri* appear more brilliant. It should also be noted that the delicate geometrical and floral designs

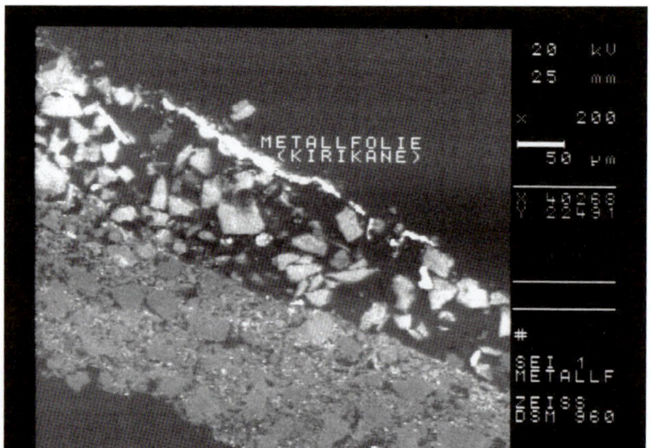

Figure 26. Bikara Taisho *(fig. 6), scanning electron micrograph of cross-section with multilayered laminated metal leaf on the bluish green layer of the breast armor (figs. 11, 20–24)*

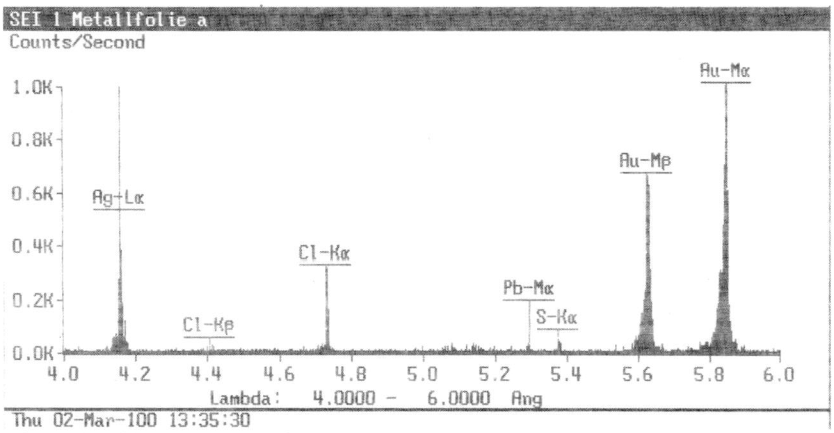

Figure 27. Bikara Taisho *(fig. 6), WDX spectrum of the top metal leaf consisting of gold. The silver content is from the silver leaf underneath.*

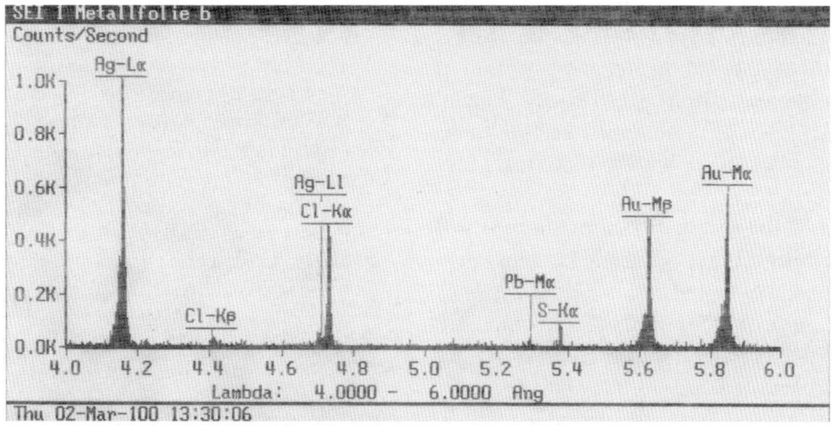

Figure 28. Bikara Taisho *(fig. 6), WDX spectrum of the middle metal leaf consisting of silver. The gold content is from the top gold leaf.*

exhibit a matte surface similar to unburnished gildings applied onto a glue size, adding subtle contrasts to the overall appearance of these areas. Designs using *kirikane* were also applied to the rough green and blue surfaces of the *Jizō-bosatsu* and the *Bikara Taisho*.

The EDX analysis of the metal foil used for *kirikane* and *kindei-nuri* on the *Miroku-bosatu* and *Jizō-bosatsu* resulted in the identification of gold. The multilayered gold leaf used on both sculptures has an approximate thickness of 0.3 µm and appears to be laminated. At least two laminae are present by all samples, with gold leaf with additional laminae not being ruled out.[12] A closer examination with SEM and UV fluorescence microscopy led to the discovery of an intermediate layer of binding media and particulate matter between the *kirikane* gilding and the layer of *kindei-nuri*. The EDX analysis of the underside of the gold leaf revealed magnesium, silicon, and calcium, which might suggest the use of talc ($Mg_3[(OH)_2/Si_4O_{10}]$). Talc was used on gilding tools (e.g., the leather cushion and bamboo knife used to cut the gold leaf) so that the metal leaf would not adhere.

An unusual composition was determined for the gold-colored metal leaf used for the *kirikane* pattern applied on the green breast armor (*muna-goh*) of the *Bikara Taisho*. EDX and WDX analyses revealed distinct peaks for both gold and silver. The detailed examination of the metal leaf with the spot analysis facility led to the identification of a multilayered laminate with alternating layers of gold and silver (figs. 26–28). In the SEM image the gold as the heavier metal appears as whitish stripes, while the silver foil sandwiched in between appears gray. Seckel (1954) describes exactly such a metal foil in his article on *kirikane*, stating that standard gold leaf is too thin to be cut for this technique and needs to be laminated first, resulting in a thicker metal foil consisting of at least three layers. The three leaves are laminated together over hot charcoal embedded in ash, resulting in a uniform and controlled source of heat. To save gold and to produce a thicker gold leaf, a silver leaf, instead of gold, may be used as an intermediate layer. Alloys or multilayered foils consisting of both metals have also been used for gilding in East Asian art. A metal leaf with such a composition was identified in fifth-century Chinese wall painting fragments (Wainwright et al. 1993).

Conclusions

The scientific examination of the samples from the three sculptures revealed a great variety of pigments and other artists' materials, including some that have rarely been identified previously, or are here identified for the first time, in Japanese polychromy and painting. Included among these identifications are the highly interesting discoveries of jarosite and fluorite, as well as the use of multilayered metal leaf on the *Bikara Taisho*. Also, the unusual zinc content in the green particles of atacamite and brochantite is puzzling, and its discovery provides more questions than answers. (As mentioned before, zinc has been previously detected in green copper pigments used on other Japanese paintings and polychrome sculptures.)

The presence of calcium and copper oxalate (weddellite and moolooite), especially in the green layers, suggests that the sculptures have in part been subject to a complex process of deterioration that may include microbiological processes. Further investigations are necessary here, since no evidence was found to allow a categorical statement or even a presumption of what agents were responsible for the occurrences of these two products in these particular layers.

It should be emphasized that in order to properly classify pigments, metals, and other compounds, it is extremely important to employ a wide range of analytical techniques. In this study the scanning electron microscope (SEM) proved to be vital for the examination and interpretation of pigment particle morphology, paint layer structure, metal leaf and metal powder gildings (*kindei-nuri* and *kirikane*) and for the identification of layers that were not recognizable with light microscopy.

Last but not least it should be noted that the complexity of the multilayered structures of the polychromy samples taken from the *Miroku-bosatsu*, *Jizō-bosatsu*, and *Bikara Taisho* is indeed comparable to those of paintings. These structures are an aspect of Japanese polychrome sculpture that has seldom been looked at in detail and demands further study.

Acknowledgments

I would particularly like to thank Vojislav Tucic, Christian Gruber (Central Laboratory, Bavarian State Department of Historical Monuments, Munich), and Klaus Rapp (mineralogist, Munich) for the pigment analysis and helpful suggestions. Special thanks to Tetsuei Tsuda (art historian, Tokyo National Research Institute for Cultural Properties) for his valuable help and providing me with photographic material. I am also grateful to Catharina Blänsdorf (Chinese-German collaboration on the conservation of the Terracotta Army of the First Chinese Emperor, Bavarian State Department of Historical Monuments) for her help and expertise during the examinations using polarized light microscopy. This study was part of the research conducted in the German-Japanese Research Project entitled Historical Polychromy, which was supported and financed by the Federal Ministry for Education and Research in Germany and by the Agency for Cultural Affairs (Bunkacho) as well as Grant-in-Aid for Scientific Research by the Ministry of Education, Culture, Sports, Science and Technology in Japan.

Photo credits

- Figures 1–7: Tetsuei Tsuda (National Research Institute for Cultural Properties, Tokyo)
- Figures 8–12, 14, 25–25: Mark Richter (Bavarian State Department of Historical Monuments, Munich)
- Figures 19–20, 23: Vojislav Tucic (Central Laboratory, Bavarian State Department of Historical Monuments, Munich)
- Figures 15, 18, 21–22: Christian Gruber (Central Laboratory, Bavarian State Department of Historical Monuments, Munich)
- Figures 13, 16–17, 26–28: Klaus Rapp (freelance mineralogist, Munich)

Notes

1. This unique chance to study Buddhist sculpture polychromy in Japanese temples would not have been possible without the unceasing efforts of Tetsuei Tsuda and Sadatoshi Miura, National Research Institute for Cultural Properties in Tokyo.
2. The word *miroku* comes from a Sanskrit word that means "one born of compassion." There are three fundamental Buddhist writings that provide explanations about Miroku. According to these writings, Miroku now preaches in the sacred world as a bodhisattva. When the teachings of the Buddha disappear completely from this world 5 billion 670 million years after the death of Shakyamuni, he will descend to the present world and become enlightened, in other words, become a Buddha. After that he will preach three times and save those who had not yet been saved by the teachings of the Buddha.
3. Jizō-bosatsu (Sanskrit, *Ksitigarbha*) appears as a gentle monk with a shaven head and is depicted as a *sravaka* disciple who has left his home. In his right hand Jizō holds a *shakuujō* stick with six rings, and by shaking this he awakens us from our deluded dreams. In his left hand he holds a jewel, which signifies that he bestows treasures and wealth on all beings. There are very few Buddhist deities as popular as Jizō in Japan. The six forms of Jizō are each responsible for the Six Paths of Transmigration: hells, hungry ghosts, beasts, demons, human beings, and heavenly beings. They all vow to save people from suffering and disaster.
4. All Twelve Divine Generals are probably originally from the Johruri-ji, Kyoto. Today they are found in various collections including the Tokyo National Museum, which owns four. The polychromy of these sculptures is very well preserved and of exceptional quality despite the numerous black areas caused by surface dirt from the atmosphere and incense sticks.
5. In Japan, around the Kamakura period (1185–1333), the Twelve Divine Generals of Yakushi Nyorai were sometimes confused or associated with the twelve animals of the twelve-year cycle based on the twelve divisions of heaven in ancient Chinese astronomy. As a result, in Japan, it is not uncommon to see depictions of the Twelve Generals with the astrological animals in their headpieces.
6. The Shōsō-in was constructed in the Nara period (710–94) as a repository belonging to the Todai-ji. About nine thousand objects from Japan, China, and Korea from the eighth to the eleventh century are stored there. They include paintings, sculptures, lacquerwares, glasswares, ceramics, textiles, furniture and interior ornaments, musical instruments, game sets, *Gigaku* masks, Buddhist ritual objects, weaponry, tableware, stationery, documents, and medicine. Materials used to make them have also been stored. They include inorganic substances, such as stones, pigments, ceramics, metals, and organic substances taken from animals and plants. The pigments identified by Yamasaki in the paintings and decorative artworks

of the Shōsō-in are kaolinite, lead white, cinnabar, red ocher, red lead, Sappanwood lake, yellow ocher, gamboge, malachite, azurite, indigo, unidentified violet colorant, Chinese ink, gold, and silver. Naruse later identified further pigments: lead chloride pigments (cotunnite, laurionite, blixite), calcite, apatite, atacamite, green earth, and the yellow orpiment.

7. The cross-sections were prepared with the embedding resin Technovit 2000 LC and were polished with Micro-mesh. A Zeiss Axiophot microscope with Epiplan-Neofluar HD and a Zeiss Axiolab polarized light microscope were used to examine cross-sections and pigment samples. The mounting medium for microscope slides was Aroclor Meltmount (n_D = 1.662).
8. The investigations were carried out with the scanning electron microscope DSM 960A from Zeiss in combination with the energy-dispersive spectrometer LINK QX 2000 (with Be-Window). Chemical elements of an atomic number greater than or equal to 5 (boron) can be detected. The WDX analyses were carried out with the wavelength-dispersive spectrometer from MICROSPEC WDX-3PC.
9. The x-ray diffraction analyses (XRD) were carried out with the Philips PW1800 powder diffractometer system. The samples were exposed for two hours with CuK[al] radiation at 40 kV and 50 mA.
10. *Asa-nuno* is a hemp cloth used for fixing. *Nunokise* is the process in which the cloth is fixed on the surface of a wooden substrate or foundation layers, for example, with *nori urushi* (*urushi* and rice paste used as an adhesive). This process is necessary to reinforce the foundation layers or wooden substrate and provide an even surface.
11. Lead sulfate was identified in a Japanese painting from 1053. Sadatoshi Miura, private communication.
12. It is not known when and where the lamination technique of metal leaf began and if the examples of the thirteenth century (Kamakura period) are comparable to those of modern times. Further studies are necessary.

References

Akai, J., K. Akai, M. Ito, et al. 1999. Biologically induced iron ore at Gunma iron mine, Japan. *American Mineralogist* 84:171–82.

Bauer, L. H., and H. Berman. 1933. Barium-muscovite from Franklin, New Jersey. *American Mineralogist* 18:30.

Colinart, S., and S. Pàgas-Camagna. 2001. Egyptian polychromy: Pigments of the "Pharaonic Palette." In *The polychromy of antique sculptures and the Terracotta Army of the First Chinese Emperor*, ed. W. Yongqi, Z. Tinghao, M. Petzet, E. Emmerling, and C. Blänsdorf. Arbeitshefte des Bayerischen Landesamtes für Denkmalpflege, 111:85–86. Munich: Bayerischen Landesamtes für Denkmalpflege. (in English and Chinese)

de Visser, M. W. 1935. *Ancient Buddhism in Japan*, vol. 2. Leiden.

Duffy, K. I., and J. A. Elgar. 1995. An investigation of palette and color notations used to create a set of Tibetan thangkas. In *Preprints: Historical painting techniques, materials, and studio practice*, ed. A. Wallert, E. Hermens, and M. Peek, 78–84. Symposium, University of Leiden, The Netherlands, 26–29 June 1995. Los Angeles: Getty Conservation Institute.

Emmerling, E., F. Juan, R. Karbacher, et al. 1996. On the formation, painted decoration, and stabilization of the figures in the cave temple of the Great Buddha of Dafosi. In *The Great Buddha of Dafosi*. Arbeitshefte des Bayerischen Landesamtes für Denkmalpflege, 82:248–79. Munich: Bayerischen Landesamtes für Denkmalpflege. (in German, Chinese, and English)

Gettens, R. J., H. Kühn, and W. T. Chase. 1994. Lead white. In *Artists' pigments: A handbook of their history and characteristics*, vol. 2, ed. A. Roy, 67–81. Washington, D.C.: National Gallery of Art.

Guoxin, Z., Z. Jianquan, and C. Huaiwan. 1993. Pigment analysis of polychrome statuary and wall paintings of the Tiantishan grottoes. In *Conservation of ancient sites on the Silk Road*, 363–67. Conference, Dunhuang, 3–8 October. Los Angeles: Getty Conservation Institute.

Hayakawa, Y., S. Miura, H. Yotsutsuji, et al. 2002. Analysis of the painting materials used in the *Scroll Paintings of the Tale of Genji*, National Treasure. *Science for Conservation* 41:1–14.

Miura, Sadatoshi. 2004. Polychromy in Japan: A chronological compilation of painting materials. In *Historical polychromy*, ed. M. Kühlenthal and S. Miura, 243–50. Munich: Hirmer Verlag.

Moffat, E. A., N. T. Adair, and G. S. Young. 1985. The occurrence of oxalates on three Chinese wall paintings. In *Application of science in examination of works of art*, ed. P. A. England and L. van Zelst, 234–38. Seminar, Boston, 7–9 September 1983. Boston: Research Laboratory, Museum of Fine Arts.

Mogi, A., T. Nakasato, and Y. Emoto. 1975. Conservation treatments with reference to color paintings on the ceiling of the Main Hall, National Treasure, of Tōshōdaiji Temple. *Science for Conservation* 14:55–69.

Naruse, Masakazu 1996. Investigation of inorganic pigments used on the Shōsōin object by x-ray analyses. In *18th International Symposium on the Conservation and Restoration of Cultural Property on Spectrometric Examination in Conservation*, 143–51. Tokyo, 31 October–2 November 1994. Tokyo: Tokyo National Research Institute for Cultural Properties.

Noll, W. 1991. *Alte Keramiken und ihre Pigmente: Studien zu Material und Technologie*. Stuttgart: E. Schweizerbart'sche Verlagsbuchhandlung.

Richter, M. 2004. Two thirteenth-century polychrome sculptures in the Shomyo-ji and Ensho-ji: Examination results of pigments and layer structures. In *Historical polychromy*, ed. M. Kühlenthal and S. Miura, 276–91. Munich: Hirmer Verlag.

Richter, M., O. Hahn, and R. Fuchs. 2001. Purple fluorite: A little known artists' pigment and its use in late gothic and early renaissance painting in northern Europe. *Studies in Conservation* 46:1–13.

Rösch, P. 2001. Color schemes on wooden Guanyin sculptures of the 11th to 13th centuries with special reference to the Amsterdam Guanyin and its cut gold-foil applications on a polychrome ground. In *The polychromy of antique sculptures and the Terracotta Army of the First Chinese Emperor*, ed. W. Yongqi, Z. Tinghao, M. Petzet, E. Emmerling, and C. Blänsdorf. Arbeitshefte des Bayerischen Landesamtes für Denkmalpflege, 111:75–84. Munich: Bayerischen Landesamtes für Denkmalpflege.

Sano, C. 2004. The examination of ground materials used on *Niozo* in the Nandai-mon of the Todai-ji. In *Historical polychromy*, ed. M. Kühlenthal and S. Miura, 322–25. Munich: Hirmer Verlag.

Scott, D. A. 2002. *Copper and bronze in art: Corrosion, colorants, conservation*. Los Angeles: Getty Conservation Institute.

Seckel, D. 1954. Kirikane: Die Schnittgold-Dekoration in der japanischen Kunst, ihre Technik und ihre Geschichte. *Oriens Extremus* 1:71–88.

Spring, M. 2000. Occurrences of the purple pigment fluorite on paintings in the National Gallery. *National Gallery Technical Bulletin* 21:20–27.

Tsuda, Tetsuei. 1999. Honzon Miroku Bosatsu ryuzo: Tokubetsu kokai, shomyji nohiho [On the statue of *Miroku-bosatsu* at Shomyo-ji]. In *The Statue of Miroku-bosatsu at Shomyo-ji*. Kanagawa Prefectural Kanazawa Bunko Museum. Yokohama: Kanagawa Kenritsu Kanazawa Bunko. (in Japanese)

Tsuda, Tetsuei. 2004. The statue of *Miroku-bosatsu* at Shomyo-ji. In *Historical polychromy*, ed. M. Kühlenthal and S. Miura, 268–76. Munich: Hirmer Verlag.

Wainwright, I., E. Moffat, J. Sirois, et al. 1993. Analysis of wall painting fragments from the Mogao and the Bingling Temple grottoes. In *Conservation of ancient sites on the Silk Road*, 334–40. Dunhuang, 3–8 October. Los Angeles: Getty Conservation Institute.

Wallert, A. 1995. Unusual pigments on a Greek marble basin. *Studies in Conservation* 40:177–88.

Winter, J. 1981. "Lead white" in Japanese paintings. *Studies in Conservation* 26:89–101.

Winter, J. 1985. Gold in Japanese paintings: A case history involving the *kirikane* technique. In *Application of science in examination of works of art*, ed. P. A. England and L. van Zelst, 46–54. Seminar, Boston, 7–9 September 1983. Boston: Research Laboratory, Museum of Fine Arts.

Winter, J. 1996. Ink: I, Types and properties; II. Uses; III, Conservation. In *The dictionary of art*, ed. J. Turner, 849–50. New York: Grove.

Yamamoto, T. 2004. The significance of *Kindei-nuri*: A special gilding technique on Buddhist sculptures. In *Historical polychromy*, ed. M. Kühlenthal and S. Miura, 326–35. Munich: Hirmer Verlag.

Yamasaki, Kazuo, and Kyotaro Nishikawa. 1970. Polychromed sculptures in Japan. *Studies in Conservation* 15:278–93.

Yamasaki, Kazuo, and Yoshimichi Emoto. 1979. Pigments used on Japanese paintings from the protohistoric period through the 17th century. *Ars Orientalis* 11:1–14.

The Identification and Light Sensitivity of Japanese Woodblock Print Colorants: The Impact on Art History and Preservation

Sandra A. Connors, Paul M. Whitmore, Roger S. Keyes, and Elizabeth I. Coombs

ABSTRACT Seventy-two eighteenth- and nineteenth-century Japanese woodblock prints from the collections of the Carnegie Museum of Art and the Library of Congress were surveyed to determine their sensitivity to visible light using a micro-fading tester developed at the Research Center on the Materials of the Artist and Conservator. When possible, the reflectance data collected by the micro-fading tester were also used to identify the natural organic colorants used on these objects. Japanese woodblock prints are considered to be light-sensitive objects, but surprisingly wide ranges of fading behavior were found. Yellow areas were created with the most light-sensitive colorants, while blue areas proved to be relatively stable to visible light. Further examination of the data showed that medium values of a colorant will fade faster than their light and dark counterparts and that the prior fading history of a colorant has little effect on its light sensitivity. The reasons for specific colorant/color use by Japanese woodblock printers are also explored.

Introduction

The light sensitivity of many Japanese woodblock print colorants is widely recognized, mainly from witnessing the damage that prolonged exhibition can produce. Despite this understanding, questions still remain about the care of these objects. The light sensitivity of individual prints is not usually known, especially for those in very good states of preservation. It is also difficult to predict how much damage these prints will suffer when displayed in low-intensity, UV-free lighting environments. Many of the sensitive colorants on these objects are natural organic materials, and it is unknown whether variations in the growth, manufacture, or aging history of those materials could affect their current light sensitivity. How much of the apparent fading damage on a print is the result of light damage, or how much may be caused by other aging factors or by intentional printing of dilute colors, is also poorly understood.

In the past few years, similar questions have been addressed for specific artifacts by studying them with a micro-fading tester developed at the Research Center on the Materials of the Artist and Conservator. This device allows the direct determination of light sensitivity and provides information that may sometimes be used to identify the colorants present. While the main use of this instrument has been to determine the preservation needs of particular artifacts, this project is its first use to survey a group of related objects—Japanese woodblock prints from the eighteenth and nineteenth century containing natural organic colorants—in order to explore the trends and variations of light sensitivity for specific colors and colorants. The relative sensitivity of colorants applied at different depths of shade and whether a colorant's prior fading may affect its current light sensitivity have also been explored. Further, the implications of these findings are considered from an art-historical perspective by discussing the specific color and colorant choices made throughout the development of this art form.

Scope of Survey and Analytical Methods Used

Seventy-two Japanese woodblock prints were chosen for this survey, fifty-six from the James B. Austin Collection at the Carnegie Museum of Art and sixteen from the Japanese Print Collection at the Library of Congress. The prints selected range in date from the early 1720s to 1863 and provide examples of both well-preserved and poorly preserved color. Each color on a print was tested for sensitivity to visible light by performing *in situ* micro-fading tests. The instrument used has been described in detail previously (Whitmore et al. 1999). Tests were performed with high-intensity visible light (400–700 nm) focused to a spot 0.4 mm in diameter. Reflectance spectra of each test spot were collected during a five minute interval, and color differences (CIE ΔE values) were calculated from them using the initial spectrum as the basis for the comparison. Large values of ΔE indicate a greater measured color change in the test area. The ΔE value from each test area was compared to the ΔE values produced in Blue Wool standards exposed under the same conditions. The Blue Wool standards are designed to fade at specific rates, with Blue Wool standard 1 more sensitive than Blue Wool 2 and Blue Wool 2 more sensitive than Blue Wool 3. For colored materials used to make works of art, fading rates comparable to Blue Wool 1–3 are considered to be "light sensitive." All fading tests on the prints were performed so that the resulting color changes never exceeded five ΔE units. Such a small change in the color of the test areas is not visibly perceptible.

The reflectance spectra recorded at the start of each fading test were also used to identify those natural organic colorants having characteristic spectral profiles. These included dayflower blue (*tsuyukusa* or *aigami*), indigo (*ai*), and safflower red. The method for identification of these colorants using reflectance data is described elsewhere (Feller et al. 1984; Leona and Winter 2001; Schweppe and Roosen-Runge 1986; Schweppe and Winter 1997). Prussian blue, a colorant used in the nineteenth century, does not have a distinctive visible reflectance spectrum, but it was identified by the absence of spectral features indicating either dayflower blue or indigo (Leona and Winter 2001). Colorants were identified on several prints. A list of identifications appears as an appendix.

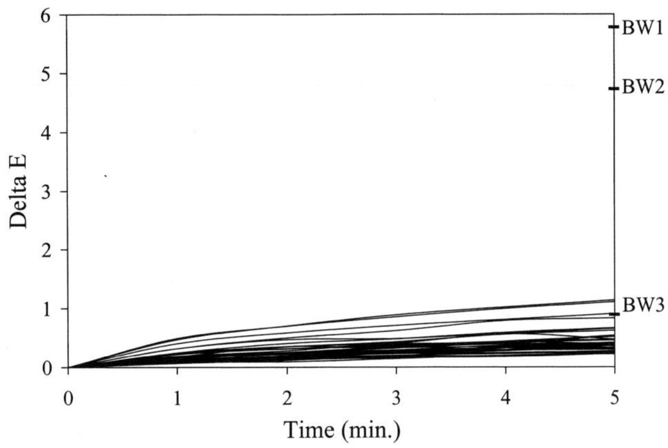

Figure 1. *Fading test results for all dayflower blue areas tested, thirty-two areas total*

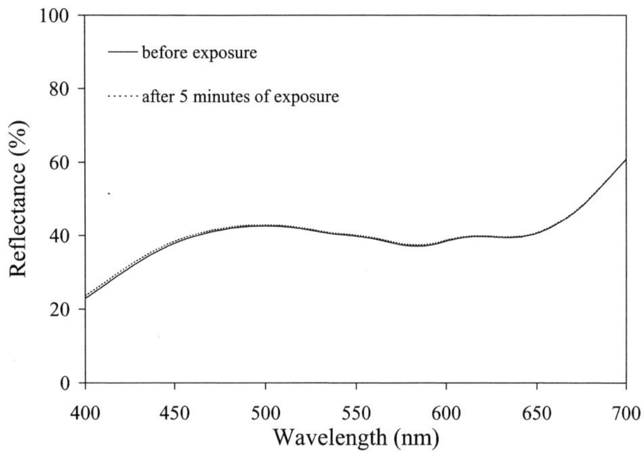

Figure 2. *Reflectance spectra for dayflower blue area on the print* Ochanomizu *by Shôtei Hokuju, ca. 1810s*

Fading Test Results: Variations in Light Sensitivity among All Prints

Blue Colorants

Of the 110 blue areas tested, all but nine (which could not be identified by their reflectance spectra) were found to be one of the three blue colorants mentioned above (dayflower blue, indigo, and Prussian blue). Dayflower blue was found in one-third of the areas tested, and figure 1 shows the fading changes produced in all areas of dayflower blue, with the final ΔE values for the Blue Wool standards 1–3 indicated on the right axis for reference. Dayflower blue showed fairly uniform light sensitivity, with fading rates similar to that of Blue Wool 3, making it one of the most light-stable organic colorants tested on these prints. Interestingly, the very slight color change produced during these fading tests resulted more from a loss of yellow (reflectance increase between 400 nm and 500 nm) than from a loss of blue color. Figure 2 shows the reflectance spectra for dayflower blue on a print entitled *Ochanomizu* by Shôtei Hokuju in the collection of the Carnegie Museum of Art, before and after the fading test. Longer fading tests on laboratory samples showed that dayflower blue begins fading by this bleaching of yellow color, followed by a gradual reflectance increase at all wavelengths, including those at the absorbance peaks associated with the blue color (unpublished results).

Indigo was found in fewer than one-quarter of the blue areas tested and produced similar results to those found with dayflower blue. Figure 3 shows the fading results for all areas of indigo tested. The final ΔE values are grouped around the value produced from the fading of Blue Wool 3, making this colorant another relatively light-stable material. Similar to dayflower blue, indigo experienced a reflectance increase in the 400–500 nm region during the fading test,

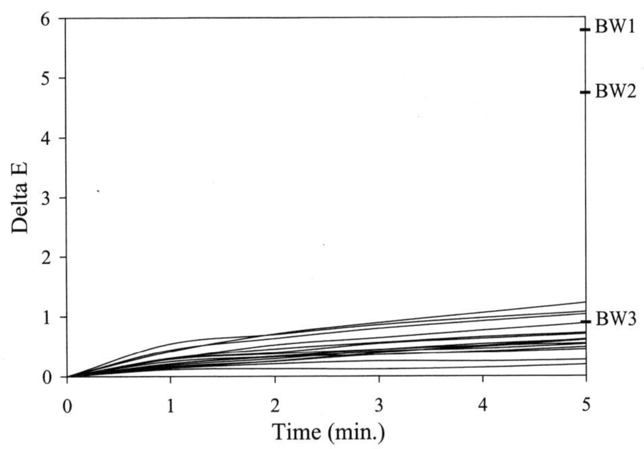

Figure 3. *Fading test results for all indigo areas tested, sixteen areas total*

rather than in the 660–700 nm region of the main indigo absorption band.

The fading tests of Prussian blue areas produced ΔE values from 0.5 to 5.4 after five minutes of exposure, which is equivalent to a range in lightfastness from Blue Wool 3 to Blue Wool 1. This result is seemingly at odds with the known light stability of this colorant (ASTM standard D5067-98). However, an earlier study of Prussian blue on cyanotypes (Whitmore et al. 2000) showed that these fading tests cause Prussian blue (ferric ferrocyanide) to undergo the well-known reversible conversion to a white compound (ferrous ferrocyanide), which converts back to the blue once the light exposure has ended. If these applications of Prussian blue behave in this typical way, then the micro-fading tests have overestimated the light sensitivity of those areas. The confirmation that these fading changes are reversible would further support the identification of these passages as Prussian blue, the only available pigment known to display such behavior.

Red Colorants

Most of the 135 red areas tested faded very slightly, showing an increase in reflectance between 400 nm and 600 nm, the main absorbance region for most red colorants. Figure 4 shows the fading test results for all red areas studied. Most of the red areas faded at rates between those of Blue Wool 2 and Blue Wool 3 and should therefore be considered moderately sensitive to visible light.

Safflower red was identified on a number of prints, so the light sensitivity of these passages can be examined separately. The fading results of all the safflower red passages lie in a range only slightly narrower than that of all the red areas combined (fig. 5), with fading rates between Blue Wool 3 and halfway between Blue Wool 2 and Blue Wool 3.

Purple Colorants

Figure 6 shows the fading results for all purple areas tested. A relatively narrow range of fading behavior centered around Blue Wool 3 was found, with the exception of two purple areas showing faster fading rates (around Blue Wool 2).

The literature describes overprinting or mixtures of blue and red colorants being used to create purple areas of color (Keyes 1988; Feller et al. 1984). Just under half of the sixty-six purple areas studied showed spectral features typical of dayflower blue and safflower red, while only six areas showed the spectral characteristics of indigo and safflower red. The composition of the remaining purple passages could not be determined from their reflectance spectra. In the mixtures of safflower red and dayflower blue, the more fugitive colorant seems to be safflower red. This finding is consistent with the observed greater light sensitivity of safflower red compared to dayflower blue measured in the passages of unmixed color. Figure 7 shows a difference spectrum describing the spectral changes produced during

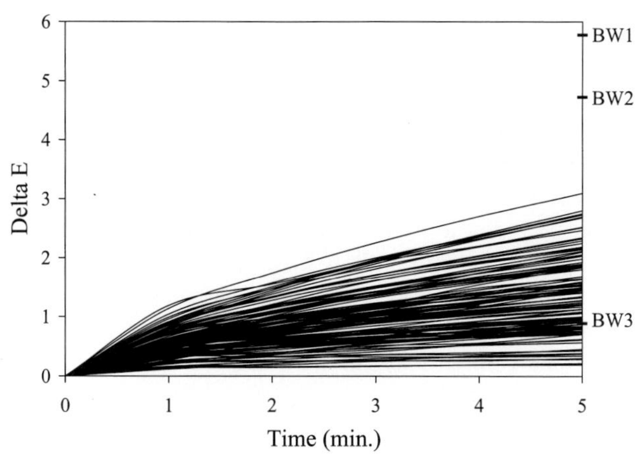

Figure 4. *Fading test results for all red areas tested, 135 areas total*

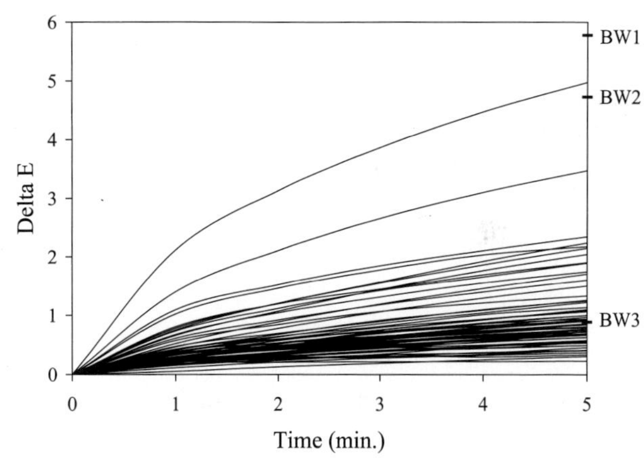

Figure 6. *Fading test results for all purple areas tested, sixty-six areas total*

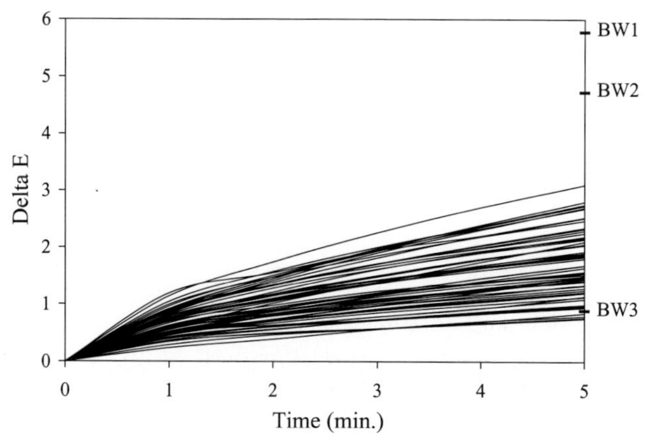

Figure 5. *Fading test results for all safflower red areas tested, sixty-one areas total*

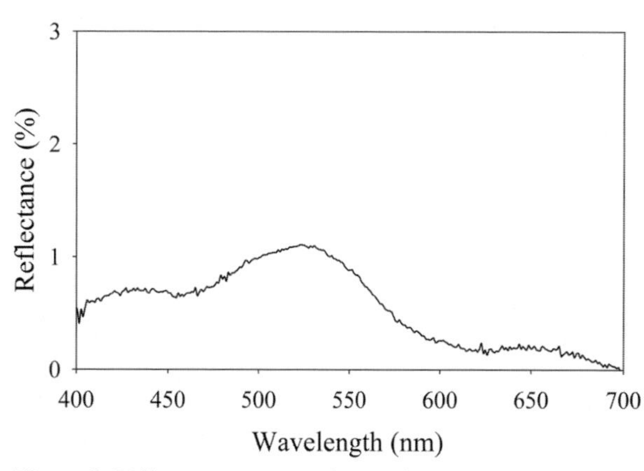

Figure 7. *Difference spectrum of a purple area on* Actor as Kintoki Hambei *by Utagawa Kunisada, ca. 1850*

the fading of the purple area on *Actor as Kintoki Hambei* by Utagawa Kunisada, in the collection of the Carnegie Museum of Art. The greatest increase in reflectance occurred around 540 nm, the absorbance peak for safflower red, indicating that the primary effect of the light exposure caused was the destruction of the safflower red in the purple mixture.

The two purple areas showing the remarkably rapid fading were also mixtures of safflower red and dayflower blue on two prints from an untitled series of eight views of Edo by Shôrinsai Hidemaro, in the collection of the Carnegie Museum of Art. Like the typical safflower-dayflower purple mixtures described in figure 7, these also faded mainly by loss of the safflower red from the mixture. On these two prints, however, the safflower red faded more rapidly from the purple mixture than was observed in the fading tests of the unmixed safflower red areas on these prints. The reason for this finding is not known, but it suggests the safflower red used for the purple mixture might have been different from that used for the applications of unmixed colorant in the red passages. It should also be noted that there were only a few eighteenth-century prints in this study that included well-preserved purple areas. Early methods of preparing safflower red or of printing purple may have resulted in these unusually light-sensitive mixtures.

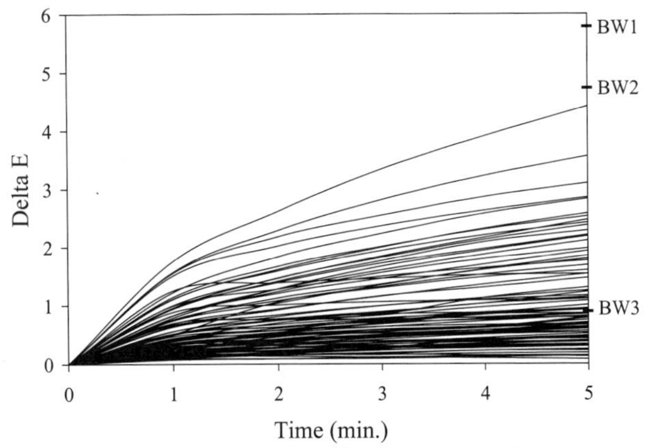

Figure 8. *Fading test results for all green areas tested, seventy-nine areas total*

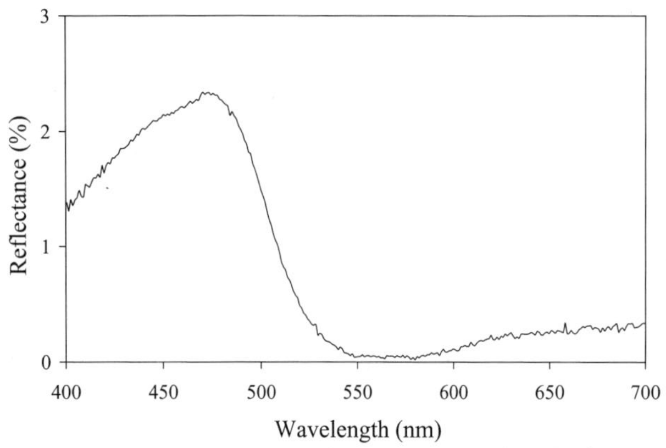

Figure 9. *Difference spectrum of a green area on the print* Actor as Oguri Hangan at Fujisawa, *from* Fifty-Three Stations of the Tôkaidô Road *by Utagawa Kunisada, 1852*

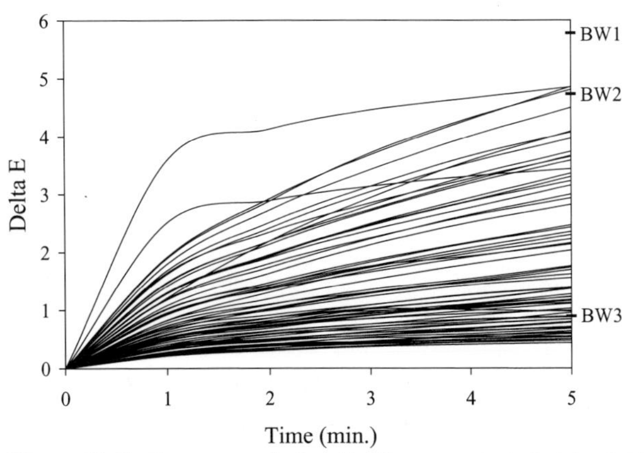

Figure 10. *Fading test results for all yellow areas tested, sixty-eight areas total*

Green Colorants

Various green passages were studied, ranging in hue from yellow-green to blue-green. Figure 8 shows the fading results from all seventy-nine green areas tested. The range of fading behavior is wider than that seen for either red or blue colorants, with ΔE values between 0.1 and 4.4 after five minutes of exposure. The majority of green areas showed fading rates near Blue Wool 3, but some areas faded at rates near Blue Wool 2, indicating their greater light sensitivity. The color change that occurred during the fading of most of these green passages was a loss of yellow. Figure 9 shows the difference spectrum after the fading of a green area on *Actor as Oguri Hangan at Fujisawa* by Utagawa Kunisada, in the collection of the Carnegie Museum of Art. The greatest change was produced between 400 nm and 500 nm, indicating fading of a yellow component in a mixture. While dayflower blue and indigo also showed a reflectance increase between 400 nm and 500 nm during the fading tests, the increase experienced by a mixture of yellow with either of these blue colorants was greater than that experienced by the blue colorant alone.

Green passages were typically created using yellow-blue colorant mixtures, by overprinting yellow and blue colorants, or by using the green mineral pigment malachite (Keyes 1988; Feller et al. 1984). Half of the green areas studied showed reflectance spectra indicating the presence of indigo, while only one print contained green passages created from a mixture of dayflower blue and a yellow colorant. The composition of the remaining green passages could not be identified because of the lack of distinctive features in their reflectance spectra.

Yellow Colorants

The yellow areas studied showed the widest range of fading behavior of any color category. Figure 10 shows the fading results observed in all yellow areas studied. The light sensitivity of these colorants ranged from relatively light-stable areas, with fading rates less than that of Blue Wool 3, to very fugitive areas, with fading rates greater than that of

Blue Wool 2. The yellows were frequently the most light-sensitive colors on a given print.

A variety of organic and inorganic yellow colorants was available for use on Japanese woodblock prints (Keyes 1988; Feller et al. 1984), and distinction among these colorants is difficult because of the lack of characteristic features in their visible reflectance spectra. However, several of the yellow areas tested showed unusual fading behavior that could be distinctive. For example, some areas darkened (experienced a decrease in reflectance) as a result of light exposure. Exploring whether such a light reaction can be used to identify yellow colorants is an area for future research.

Paper Substrates

Along with areas of color, reference measurements were made on unpigmented areas of each print. It was found that the paper substrate of these prints may experience some slight color change as a result of these light exposure tests. When color change occurred, there was an increase in percent reflectance between 400 nm and 450 nm—bleaching of yellow discoloration in the paper—at a rate similar to that of Blue Wool 3. Typically, it is easy to distinguish between the fading of most colors and the spectral changes caused by bleaching; however, paper bleaching may interfere with the interpretation of data in areas of very light color, or colorants on strongly discolored paper, or colorants that experience similar spectral changes to those experienced by the paper itself. The majority of prints in this study did not have strongly discolored paper substrates, and an effort was made to avoid areas of very light color where bleaching of the paper may be confused with colorant fading. However, with colorants that experience a loss of yellow color as they fade (dayflower blue, indigo, green or yellow colorants), it is possible that a slight contribution to the overall color changes measured came from the bleaching of the paper substrate.

Light Sensitivity for Different Concentrations of the Same Colorant

To produce a variety of shades for a particular color, colorants were often used in different concentrations, were overprinted, or were applied as gradations of color during the printing. *The Great Bridge at Senju* by Utagawa Hiroshige, in the collection of the Library of Congress, exhibits a gradation from medium to light shades of safflower red along the horizon, and a dark shade of safflower red was used for the cartouche in the upper right corner of the print. When tested, each of these areas showed different light sensitivities. Figure 11 shows the fading results for light, medium, and dark applications of safflower red on *The Great Bridge at Senju*. Greater fading was observed for the medium-value safflower red area compared to either the very light or very dark areas. This finding is consistent with the fading behavior observed in light, medium, and dark paint glazes, where the greater sensitivity of the medium-value paints is well known (Whitmore and Bailie 1997).

Fading Behavior of Well-Preserved and Poorly Preserved Prints

These results can also be examined to determine whether prior fading of a print has any effect on the current light sensitivity of its colorants. Equivalent shades (reflectance minimum around 40 percent) of safflower red were compared from *Courtesans Enjoying Cherry Blossoms at the Yoshiwara* by Torii Kiyomine (a print with well-preserved color) and *The Actors Ôtani Hiroji and Sanogawa Ichimatsu in a Soga Play* by Okumura Masanobu (a print with poorly preserved color), both in the collection of the Carnegie Museum of Art. The fading results for the areas tested are seen in figure 12, where it can be observed that the well-preserved and previously faded areas of safflower red have essentially the same light sensitivity. This finding suggests that safflower red is probably not composed of a mixture of components of varying lightfastness. It may also explain why the various safflower red passages on all the prints had

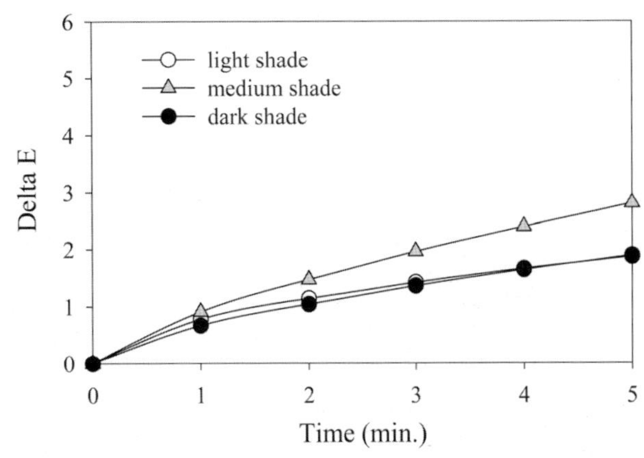

Figure 11. *Fading test results for light, medium, and dark applications of safflower red on the print* The Great Bridge at Senju, *from* One Hundred Famous Views of Edo *by Utagawa Hiroshige, 1856*

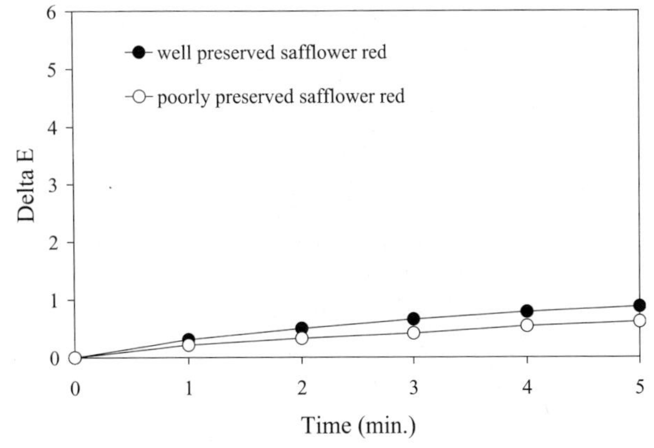

Figure 12. *Fading test results for well-preserved and poorly preserved areas of safflower red*

reasonably similar light sensitivity. While this behavior has not been tested with other colorants because of a lack of suitable prints for comparison, one may reasonably expect that the other colorants found in this survey are similarly homogeneous because they also tended to have relatively constant light sensitivity among the prints in this group.

Art-Historical Implications

Let us now consider these data from a historical perspective. The earliest print in this study is a hand-colored picture of kabuki actors by Okumura Toshinobu that was published in the early 1720s (fig. 13). Print artists often combined colors in characteristic ways to produce particular effects. The colorist here emphasized the three primary colors—red, yellow, and blue—to produce a lively effect by contrast of hue. When Japanese craftsmen developed techniques of color printing in the middle of the eighteenth century, many favored the same palette of three primary colors that dominate the earlier prints, as in the Torii Kiyomitsu actor portrait from the early 1760s seen in figure 14. One important result of this study is the discovery that *beni*, or safflower, produces most of the common reds in eighteenth- and early nineteenth-century prints and that dayflower, or *tsuyukusa*, produces nearly all the blue and gray-blue colors in these prints. This limitation is surprising, considering that other red and blue colorants were available and occasionally used on prints. All three primary colors are moderately light sensitive, but we should not assume that light hues of organic colorants on Japanese prints are always the result of fading, because printers often diluted the colorants. Diluted colorants were typical of inexpensive color prints of the period and may have been a cost-cutting measure.

Later artists continued to produce prints using this same color contrast. Figure 15 shows the unfaded courtesan portrait by Torii Kiyomine, which was published in 1807. The printer has added some green, but the main colors are still red, blue, and yellow, a combination now associated with inexpensive color prints. Patronage allowed artists to experiment with color and encouraged printers to introduce new colorants. Suzuki Harunobu's *Night Rain on the Tea Stand* (fig. 16), published in 1766, is clearly much more colorful and complex than Torii Kiyomitsu's simple three-color print.

Commercial two-color printing developed in the city of Edo in the mid-1740s, and for more than a decade the two colors of choice were green and red, as in the actor portrait by Okumura Masanobu in figure 17. This palette achieved its effect by the contrast of the cool green with the warm red. Another of the surprising results of this study is the discovery that virtually all the greens in eighteenth-century prints are mixtures of indigo blue with yellow. Many hues of green appear on the prints, as in the picture by Katsukawa Shunchô from ca. 1790 (fig. 18). It seems likely that this variation results from mixing the indigo blue with different yellows that could not be identified in this study. It is all the more surprising that indigo was so rarely used as a print

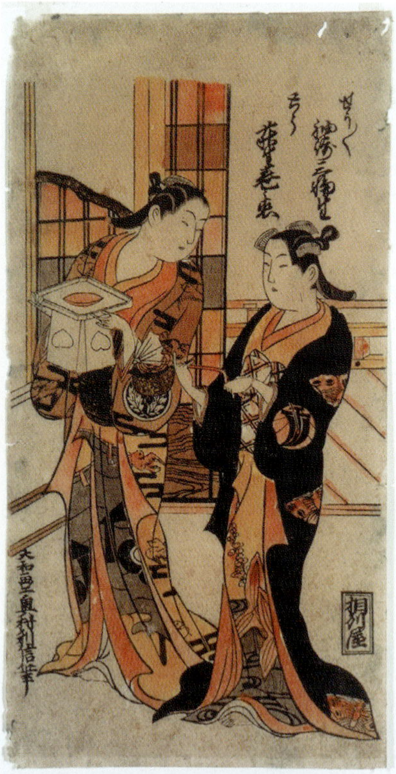

Figure 13. *Okumura Toshinobu,* The Actors Sodezaki Miwano and Ogino Makinojô as Tora and Shôshô in a Soga Play, *early 1720s, hand-colored* hosoban. *Carnegie Museum of Art, Pittsburgh, Bequest of Dr. James B. Austin, 89.28.1363*

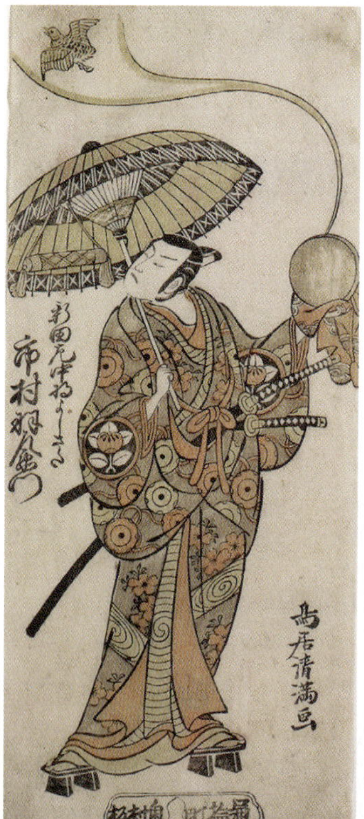

Figure 14. *Torii Kiyomitsu,* The Actor Ichimura Uzaemon IX as Nitta Yoshisada, *early 1760s, three-color* hosoban. *Carnegie Museum of Art, Pittsburgh, Bequest of Dr. James B. Austin, 89.28.816*

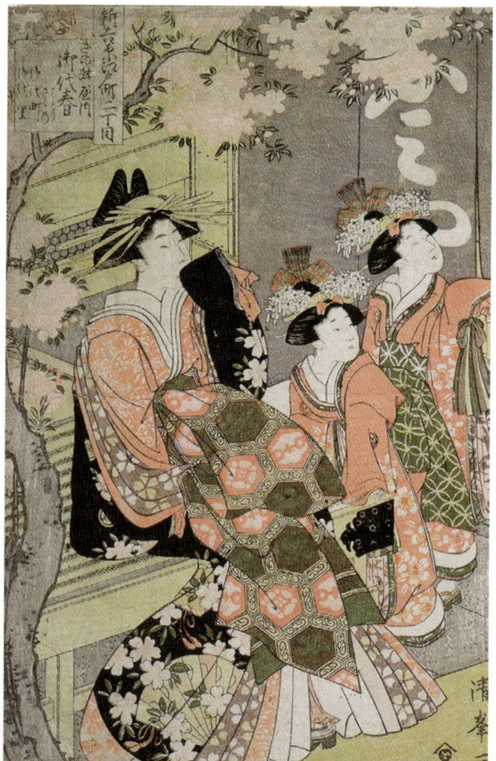

Figure 15. *Torii Kiyomine,* Courtesans Enjoying Cherry Blossoms at the Yoshiwara, *1807,* ôban *triptych. Carnegie Museum of Art, Pittsburgh, Bequest of Dr. James B. Austin, 89.28.812*

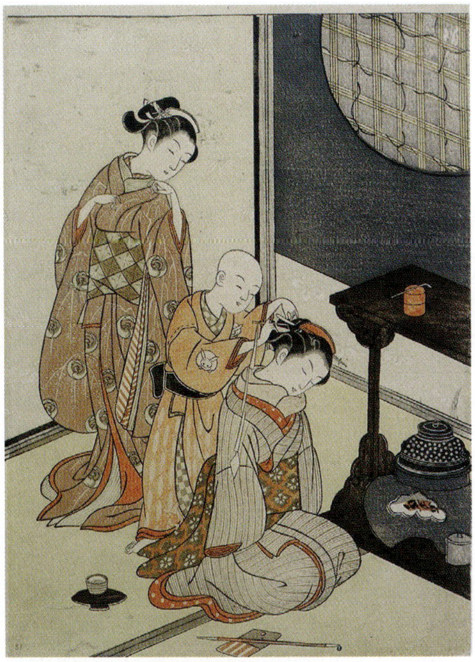

Figure 16. *Suzuki Harunobu,* Night Rain on the Tea Stand, *from* Eight Parlor Views, *1766,* chûban. *Carnegie Museum of Art, Pittsburgh, Carnegie Institute Purchase, 18.14.8*

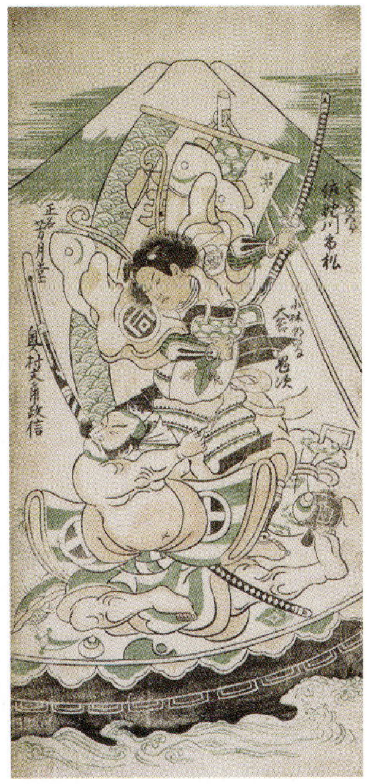

Figure 17. *Okumura Masanobu,* The Actors Ôtani Hiroji and Sanogawa Ichimatsu in a Soga Play, *1743, two-color* hosoban. *Carnegie Museum of Art, Pittsburgh, Bequest of Dr. James B. Austin, 89.28.1002*

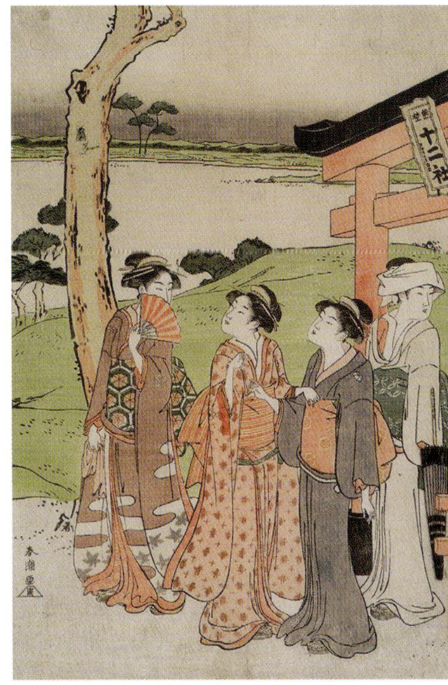

Figure 18. *Katsukawa Shunchô,* Women Visiting the Jûnisô Shrine at Tsunohazu, *ca. 1790,* ôban. *Carnegie Museum of Art, Pittsburgh, Bequest of Dr. James B. Austin, 89.28.1256*

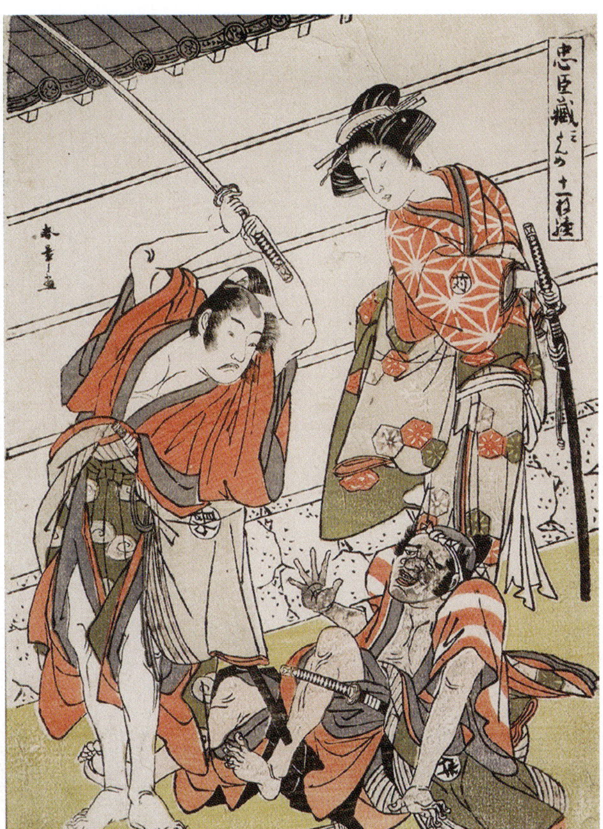

Figure 19. *Katsukawa Shunshô,* Act 3, *from the play* Chûshingura, *ca. 1780,* chûban. *Carnegie Museum of Art, Pittsburgh, Bequest of Dr. James B. Austin, 89.28.1286*

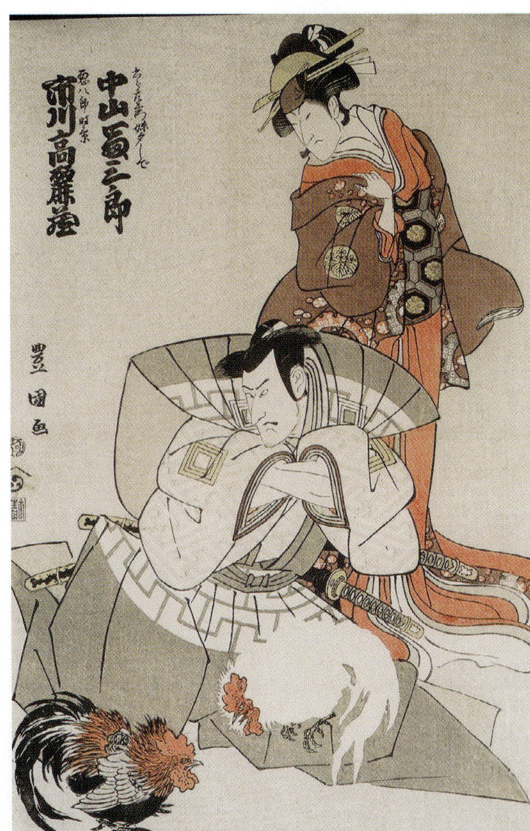

Figure 20. *Utagawa Toyokuni,* The Actors Nakayama Tomisaburô and Ichikawa Komazô II as Yûshide and Takakage, *ca. 1798,* ôban. *Carnegie Museum of Art, Pittsburgh, Bequest of Dr. James B. Austin, 89.28.1406*

colorant in its own right during this period. The study found indigo as blue on only one eighteenth-century print. By contrast, the print in figure 19 by Shunshô of around 1780 contains an uncommon yellowish green produced by mixing dayflower blue with yellow.

The unfaded 1795 actor portrait by Utagawa Toyokuni, in figure 20, is an excellent example of dayflower blue, safflower red, and one of the purples that is produced by mixing them. Dayflower was a versatile color, and the dilute blue here looks like a warm gray, in deliberate contrast to the cooler dilute *sumi* gray of the background.

A dramatic change occurred in the late 1820s when imported Prussian blue became more readily available to printers. The vivid, intense blues in the Tsukioka Yoshitoshi print published in 1863, shown in figure 21, are Prussian blue, and it is easy to understand its appeal and its swift replacement of both dayflower and indigo. Unlike dayflower, Prussian blue was also stable in the presence of moisture. The current study was not able to confirm the presence of Prussian blue in mixtures because of its lack of characteristic spectral features, but Prussian blue easily mixed with available reds and yellows and may be responsible for a variety of greens and purples on mid- and late nineteenth-century prints. This is an area that must be explored in future research.

This study shows that many Japanese print artists, like Chôbunsai Eishi in the 1790s (fig. 22) and Katsushika Hokusai in the 1830s (fig. 23), often used a very limited range of familiar colorants. No doubt publishers had an economic incentive to minimize their capital outlay by limiting the number of blocks in a given print and restricting the colorants. Audiences during certain periods also seem to have been conservative in their taste, preferring familiar color combinations. The artists satisfied both groups by producing both extremely bold and extremely subtle effects with a very limited range of colorants. Of course, tastes changed, and private patronage was one force behind these changes. In 1822, Hokusai designed a privately commissioned series of thirty prints, including the one in figure 24, for a group of poets, utilizing indigo as blue and metallic powders. The designers of *surimono*, as these prints are called, pioneered the use of highly saturated colors and other printing effects that became prevalent in commercial prints of the mid-1840s, like the fan print by Hashimoto Sadahide in figure 25. This study indicates, however, that commercial publishers at the end of this long tradition of color printing still encouraged artists to limit their palettes to readily available colorants, although the printers' increasing skills produced an ever-widening range of color mixtures, color contrasts, and special visual effects.

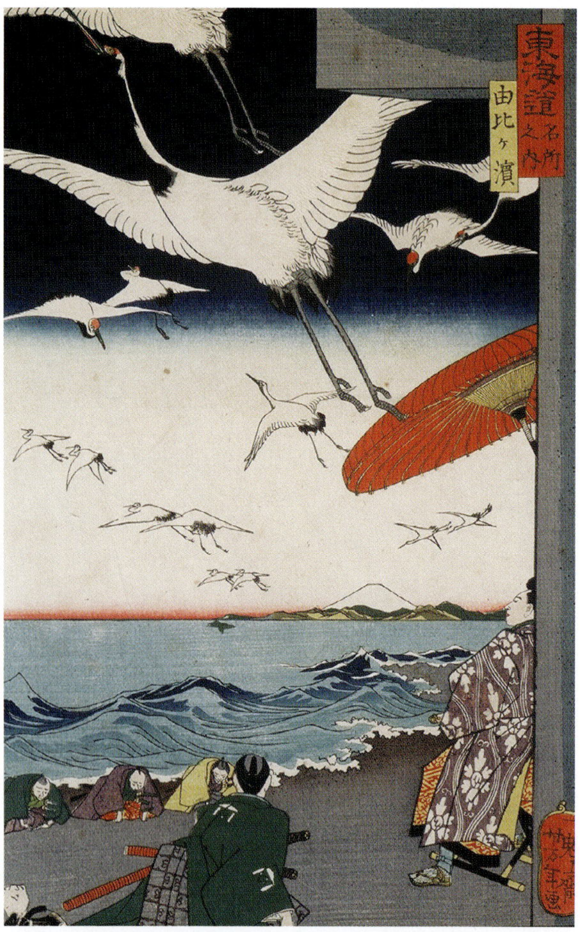

Figure 21. *Tsukioka Yoshitoshi,* Yuigahama, *from* Famous Places on the Tôkaidô Road, *1863,* ôban. *Carnegie Museum of Art, Pittsburgh, Bequest of Dr. James B. Austin, 89.28.1516*

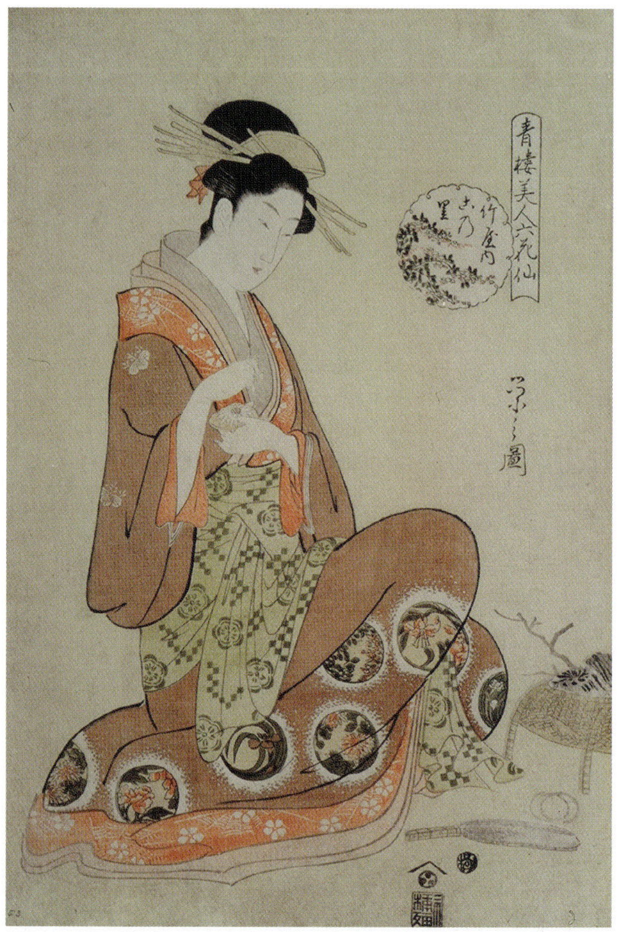

Figure 22. *Chôbunsai Eishi,* The Courtesan Konosato of Takeya, *from* Six Beauties of the Green Houses, *ca. 1795,* ôban. *Carnegie Museum of Art, Pittsburgh, Gift of Andrew Carnegie, 16.29.16*

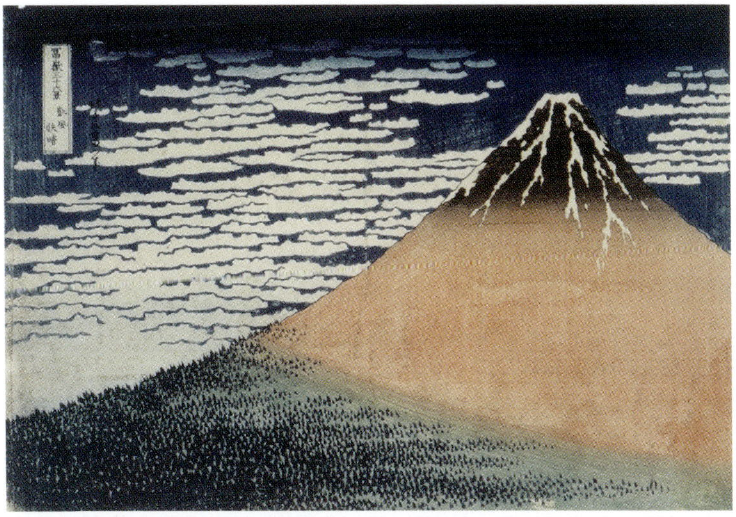

Figure 23. *Katsushika Hokusai,* Fuji in Clear Weather (Red Fuji), *from* Thirty-Six Views of Mount Fuji, *ca. 1832,* ôban. *Carnegie Museum of Art, Pittsburgh, Bequest of Dr. James B. Austin, 89.28.408*

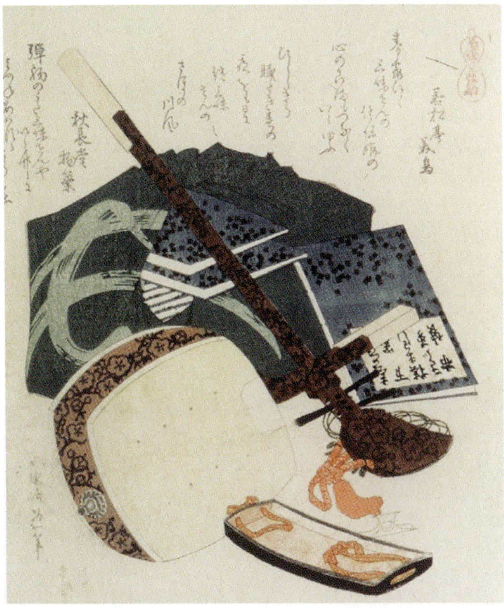

Figure 24. *Katsushika Hokusai,* Samisen 'Ponies', *from* Horses, *1822, square* surimono. *Carnegie Museum of Art, Pittsburgh, Bequest of Dr. James B. Austin, 89.28.411*

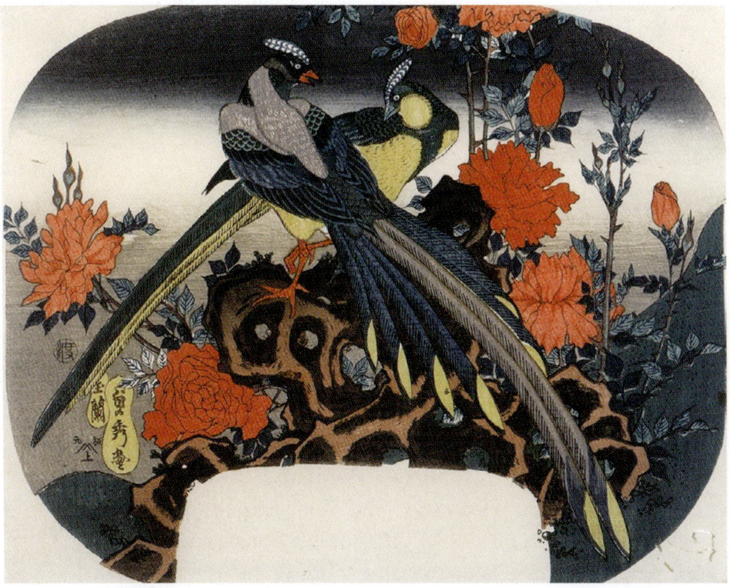

Figure 25. *Hashimoto Sadahide,* Long-tailed Birds and Roses, *ca. 1845, fan print. Carnegie Museum of Art, Pittsburgh, Bequest of Dr. James B. Austin, 89.28.1148*

Conclusions

This study successfully used a micro-fading tester to evaluate seventy-two Japanese woodblock prints. The results give a good prediction of visible light sensitivity when compared to the Blue Wool standards used for reference and will therefore be useful for the future exhibition of these prints. The light source used for these tests, however, is not likely to be the same light source used for museum exhibition conditions. To further understand how such prints will react to specific lighting conditions, the Blue Wool reference strips can be used.

The natural organic colorants on these prints exhibited a range of light sensitivity. The fading rates of most of the areas tested were between Blue Wool 3 and Blue Wool 2. For the five color categories tested, the colors ranked from relatively stable (fading rates slower than Blue Wool 3) to fugitive (fading rates faster than Blue Wool 2) in the following way: blue < red < purple < green < yellow. The result that blue colorants are the most stable is somewhat surprising because at least one blue colorant—dayflower blue—was previously thought to be extremely light sensitive, a perception arising from the observation of apparent fading damage to dayflower areas. The findings of this study suggest that the "damage" to dayflower blue passages may not have been caused by visible light exposure. Exposure to ultraviolet light (which will be limited in future museum exhibition) or other aging processes besides light fading may be to blame. Exploration into some of these factors is an area for future research.

Further examination of the data showed that medium shades of a colorant had faster rates of fading compared to light and dark areas of the same colorant. It is important to note that dark shades may experience an increase in their rate of appearance change over time as they fade toward these medium shades. Therefore, the results of light-sensitivity testing should be considered an evaluation of a dark color's current stability and should not be extrapolated to light exposure doses greater than those used in the test.

It was also shown that the prior fading history of a print does not seem to affect the light sensitivity of safflower red. This finding suggests that safflower red (or the natural organic materials that make up safflower red) is not a mixture of materials with different light sensitivities. Instead, the color is produced by a single colorant or colorants with very similar light sensitivities. Therefore, faded areas of color will not achieve greater light stability from loss of a light-sensitive component.

The tests performed during this study do give a good prediction of a color's light sensitivity; however, they do not indicate how much color change has already occurred. Curators, conservators, and scholars should be cautious in their assessment of color on traditional Japanese prints. Many colors become paler as they fade, but some printers intentionally dilute colors, so pale prints are not necessarily faded, although this is often the case.

Acknowledgments

Experimental work was performed at the Carnegie Museum of Art and the Library of Congress and was supported by grants from the Andrew W. Mellon Foundation and the United States Japan Foundation. The authors would like to thank Shiho Sasaki for many informative discussions regarding Japanese colorants and for preparation of dayflower blue samples used in laboratory analysis. This work would not have been possible without the assistance and consultation of Linda Batis and Ellen Baxter from the Carnegie Museum of Art. The sixteen prints from the Library of Congress were chosen in consultation with library staff Linda Stiber Morenus, senior paper conservator; Katherine Blood, curator for fine prints; and Betty Fiske, contract paper conservator specializing in Japanese prints. The materials selected by this team for study were representative of a historical cross-section of the library's Japanese print collection, dating from the early eighteenth century through the nineteenth century. The authors would also like to thank Dianne van der Reyden, Maria Nugent, Jeremy Adamson, and Harry Katz from the Library of Congress for their contribution to this project.

References

Feller R., M. Curran, and C. Bailie. 1984. Identification of traditional organic colorants employed in Japanese prints and determination of their rates of fading. In *Japanese woodblock prints: A catalogue of the Mary A. Ainsworth Collection*, ed. R. Keyes, 253–66. Oberlin, Ohio: Oberlin College.

Keyes, K. 1988. Japanese print conservation: An overview. In *The conservation of Far Eastern art*, ed. J. Mills, P. Smith, and K. Yamasaki, 30–36. London: International Institute for Conservation.

Leona, M., and J. Winter. 2001. Fiber optics reflectance spectroscopy: A unique tool for the investigation of Japanese paintings. *Studies in Conservation* 46:153–62.

Schweppe, H., and H. Roosen-Runge. 1986. Carmine: Cochineal carmine and kermes carmine. In *Artists' pigments: A handbook of their history and characteristics*, vol. 1, ed. R. Feller, 255–83. Cambridge: Cambridge University Press.

Schweppe, H., and J. Winter. 1997. Madder and alizarin. In *Artists' pigments: A handbook of their history and characteristics*, vol. 3, ed. E. W. FitzHugh, 109–42. Washington, D.C.: National Gallery of Art.

Whitmore, P., and C. Bailie. 1997. Further studies on transparent glaze fading: Chemical and appearance kinetics. *Journal of the American Institute for Conservation* 36:207–30.

Whitmore, P., Xun Pan, and C. Bailie. 1999. Predicting the fading of objects: Identification of fugitive colorants through direct nondestructive lightfastness measurements. *Journal of the American Institute for Conservation* 38:395–409.

Whitmore, P., C. Bailie, and S. Connors. 2000. Micro-fading to predict the result of exhibition: Progress and prospects. In *Tradition and innovation: Advances in conservation*, ed. A. Roy and P. Smith, 200–205. London: International Institute for Conservation.

Appendix. A list of all the colorants identified.

All distinct areas of color were measured on each print; however, only those measurements that resulted in the identification of a colorant are listed here.

CMA: Carnegie Museum of Art; LC: Library of Congress; D: dayflower blue; I: indigo; S: safflower red; P: Prussian blue

Artist	Title, institution, format, number	Date	Colorant and location on print
Okumura Toshinobu (act. late 1710s–ca. 1750)	*The Actors Sodezaki Miwano and Ogino Makinojô as Tora and Shôshô in a Soga Play*, hand-colored *hosoban*, CMA, 89.28.1363	Early 1720s	D: blue square on screen behind figure
Nishimura Shigenobu (act. ca. 1730–1735)	*The Actor Ichimura Uzaemon VIII*, hand-colored *hosoban*, LC	Ca. early 1730s	S: red inside collar of figure's robe
Okumura Masanobu (1686–1764)	*The Actors Ôtani Hiroji and Sanogawa Ichimatsu in a Soga Play*, two-color *hosoban*, CMA, 89.28.1002	1743	I: green "ruffle" detail on boat

Artist	Title, institution, format, number	Date	Colorant and location on print
Torii Kiyomasu II (1705–1763)	*The Actors Onoe Kikugorô and Sanogawa Ichimatsu*, two-color *hosoban*, CMA, 18.14.12	Ca. early 1750s	I: green areas throughout the print
Torii Kiyomitsu (1735–1785)	*The Actor Ichimura Uzaemon IX as Nitta Yoshisada*, three-color *hosoban*, CMA, 89.28.816	Early 1760s	S: red fabric under drum in figure's hand
Suzuki Harunobu (act. ca. 1760–1770)	*Night Rain on the Tea Stand*, from *Eight Parlor Views*, *chûban*, CMA, 18.14.8	1766	D: stripes on seated figure's robe; edge of floor mat along bottom edge of print S: dark red cuff of seated figure's robe
Suzuki Harunobu	*Mid-Autumn*, from *Scenes from Daily Life with Poems for the Seasons*, *chûban*, CMA, 18.14.13	Late 1760s	D: blue stripes on robe of reclining figure I: leaf detail on fan held by seated figure S: red blanket on bench
Suzuki Harunobu	*Lovers*, plate 14, from *Mane'emon*, *chûban*, CMA, 89.28.121	Ca. 1770	I: green floor next to kneeling figure S: red mattress next to screen
Unidentified artist	*Poppies and Flying Bird*, *chûban*, CMA, 89.28.705	Ca. 1770	I: blue sky
Katsukawa Shunshô (1726–1793)	*Couple by a Well*, from *Tales of Ise*, *chûban*, CMA, 89.28.1280	Early 1770s	I: green tree trunk D: blue collar on robe of figure pointing into well; purple robe of lower figure S: red robe of figure pointing into well; purple robe of lower figure
Katsukawa Shunshô	*A Fabric Merchant*, plate 12, from *The Cultivation of Silkworms*, *chûban*, CMA, 89.28.1289	Mid-1770s	D: blue robe on seated figure in foreground; purple collar of robe on seated figure at left S: purple collar of robe on seated figure at left; background color of robe on seated figure at left
Katsukawa Shunshô	*The Actor Bandô Mitsugorô as Soga no Jûrô*, *hosoban*, CMA, 89.28.1281	Ca. late 1770s	D: purple on lower half of figure's robe S: dark and light red areas on figure's robe; purple on lower half of figure's robe
Katsukawa Shunshô	*Act 3*, from the play *Chûshingura*, *chûban*, CMA, 89.28.1286	Ca. 1780	D: inside cuff on robe of left standing figure; green detail at waist on left standing figure S: exterior of robe of left standing figure
Isoda Koryûsai (act. late 1760s–1807)	*The Courtesan Nishikigi of Yotsumeya*, pillar print, LC, LC-USZC4-8534	Ca. late 1770s	S: red collar of robe on seated figure
Kubo Shumman (1757–1820)	*Entertainment at the Shikian Restaurant*, *ôban*, CMA, 89.28.1272	Ca. late 1780s	D: blue plaid robe on seated figure on left side of print I: dark green plant next to seated figure on left side of print S: dark red back of robe on center seated figure in foreground
Katsukawa Shunchô (act. 1780s–1790s)	*Women Celebrating Boy's Day*, *ôban*, CMA, 89.28.1255	Ca. 1790	D: blue plaid robe on kneeling figure; blue robe on center standing figure; purple detail on mask in foreground S: dark and light red on fabric on kneeling figure's robe; purple detail on mask in foreground
Katsukawa Shunchô	*Women Visiting the Jûnisô Shrine at Tsunohazu*, *ôban*, CMA, 89.28.1256	Ca. 1790	I: dark green leaves in trees; light green in grass D: blue sky at horizon; purple robe of standing figure on left; blue robe of standing figure second from right S: purple robe of standing figure on left; dark red cuff on robe of standing figure on left; gate in background; light red on robe of standing figure second from left
Katsushika Hokusai (1760–1849)	*The Actor Segawa Kikunojô III as the Geisha Kashiku*, *hosoban*, CMA, 89.28.415	1790	D: blue background S: red background color on figure's robe
Chôbunsai Eishi (1756–1829)	*The Courtesan Konosato of Takeya*, from *Six Beauties of the Green Houses*, *ôban*, CMA, 16.29.16	Ca. 1795	I: dark green floral medallion on lower half of robe S: red collar on figure's robe
Kitagawa Utamaro (1754–1806)	*The Hour of the Dragon*, from *Twelve Hours of the Green Houses*, *ôban*, CMA, 82.52.12	Ca. 1794	D: cuff on sleeve of lying figure's robe I: green blanket S: red mattress under lying figure; light red in upper right corner near writing
Katsukawa Shun'ei (1762–1819)	*A Sumô Wrestling Tournament*, *ôban* triptych, CMA, 89.28.1259, A–C	1794	D: blue sky near horizon I: dark green in trees; light green in grass S: red flowers in trees
Katsukawa Shun'ei	*Bust Portrait of the Actor Arashi Ryûzô II*, *chûban*, CMA, 89.28.1262	Ca. 1795	I: green stripe on shirt
Kabukidô Enkyô (act. 1796)	*Bust Portrait of the Actor Nakayama Tomisaburô*, *ôban*, LC, LC-USZC4-8439	1796	S: red collar on figure's robe near right shoulder; purple cloth over figure's forehead
Rekisentei Eiri (act. late 1780s–early 1800s)	*The Whale Caught in Edo Bay near Shinagawa*, *aiban* diptych, CMA, 89.28.41	1798	S: (left panel of diptych) bottom of robe on seated figure on right

Artist	Title, institution, format, number	Date	Colorant and location on print
Ichirakutei Eisui (act. late 1790s–early 1800s)	*Osono Watching the Lovers Sankatsu and Hanshichi*, *ôban*, CMA, 89.28.57	Late 1790s	D: blue cuff on sleeve of male figure's robe; purple robe on female figure I: green on female figure's robe S: red collar on female figure's robe; purple robe on female figure
Utagawa Toyokuni (1769–1825)	*The Actors Nakayama Tomisaburô and Ichikawa Komazô II as Yûshide and Takakage*, *ôban*, CMA, 89.28.1406	Ca. 1798	D: purple background color of standing figure's exterior robe; blue background color on lower half of kneeling figure's robe S: red collar of standing figure's robe
Utagawa Toyokuni	*The Actors Arashi Hinasuke II and Iwai Kumesaburô as Yorimitsu and Yosooi-hime*, *ôban*, CMA, 06.1.2	Ca. 1798	D: bottom half of robe, standing figure S: red cuff at bottom of robe, seated figure
Utagawa Toyokuni	*The Actors Matsumoto Kôshirô V and Ichikawa Komazô as Ashiya Dôman and His Son*, *ôban*, CMA, 89.28.1394	Ca. 1801–2	D: purple cuff on sleeve of standing figure's robe I: blue background color on standing figure's robe S: purple cuff on sleeve of standing figure's robe; red color of seated figure's robe
Kitagawa Utamaro (1754–1806)	*Net Fishing at Night*, *ôban* triptych, CMA, 80.55.2	Ca.1800	D: (center panel of triptych) water around boat; (right panel of triptych) purple robe on standing figure I: (center panel of triptych) green pole of fishing net S: (center panel of triptych) light red bucket behind seated figure; (right panel of triptych) purple robe on standing figure
Eishôsai Chôki (act. late 1780s–early 1800s)	*Evening Bell for Kumenosuke*, from *Eight Views for Chanted Plays*, long horizontal format, CMA, 89.28.1236	Ca. 1800	I: dark green bushes along right edge of print
Katsukawa Shun'ei	*A Benefit Sumô Wrestling Tournament*, *ôban* triptych, CMA, 89.28.1265	Ca. 1800	D: (center panel of triptych) purple robe of seated figure on right side of ring I: (center panel of triptych) sky in background; green robe of figure in crowd in the foreground S: (center panel of triptych) red stripe on balcony; purple robe of seated figure on right side of ring
Katsukawa Shun'ei	*The Wrestlers Raiden and Sadogatake*, *ôban*, CMA, 78.43.3	1803	D: blue fabric on inside of seated figure's robe I: green stripes on seated figure's robe S: red fabric inside seated figure's sleeve
Kitagawa Utamaro	*The Taikō Hideyoshi and His Wives Viewing Cherry Blossoms*, *ôban* triptych, LC	Ca. 1804	I: green robe on seated figure at far right S: purple tie on curtain in background; red background color on floral print robe worn by standing figure
Shôrinsai Hidemaro (act. 1800s–mid-1810s)	*Evening Glow at Ryôgoku Bridge*, from an untitled series of eight views of Edo, *chûban*, CMA, 89.28.193.7	Ca. 1805	D: purple foreground; water S: purple foreground; red border around print
Shôrinsai Hidemaro	*Night Rain at Massaki*, from an untitled series of eight views of Edo, *chûban*, CMA, 89.28.193.8	Ca. 1805	D: purple foreground; water S: purple foreground; red border around print
Torii Kiyomine (1787–1868)	*Courtesans Enjoying Cherry Blossoms at the Yoshiwara*, *ôban* triptych, CMA, 89.28.812	1807	I: (center panel of triptych) green on bottom of left figure's robe D: (center panel of triptych) purple in center of left figure's robe; blue screen behind figures S: red on bottom of left figure's robe; red flowers in trees; purple in center of left figure's robe
Shôtei Hokuju (act. late 1790s–1824)	*Ochanomizu*, from an untitled series of views of Edo, *ôban*, CMA, 89.28.374	Ca. 1810s	D: dark and medium blue tones in sky I: light blue around clouds in sky; green bushes and trees
Utagawa Kuniyasu (1794–1832)	*The Actor Bandô Mitsugorô III*, square *surimono*, LC	Ca. late 1810s	D: blue front of figure's robe I: green hat behind figure S: red frame around mirror
Katsushika Hokusai (1760–1849)	*Samisen 'Ponies'*, from *Horses*, square *surimono*, CMA, 89.28.411	1822	I: purple detail around instrument; light green area on cloth; blue book S: red cloth between instrument and tray; purple detail around instrument; handle on tray
Totoya Hokkei (1780–1850)	*Feeding the Salt Dragon*, square *surimono*, CMA, 89.28.371	1832	P: dark blue around base of tank; medium and light blue inside tank
Katsushika Hokusai	*Fuji in Clear Weather (Red Fuji)*, from *Thirty-Six Views of Mount Fuji*, *ôban*, CMA, 89.28.408	Ca. 1832	I: green grass at base of mountain; blue writing in upper left corner P: light blue in clouds
Katsushika Hokusai	*Fuji from Fujimigahara*, from *Thirty-Six Views of Mount Fuji*, *ôban*, CMA, 18.14.5	Ca. 1832	I: blue outline of planks in barrel; tool on left side of barrel P: light blue in sky along right edge of print

Artist	Title, institution, format, number	Date	Colorant and location on print
Katsushika Hokusai	*Yatsuhashi Bridge in Mikawa Province*, from *Famous Bridges in the Provinces*, *ôban*, CMA, 89.28.407	Ca. 1833	P: water
Katsushika Hokusai	*Bun'ya no Asayasu*, from *One Hundred Poems as Explained by the Nurse*, *ôban*, CMA, 78.43.4	Ca. 1835	P: blue in water I: blue edge at very top of print; bottom half of robe on center standing figure in boat
Utagawa Kunisada (1786–1864)	*Kambara*, from *Fifty-Three Stations of the Tôkaidô Road*, *chûban*, LC, LC-USZC4-8431	Ca. 1838	P: blue on inside of figure's robe I: purple background under bull's tail S: circular stamp at bottom of print; light red background around writing in upper left corner of print; purple background color under bull's tail
Hashimoto Sadahide (1807–1873)	*Long-tailed Birds and Roses*, fan print, CMA, 89.28.1148	Ca. 1845	P: light blue leaves around flowers I: purple on bird's back; purple in upper sky S: light red on flower along top edge of print; purple in upper sky
Utagawa Kunisada	*Actor as Kintoki Hambei*, from a series of eighteen actor portraits, *ôban*, CMA, 89.28.907	Ca. 1850	D: dark purple on top half of figure's robe I: dark blue cuff on bottom of figure's robe; sky P: blue background near figure's feet S: red background around writing; dark purple on top half of figure's robe
Utagawa Kunisada	*Actors as Saitôgo Kunitake, Tada Kurando Yukitsuna, Naruto no Mae, and Akugenta Yoshihira*, four *ôban*, LC, LC-USZC4-8521, 8522, 8523, 8524	Ca. 1850	P: (left panel) light blue straps on figure's shoes I: (left panel) dark blue background color of figure's robe S: (left panel) light red floral detail on figure's robe; narrow purple stripe on figure's pants
Utagawa Kunisada	*Actor as Gofukuya Jûbei at Hara*, from *Fifty-Three Stations of the Tôkaidô Road*, *ôban*, CMA, 89.28.899.2	1852	D: purple cloud behind figure P: water in background S: purple in cloud behind figure; sky behind mountain
Utagawa Kunisada	*Actor as Oguri Hangan at Fujisawa*, from *Fifty-Three Stations of the Tôkaidô Road*, *ôban*, CMA, 89.28.899.4	1852	D: dark purple on figure's robe in floral detail; purple clouds behind figure P: light blue detail in figure's robe S: red horizon; dark purple on figure's robe in floral detail; purple clouds behind figure
Utagawa Kunisada	*Actor as Kanae Tanigorô at Kambara*, from *Fifty-Three Stations of the Tôkaidô Road*, *ôban*, CMA, 89.28.899.3	1852	P: blue detail on robe (border around flower) S: light purple detail on robe (inside flower)
Utagawa Kunisada	*The Floating Bridge of Dreams*, from an untitled Genji series, *ôban*, LC, LC-USZC4-8432	1854	D: purple sleeve of seated figure's robe P: light blue detail on mat near bottom of print S: dark red background around writing in upper right corner of print; purple sleeve of seated figure's robe
Utagawa Hiroshige (1797–1858)	*Distant View of Kinryûzan from Azuma Bridge*, from *One Hundred Famous Views of Edo*, *ôban*, LC, LC-USZC4-8423	1857	P: sky at top of print; water S: dark red background color around writing in upper right corner; red in sky at horizon; purple sky at top of print
Utagawa Hiroshige	*The Great Bridge at Senju*, from *One Hundred Famous Views of Edo*, *ôban*, LC, LC-USZC4-8425	1856	P: water S: red in sky at horizon; dark red background color around writing in upper right corner
Utagawa Kunisada	*The Actor Ichikawa Danjûrô VIII as Matsuri Sashichi*, *ôban*, CMA, 89.28.900	1856	P: blue sky
Ochiai Yoshiiku (1833–1904)	*Foreigners of the Five Nations Partying at the Gankirô Restaurant*, *ôban* triptych, LC, LC-USZC4-8479, 8480, 8481	1860	P: (left panel of triptych) dark blue cuff on bottom of central standing figure's robe; light blue floral decoration on figure's robe I: (left panel of triptych) brown floor next to writing along left side of print; blue background behind bridge at top of print
Hashimoto Sadahide	*Commerce at Yokohama: Western Traders Transporting Goods*, *ôban*, LC, LC-USZC4-8538	1861	P: water
Utagawa Kunisada	*The Actor Kawarazaki Gonjûrô Applying Makeup*, *ôban*, CMA, 89.28.908	1861	P: blue background in mirror
Toyohara Kunichika (1835–1900)	*Prince Genji and Companions by a Porcelain Water Basin*, two *ôban*, LC, LC-USZC4-8451	1862	P: (right panel of diptych) light and dark blue areas in water outside window S: (right panel of diptych) light red flowers in trees; dark red background color around writing in lower right corner of print; purple robe of standing figure
Tsukioka Yoshitoshi (1839–1892)	*Yuigahama*, from *Famous Places on the Tôkaidô Road*, *ôban*, CMA, 89.28.1516	1863	P: dark blue in sky; water D: purple exterior robe of seated figure along right edge of print S: light red in sky at horizon; purple exterior robe of seated figure along right edge of print

Dayflower Blue: Its Appearance and Lightfastness in Traditional Japanese Prints

Shiho Sasaki and Elizabeth I. Coombs

ABSTRACT Dayflower blue is one of the more fugitive colorants used in traditional Japanese woodblock prints of the eighteenth and nineteenth centuries, being sensitive to both light and humidity. It is an organic dye produced from petals of the hybridized dayflower. It is still sold in the traditional form of pieces of paper saturated with the colorant. The paper carrier is soaked in water to extract the dye for use. Previous research has shown that the extracted color ranges from a warm bright blue to a distinctly grayish blue depending on the length of time it has been stored in the paper carrier, and it can be printed so as to appear blue or gray. In this scientific investigation the fading rates of printed samples of dayflower blue from several different years of manufacture were measured using an accelerated light-aging tester and a micro-fading tester. The results show that different degrees of lightfastness of printed materials can be anticipated from the different colors and ages of dayflower blue in the paper carrier. The discovery that the age of the colorant and the way it is printed affect its appearance raises the prospect that some of these variations were intentional. Possible reasons, such as artistic intent and publishing history, are discussed.

Introduction

Dayflower blue has been traditionally known as one of the most fugitive colorants used in the production of traditional Japanese woodblock prints, known as ukiyo-e prints, during the eighteenth and nineteenth centuries. It was used first in the hand-colored prints known as *beni-e* (1720s–40s) and flourished with the invention of polychrome multi-block prints (*nishiki-e*) in the late 1760s.

It had been widely assumed that fading accounted for the dayflower blue color variations seen in Japanese prints and illustrated books, even when other colorants on the same sheet appeared unfaded. Yet the dayflower blue color in a number of these prints has different hues that do not seem to be a result of fading by light exposure. However, the colorant often shows a partial loss or change of the color, damage characteristic of contact with moisture.

A previous study of dayflower blue (Sasaki and Webber 2002) examined whether the color faded simply by exposure to light, as has been traditionally believed, and whether the poor lightfastness of dayflower blue might be related to differences in hue. The study revealed that the extracted dayflower color ranges from a warm bright blue to a distinctly grayish blue depending on the length of time the colorant has been stored in the paper carrier, and therefore it can be printed so as to appear blue or gray.

The gray color in ukiyo-e prints has long been attributed to fading of the dayflower blue colorant. However, the ability to print either blue or gray from the same paper carrier, depending on the time elapsed between the production of the paper carrier and the time of printing, suggests that the range of colors seen in the prints today may be due to the initial production intent as well as the effects of aging. We questioned whether dayflower blue in the printed state is actually more stable than was previously thought. In this study we investigated whether color changes occurred during the aging tests, whether a warm blue color maintains its color after aging or changes to gray as a result of fading, and whether gray results from a fading of blue or is the original printed color. We also wondered whether there is a difference in the stability to light exposure between samples printed from paper carriers of different ages. To address these questions and better understand the behavior of the colorant, printed samples of dayflower blue using a variety of ages of manufacture and elapsed time to printing were prepared, and their light stability was tested using an accelerated light-aging tester and a micro-fading tester.

Dayflower Blue: Preparation and Use

Dayflower blue is still produced today using traditional methods. The petals of the hybridized dayflower (*Commelina communis* L. var. *hortenis* Makino) are harvested in midsummer, and the juice is extracted by hand. Sheets of paper are saturated with the blue juice and stored for future use. This method of production gave rise to two common names—*aobanagami* (blue flower paper) and *aigami* (blue paper). The latter name, *aigami*, was used in Edo where ukiyo-e prints were primarily produced (Sasaki and Webber 2002).

For use, *aigami* is cut into small pieces, and a suitable amount is soaked in water, which immediately releases the colorant from the paper. *Aigami* is believed to have been used first by dyers for the "underdrawing" for certain types of textiles, since dayflower blue is extremely fugitive in water and so easily removed with cold water after dyeing without disturbing other colors. Since the dayflower blue becomes more difficult to remove from the textile when older *aigami* is used, dyers prefer to use fresh *aigami*.

Previous study (Sasaki and Webber 2002) has revealed that different tones of dayflower blue can be produced by printing from different ages of *aigami*, ranging from a warm bright blue to a distinctly grayish blue. For this study, *aigami* is described as either fresh (used within one year of manufacture) or aged (used after one year).

The Samples

Traditional materials and techniques were used for making the samples for testing. Samples of dayflower blue from *aigami* were produced in four different years (1986, 1997, 1999, and 2002), so that the elapsed time from production to printing ranged from four months to thirteen years. Samples of the colorant were obtained by cutting up a 2 cm^2 piece of *aigami* and placing it in a dish containing 3 ml of deionized water. The resulting colorant was mixed with a very small amount of rice starch paste on a woodblock and printed on pieces of *hōsho* paper that had first been sized.[1]

A range of tones of dayflower blue was produced in the printed samples (table 1 and fig. 1). Samples A and D were prepared using *aigami* less than one year old, and both show a warm blue. Samples A and E were prepared using the same sheet of *aigami*, made in 1999. Whereas the sample also printed in 1999 (sample A) has a warm bright blue tone, the sample printed nearly four years after manufacture (sample E) is grayish blue. Dayflower blue appears to lose its blue tone the longer it is stored as *aigami*.

Samples E and F were prepared using the same sheet of *aigami* manufactured in 1999. The *aigami* was cut in two and stored under two different sets of environmental conditions. Both pieces of *aigami* were kept in plastic bags in the dark, but the piece used for making sample E spent part of the time in an environment with fluctuating relative humidity, while the piece used for sample F was kept in a controlled environment. The blue color of sample F appears much warmer than that of sample E, demonstrating the colorant's sensitivity to a range of environmental factors. Since this study was attempting to use traditional materials and production in sample making and there was no environmental control used during traditional ukiyo-e printmaking, sample E, and not sample F, was used for the subsequent fading and aging tests.

Fading Tests

The first series of tests was conducted using an accelerated light-aging tester[2] in the Conservation Department at the Victoria and Albert Museum in London in 1999. The second tests were carried out by Sandra Connors at the Research Center on the Materials of the Artist and Conservator, Carnegie Mellon University, in Pittsburgh, in 2003 using a microfading tester that has been previously described in detail (Whitmore et al. 1999). Color change (ΔE) was calculated using the CIE L*a*b* system for both sets of experiments.

The light generated by the accelerated light-aging tester contained ultraviolet radiation and used a high intensity discharge (HID) 500 watt lamp that combined an incandescent light source with a high-pressure mercury source.

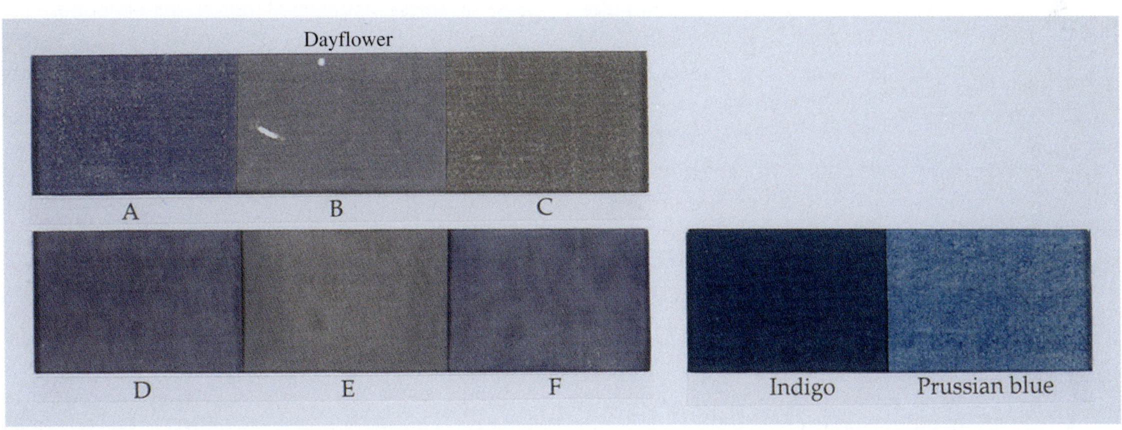

Figure 1*. Printed samples of dayflower, indigo, and Prussian blue, unexposed to light*

Table 1 Aigami *and the printed samples*

Sample	Year of *aigami* manufacture	Date of printing (elapsed time since manufacture)	Storage conditions of *aigami*	Observed color when printed
A	1999 summer	1999 December (4 months)	Uncontrolled	Warm bright blue
B	1997 summer	1999 December (2 years, 4 months)	Uncontrolled	Grayish blue
C	1986 summer	1999 December (13 years)	Uncontrolled	Distinctly grayish blue
D	2002 summer	2003 March (7 months)	Uncontrolled	Warm bright blue
E	1999 summer	2003 March (3 years, 7 months)	Uncontrolled	Grayish blue
F	1999 summer	2003 March (3 years, 7 months)	Controlled (68–70°F, 50–55% RH)	Warmer blue than sample E

Samples A, B, and C were exposed to the light for approximately 7,000,000 lux hours. The area of exposure for the samples in this tester was approximately 4 × 10 cm. Measurements of color difference were taken from the same four positions on each sample, and an average was given as the color change value in order to compensate for slight unevenness in the printing of the colors.

There were no ultraviolet or infrared components from a filtered 75 watt xenon arc lamp, the light source for the micro-fading tester. Here, the spectrum and color change were measured from samples A to E during a thirty hour exposure. Color change measurements were also taken from Blue Wool standards 1 to 4 during each test in order to compare fading rates from the two instruments, since it is difficult simply to compare fading rates by exposure time or lux hours, due to the difference in the light sources and types of instrument.

For a comparison with dayflower blue, indigo (*Polygonum tinctorium* Ait.) and Prussian blue ($Fe_4[Fe(CN)_6]_3 \cdot \chi H_2O$)—the other common blue colorants used in ukiyo-e prints—were also printed in December 1999, and their light stability was tested during the same exposure and with the same accelerated light-aging tester as used for the dayflower blue.

Results and Discussion

The given value of the color differences from testing by the two instruments is similar (figs. 2 and 3). In the results from the accelerated light-aging tests carried out in 1999, the fading rates were: sample A, printed using fresh *aigami*, close to Blue Wool 4; and samples B and C, printed from aged *aigami*, the same as Blue Wool 3. The difference of the color change value (ΔE) between the fresh and aged samples was ΔE = 5 at 1,000,000 lux hours and ΔE = 12 at 7,000,000 lux hours. However, the micro-fading tester measured little difference in the fading rates among the samples. The presence of ultraviolet radiation in the accelerated light-fading tester and not in the micro-fading tester may account for the small (less than halfway between Blue Wool 3 and Blue Wool 4) difference in the fading rates between the two tests. The difference in the fading rates between the printed samples prepared from fresh and aged *aigami* may not be a sufficient change in value to clearly define the light sensitivity of each colorant. However, after the 1999 test, it was first thought that there was a perceptible difference in the fading rates of aged and fresh samples even though botanically they are the same and have the same color. The difference of the color change value for the indigo and Prussian blue samples from the 1999 test are also close to the value for Blue Wool 4. Significantly, dayflower blue has better light stability than was previously believed.

The color changes that occurred on the samples during exposure to light are also illustrated as reflectance spectra (figs. 4 and 5). The characteristic reflectance curve of dayflower blue is seen in the spectrum: a large absorption occurs at approximately 590 nm, a small shoulder at a wavelength around 550 nm, and a large shoulder at a wavelength between 600 and 650 nm. In the samples prepared from aged *aigami*, particularly sample C, as the colorant faded with greater exposure times the absorption peaks in the spectrum lost their sharpness and overall reflectance increased (see fig. 5).

During the 1999 test, small pieces of the samples were removed for documentation on four occasions, as visible changes in color were noticed (fig. 6). When the samples from aged *aigami* (samples B and C) dramatically lost blue color, testing was stopped. At the conclusion of the test the color change values of both samples was approximately ΔE = 21. In 2003, with no actions other than the passage of time, the faded areas of the samples tested in 1999, particularly those produced using aged *aigami*, had become a brownish yellow. This change may be a result of oxidation of the colorant due to the light-induced fading that had occurred (fig.

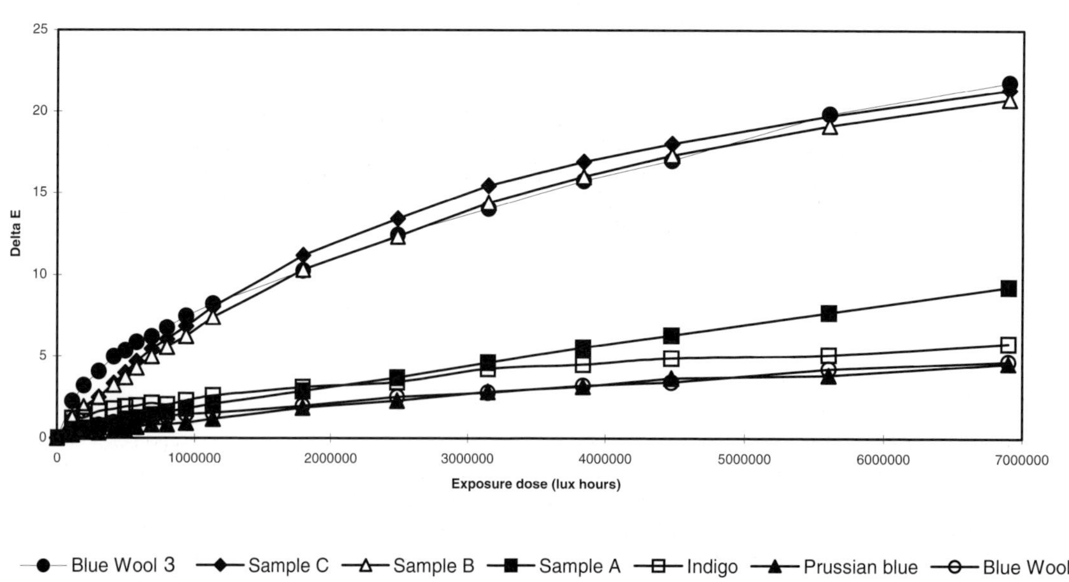

Figure 2. *Comparison of dayflower blue fading rates with Blue Wool standards 3 and 4, ΔE vs. exposure dose, accelerated light-aging tester*

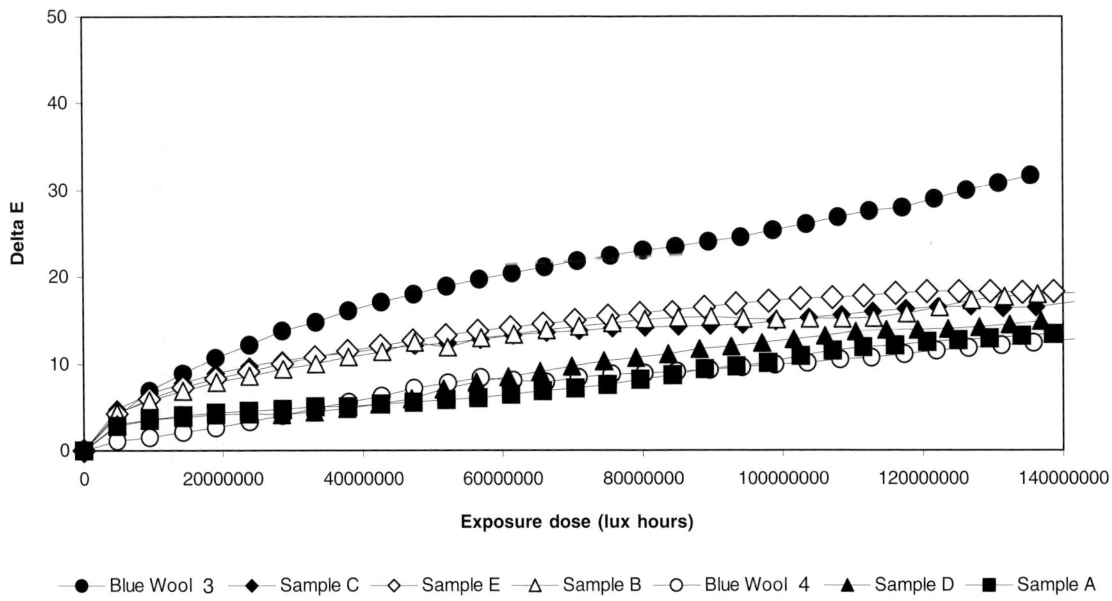

Figure 3. *Comparison of dayflower blue fading rates with Blue Wool standards 3 and 4, ΔE vs. thirty hours exposure dose, micro-fading tester*

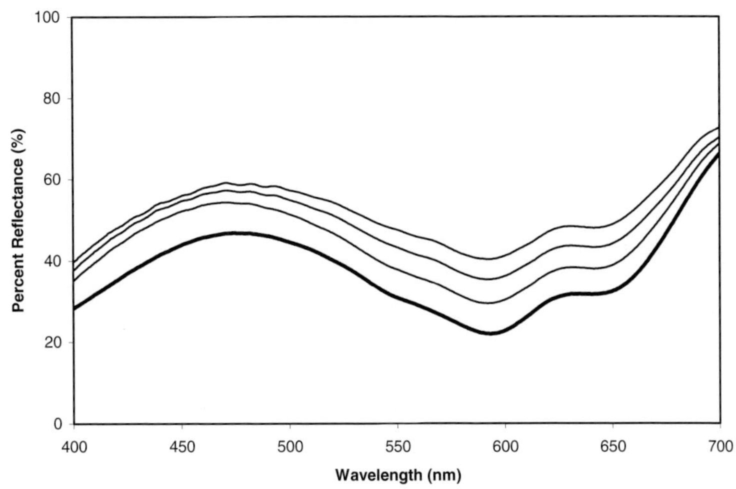

Figure 4. *Reflectance spectra of sample A (made in 1999, printed in 1999). From the bottom: unexposed, ten hours, twenty hours, and thirty hours exposure*

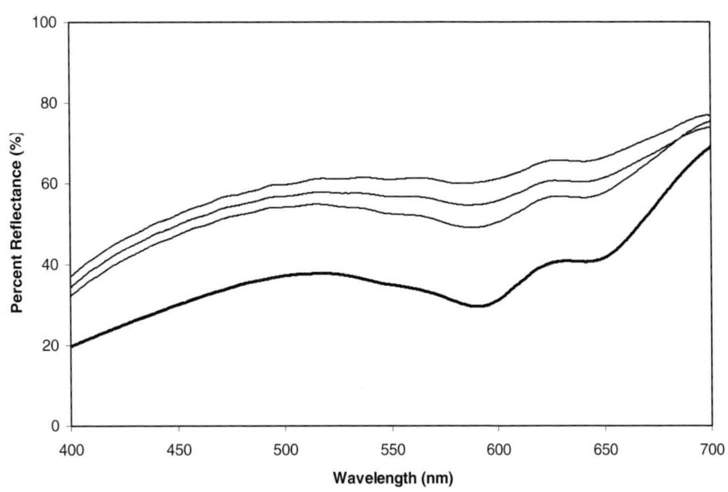

Figure 5. *Reflectance spectra of sample C (made in 1986, printed in 1999). From the bottom: unexposed, ten hours, twenty hours, and thirty hours exposure*

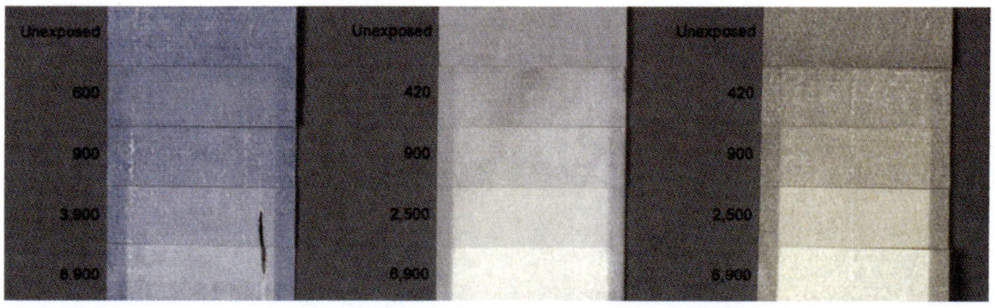

Figure 6. *The printed samples from the fading test in 1999. From left: sample A, B, and C, 1,000 lux hours*

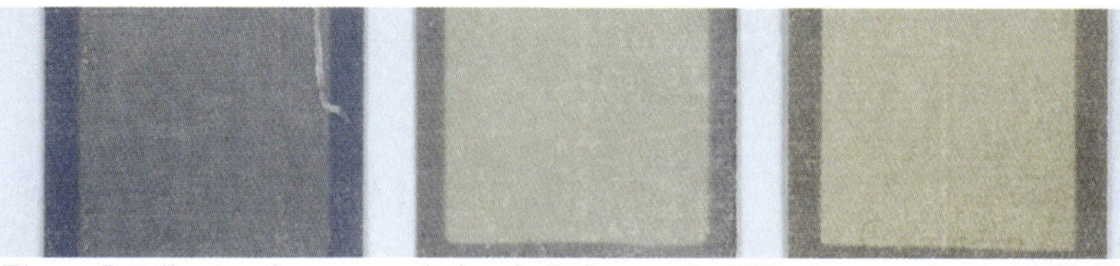

Figure 7. *Yellowing that occurred after the fading test in 1999, photographed in 2003. From left: sample A, B, and C*

7). Also of note is that there is no significant visible difference in the unexposed parts or in the unused samples. This discoloration is similar in appearance to faded dayflower blue seen in ukiyo-e prints.

The Blue Wool standards of photochemical stability are classified from unstable to stable. A material with a stability equivalent to Blue Wool 3 or less is considered unstable or fugitive, and a material with a stability between 3 and 6 is considered to have intermediate stability (Feller et al. 1984). Previous research (Feller et al. 1984) found dayflower blue to be extremely fugitive, equivalent to Blue Wool 1 or less. Our test results indicate that dayflower blue is equivalent to Blue Wool 3–4, a range that falls in the intermediate category of photochemical stability. Having samples prepared using traditional printing techniques as opposed to the brushed samples of aged *aigami* and of the freshly squeezed juice from the wild species used by the previous researchers might account for the difference in the results. In addition, differences in the thickness or types of the paper support and the age of *aigami* when the samples were prepared may have affected the results.

These tests were designed to describe the light-aging behavior of dayflower blue on prints. However, to complete the picture, the light stability of dayflower blue in relation to different levels of relative humidity should be investigated. Since the colorant is known to be sensitive to moisture, the light aging that results in environments with elevated levels of relative humidity may differ from those reported here. However, the use of the term "extremely fugitive" for dayflower blue should be redefined to indicate that the colorant is fugitive when exposed to moisture but not so fugitive when exposed to light.

The Visual Appearance of Dayflower Blue

Now that we have a better understanding of how dayflower blue responds to light, we will examine its appearance in traditional Japanese woodblock prints and printed books. There has been much confusion among those who work with and appreciate this material as to whether the appearance of the colors at present is original or whether they have been altered over time. Dayflower blue, especially, has been the object of much speculation because it appears in such a range of hues and tones. When produced in pale tones, it is often thought to have faded, and when appearing in one of its gray forms it can be mistaken for another colorant altogether (see fig. 1).

Let us consider two distinctive physical properties of dayflower blue. First, the colorant is very moisture sensitive. It can disappear when in contact with direct moisture and may shift toward a brown tone with elevated levels of relative humidity. This shift has sometimes been mistaken for light fading, but, in fact, the color is not as fugitive to light as previously thought. The fresh, bluest form of dayflower blue can be identified by visual inspection when a stray water spot (an area with pale or absent color surrounded by a tideline) is present, since the other common blues—Prussian blue and indigo—are not similarly affected (fig. 8).

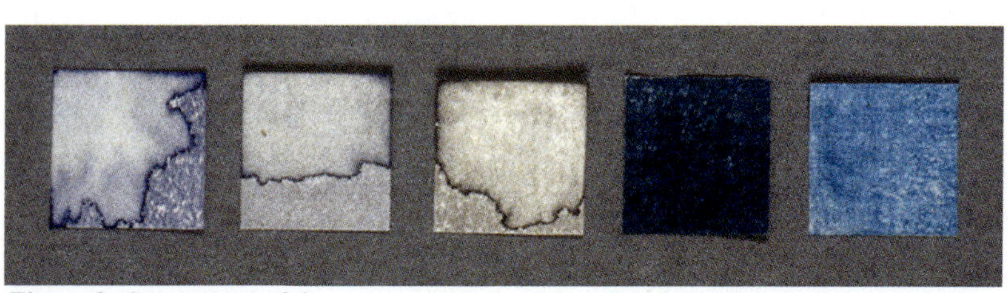

Figure 8. *Comparison of the water sensitivity of the printed samples. A water droplet was placed on each sample. From left: sample A, B, and C, indigo, and Prussian blue*

This kind of accidental spotting is not uncommon, and, in fact, a "watery" appearance for all but the most blue form of the printed color is typical.

Second, the elapsed time between making the *aigami* and printing the colorant it contains appears to be the main determinant of the hue that will be produced. The freshest *aigami* produces the bluest color, and older material produces increasingly gray variations (see fig. 1).

We have chosen woodblock printed images from five books to demonstrate the varied appearance of dayflower blue and to show how often it can be identified by visual inspection alone. The books are Kitagawa Utamaro (ca. 1754–ca. 1806), *Waka Ebisu* (Young Ebisu), ca. 1789; Utamaro, *Shioi no tsuto* (Gifts of the Ebb Tide), ca. 1790; Utamaro, *Ehon seirō nenju gyōji* (The Picture Book of Annual Events in the Green Houses), two paintings, both 1804; and Keisai Masayoshi (1764–1824), *Sansui ryakuga shiki* (How to Draw Simple Landscapes), 1797.

Books rather than prints were chosen, since the colors are more likely to have been protected from light and high humidity and we tried to choose early printings. We chose books mostly by the same artist, Utamaro, and one book by a contemporary, Masayoshi, which reveals a different sensibility. Since books often exist in multiple printings it is possible to compare differences in the choice of colors, and we do this in the two printings of Utamaro's *The Picture Book of Annual Events in the Green Houses*. Lastly, we chose books where dayflower blue was most likely to be the only blue, both to show the range of that colorant and to minimize confusion. All the books were published well before Prussian blue's debut in Japanese printmaking in the 1820s, and during this period indigo was used infrequently as an unmixed blue color (Connors et al. 2005).

The first book is Utamaro's *Gifts of the Ebb Tide* (fig. 9). Dayflower blue is used in different strengths for the sky, the umbrella, and the *obi* or kimono belt of the woman on the right, and it is barely visible in the water where the waves are beautifully embossed. The brilliance of the safflower red indicates good light preservation. Gradation printing in the sky gives a range of shades from deep to light blue. Moisture, possibly from when the album was bound, has turned the colorant a very characteristic pale brown along the centerfold. This effect is sometimes mistaken for light fading (fig. 10). Note also the curving pale lines, which are the marks from the printer's *baren*. The printer had to work carefully to avoid diminishing the color with too much pressure or by keeping the paper too damp. In the woman's *obi* we see the characteristic watery appearance of dayflower blue (fig. 11). The shore is printed with a pale mineral gray (which appears to be *sumi*, a form of carbon black, mixed

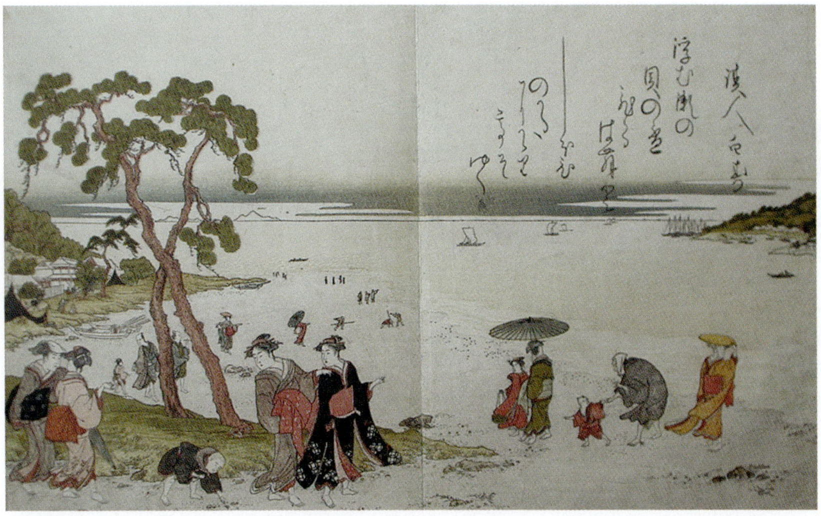

Figure 9. *Spread from Kitagawa Utamaro,* Shioi no tsuto*, ca. 1790. Collection of Arthur and Charlotte Vershbow*

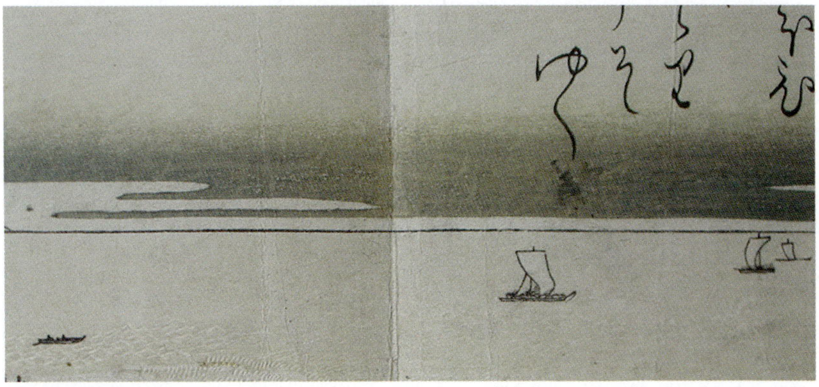

Figure 10. *Detail of spread from Kitagawa Utamaro,* Shioi no tsuto *(fig. 9), showing color gradation in the sky*

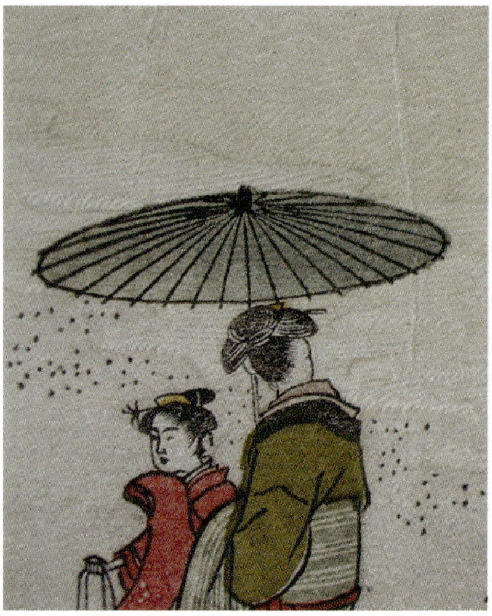

***Figure 11.** Detail of page from Kitagawa Utamaro,* Shioi no tsuto *(fig. 9), showing the woman's* obi, *or kimono belt*

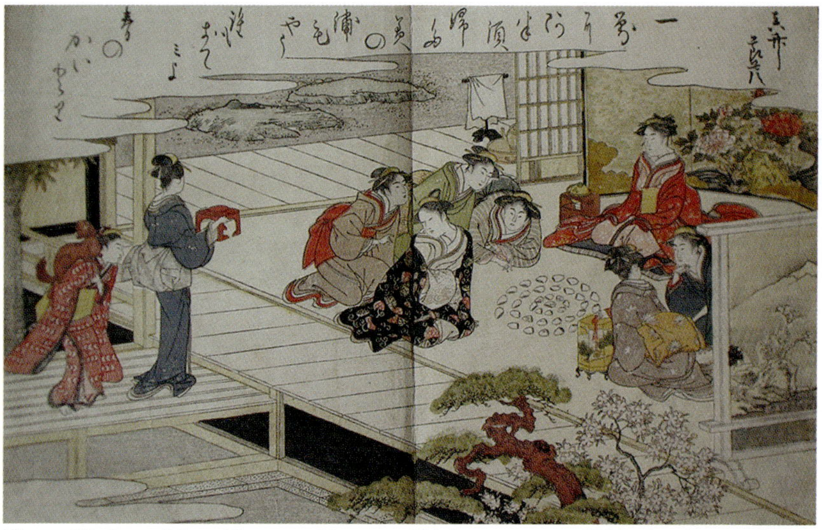

***Figure 12.** Spread from Kitagawa Utamaro,* Shioi no tsuto, *ca. 1790, showing mottling in the pale blue of screen at right and the brown in the blue river on the painted folding screen*

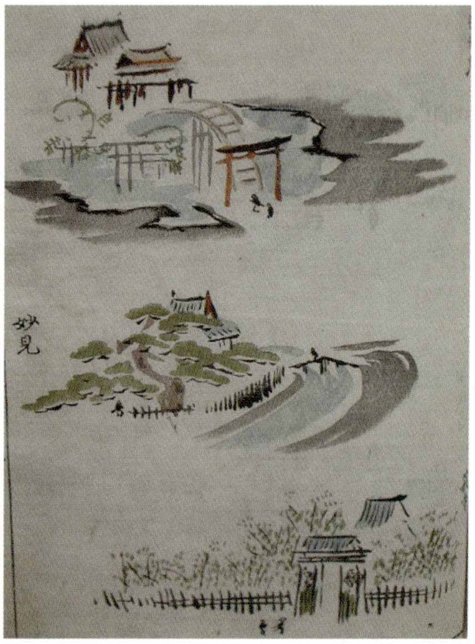

***Figure 13.** Page from Keisai Masayoshi,* Sansui ryakuga shiki, *1797. Collection of Arthur and Charlotte Vershbow*

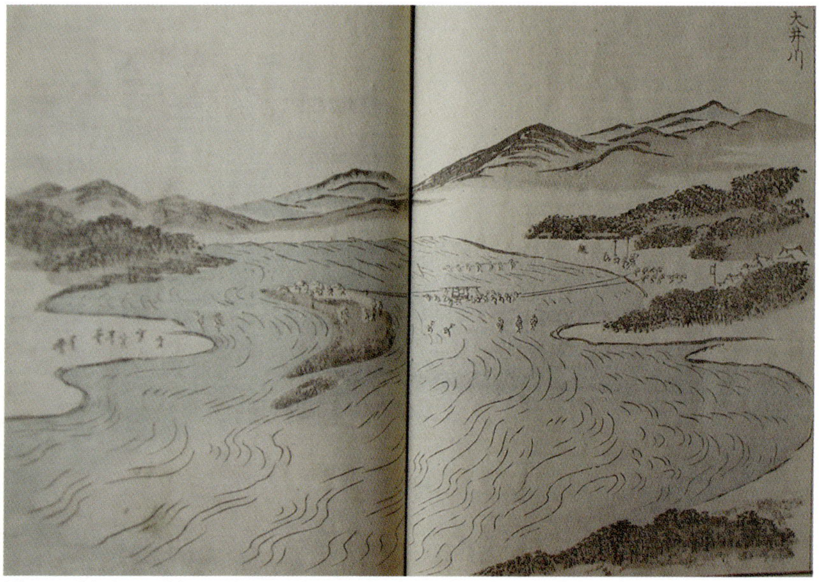

***Figure 14.** Spread from Keisai Masayoshi,* Sansui ryakuga shiki, 1797

with white), which contrasts very subtly with the water, an effect we will see in the other books by Utamaro.

In another sheet from *Gifts of the Ebb Tide* we again see a variety of blues, from vivid to pale, complemented by the different strengths of safflower red, and we also see the accidental effects of moisture on the colorant (fig. 12). The mottling in the pale blue of the screen at lower right and the brown that is part of the darker blue in the painting of the river depicted on the folding screen are both the result of water damage. To reiterate: of the three blue colorants commonly used in printmaking, only dayflower blue behaves this way.

The second book is Masayoshi's *How to Draw Simple Landscapes* (figs. 13 and 14). In contrast with the last example, this one features a palette of delicate tints. Throughout, the dayflower is printed as a very pale, but distinct, blue, and is not faded (see fig. 13). As usual, each color is printed from

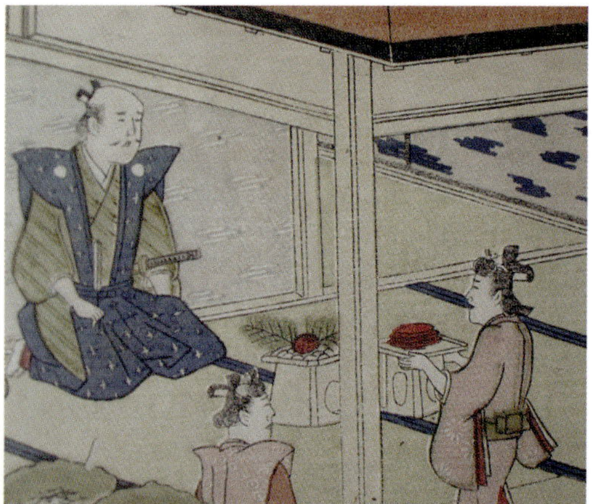

***Figure 15.** Detail of page from Kitagawa Utamaro,* Waka Ebisu, *ca. 1789, showing multiple shades of blue. Collection of Arthur and Charlotte Vershbow*

a separate block. In another sheet water is depicted by means of gradation printing of the dayflower blue (see fig. 14). Note how, at the bottom of the page, the blue curves around the pale space above it, distinguishing the printed effect from accidental fading. In the hands of skilled printers dayflower blue was beautifully suited to depict the varied hues of sky and water and the ever-changing effects of light.

Next is an example from Utamaro's *Young Ebisu*, the earliest book in the group, and it displays a palette of contrasts in which colors are shown both full strength and dilute in the same image (fig. 15). The versatility of dayflower blue is used to great advantage, both as dense blocks of brilliant color and as the palest of washes. Pale blue is used in both the vertical panels, in subtle contrast with the pale yellow of *tatami* (floor mats). The pooling of color around the edges of the blank pattern elements reveals the blue-gray hue that might otherwise be mistaken for a light mineral gray.

For the last book, Utamaro's *The Picture Book of Annual Events in the Green Houses*, we have two early printings of the same book with consistently distinct color palettes. Both palettes were also used in other copies of the book. Dayflower blue is used in its fresh *aigami* form throughout one printing (fig. 16) and in its aged form in the other (fig. 17). One cannot tell from the colors alone which version is closer to the original artistic intent or which is the earlier printing. Close examination of the black outlines of the keyblocks for signs of wear is a more reliable guide to printing history.

In figure 16 the white spot in the fresh dayflower of the sliding screen on the left is water damage. No color remains in the center, and the tideline is distinct. At first glance it could be mistaken for a design element. We also see a typical example of the purple composed of dayflower blue and safflower red in undamaged condition. It was a popular color, but, because of the dayflower blue it contains, it is also highly susceptible to moisture. In figure 17 the purple is browner in tone.

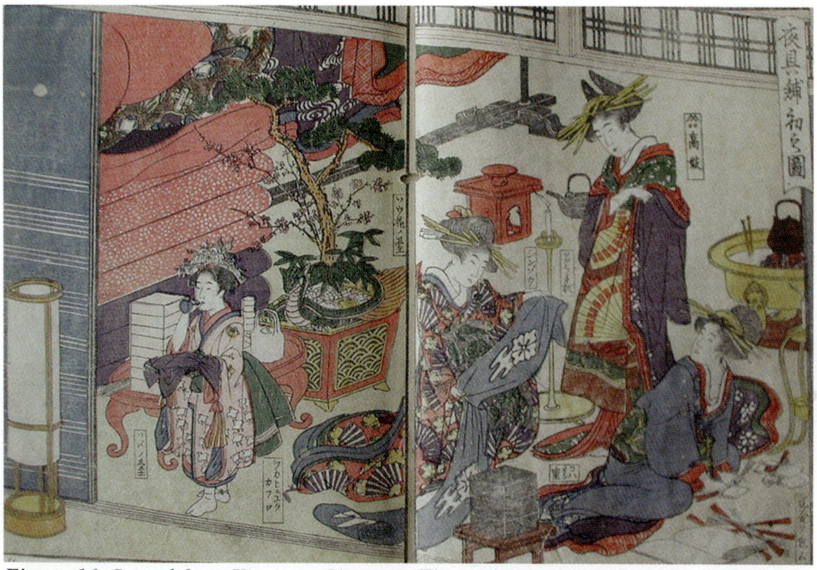

Figure 16. *Spread from Kitagawa Utamaro,* Ehon seirō nenju gyōji, *1804. Collection of Arthur and Charlotte Vershbow*

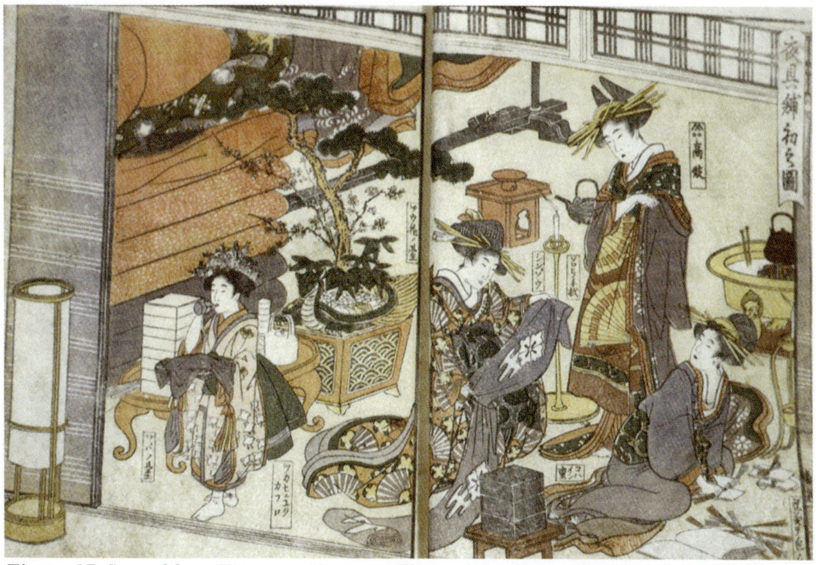

Figure 17. *Spread from Kitagawa Utamaro,* Ehon seirō nenju gyōji, *1804. Private collection*

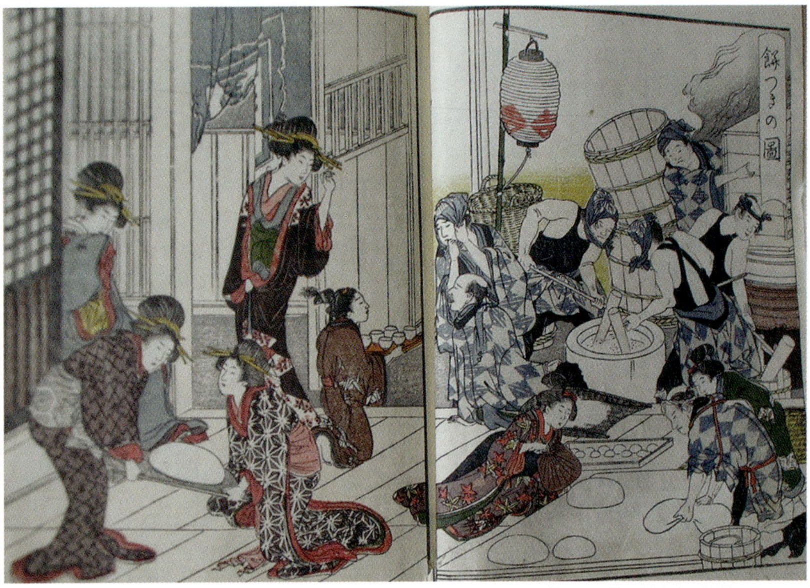

Figure 18. *Spread from Kitagawa Utamaro,* Ehon seirō nenju gyōji, *1804. Collection of Arthur and Charlotte Vershbow*

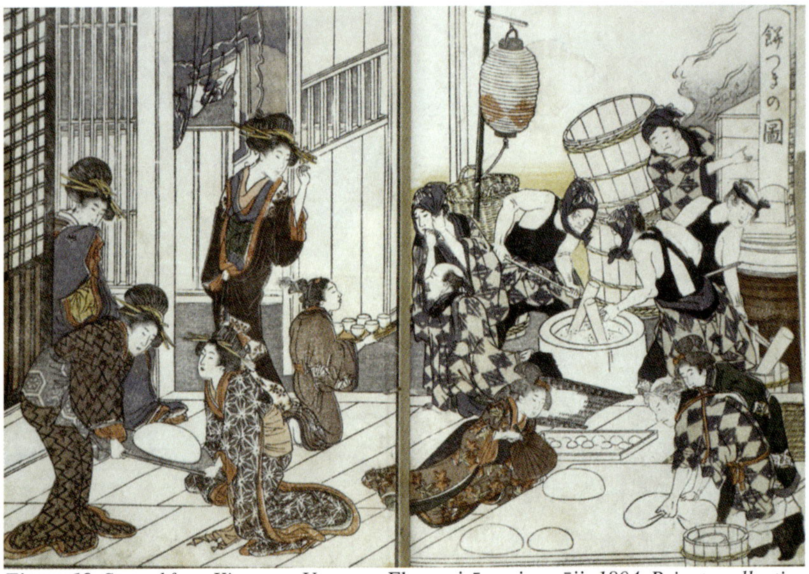

Figure 19. *Spread from Kitagawa Utamaro,* Ehon seirō nenju gyōji, *1804. Private collection*

In another sheet from *The Picture Book of Annual Events*, dayflower is used both as blue and as pale blue-gray in the pattern on the servants' robes (fig. 18). In the corresponding sheet from the other printing there is water damage characteristic of dayflower blue from aged *aigami* in the curtain on the left page (fig. 19). The color does not disappear as completely when moistened as it does with the fresh version. Although this aged color looks like a purple, the absence of pink in the water spot proves otherwise.

Conclusions

We have provided further information about the appearance and physical properties of dayflower blue. In contrast with earlier conclusions, this scientific study of light behavior concludes that the color is only moderately sensitive to light, even when ultraviolet radiation is included in the spectrum. Examples of dayflower blue in Japanese printed books of the late eighteenth and early nineteenth centuries illustrate the varied appearance of the colorant in its unfaded state and show how its presence may be identified by means of its striking sensitivity to moisture.

Acknowledgments

The authors would like to thank Pauline Webber, Boris Pretzel, and the Conservation Department staff at the Victoria and Albert Museum; Sandra Connors and Paul Whitmore at the Research Center on the Materials of the Artist and Conservator, Carnegie Mellon University; Roger

Keyes at the Center for the Study of Japanese Prints; Arthur Vershbow, whose books we use to illustrate the use of dayflower blue; the staff of the Intermuseum Conservation Association; Shinya Katsuhara, woodblock print artist; and Mark Sandy and Debbie Glynn at the Science Research Room at the Camberwell College of Arts.

Notes

1. The quantity for making a sizing agent, or *dōsa*, was water 1.8 l, *nikawa* (animal skin glue) 38g, and *myōban* (alum) 26g. The *dōsa* was applied onto both sides of *hōsho* paper using a wide brush, and the paper was hung to air-dry.
2. The tester was a Microscal Light Fastness tester with sample cells and cooling capability, which was operated using a Nestlab RTE 210 circulation water bath.

References

Connors S., P. Whitmore, R. Keyes, and E. Coombs. 2005. The identification and light sensitivity of Japanese woodblock print colorants: The impact on art history and preservation. In this volume, 35–47.

Feller R., M. Curran, and C. Bailie. 1984. Identification of traditional organic colorants employed in Japanese prints and determination of their rates of fading. In *Japanese woodblock prints: A catalogue of the Mary A. Ainsworth Collection*, ed. R. Keyes, 253–66. Oberlin, Ohio: Oberlin College.

Sasaki, Shiho, and P. Webber. 2002. A study of dayflower blue used in *ukiyo-e* prints. In *Works of art on paper—books, documents and photographs: Techniques and conservation*, ed. V. Daniels, A. Donnithorne, and P. Smith, 185–88. London: International Institution for Conservation of Historic and Artistic Works.

Whitmore, P., Xun Pan, and C. Bailie. 1999. Predicting the fading of objects: Identification of fugitive colorants through direct nondestructive lightfastness measurements. *Journal of the American Institute for Conservation* 38:395–409.

Recent Research on Historic Paper Components in East Asian Art Objects

Agnieszka Helman-Ważny

ABSTRACT This study examines the characteristics of historic paper components in art objects from Korea, China, and Japan that are preserved in Chungbuk National University Museum in Cheongju, South Korea, and the Museum of Asia and the Pacific in Warsaw, Poland. Knowledge about the fiber composition of Korean historic objects could yield information on Korean papermaking in the past. Of thirty-five samples taken from Korean paper, thirty-four contained paper mulberry fibers. In contrast, the Chinese and Japanese papers had various fiber components including paper mulberry, mulberry, cane, rice straw, and Japanese *mitsumata* and *gampi*. Observation under a scanning electron microscope revealed that the Korean specimens consisted of long fibers, very well preserved. The paper mulberry fibers observed under a light microscope and a scanning electron microscope reflected the condition one would expect of historic paper objects. Energy-dispersive x-ray analysis tests were also performed. The examined papers differed in the content of silica, potassium, aluminum, and calcium. The future condition of the paper can be predicted by taking into account the high content of some of the elements. Unfortunately, the limited number of fiber types found, primarily paper mulberry, did not allow clear differentiation of the art objects to establish their places of origin. These are only initial results, and the trace elements found in paper may or may not reflect differences in factories and/or areas of production.

Introduction

The invention of paper was a most important development as it became the basic medium for writing. Paper opened up new ways of communication, record keeping, and artistic expression. According to historical sources, the invention of paper can be traced back into Chinese history (Dąbrowski and Siniarska-Czaplicka 1991; Dąbrowski 2001). Previously, various materials were used for writing, including stone, animal bones, wooden tablets, bark, bamboo, and silk cloth. Although written sources indicate otherwise, the origin of paper is still not clear. The introduction of papermaking is commonly attributed to Cai Lun of China, but we know that paper existed in China before his time, from approximately 200 B.C.E. (Pan 1981). It is not recorded when paper spread from China into Korea and Japan, but scholars believe that paper may have come to Korea along with Buddhism, in the fourth century (Jelisejev 1988). The origin of traditional Japanese handmade paper, *washi*, is said to go back to the seventh century (Hunter 1978).

Particularly in East Asian countries (Korea, China, Japan), bast fibers are used for traditional papermaking. These bast fibers are derived from living plants. In China, these plants include mulberry (*Morus alba* and *Morus lhoe*, which is used during the process of silk production), paper mulberry (*Broussonetia papyrifera*), bamboo, and rice straw. Sometimes the Chinese added flax, hemp, ramie, or climbing rattan as components (Winter 1985). The best-known Chinese paper for calligraphy and other works of art is *xuan* paper (*xuan zhi*). To make this paper, depending on the factory, bamboo, mulberry, and rice straw were used. The papermaking process was also described as specific (Koretsky 1986). In Korea, paper mulberry (*Broussonetia papyrifera* and *Broussonetia kazinoki*) is the most popular for traditional paper production (Kim 1983–84; Dam 1994; Kim 2001). The Korean name for paper mulberry is *dak*. Traditional paper in Korea is called *hanji*, which means, literally, "Korean paper" (*han* = Korea, *ji* = paper). The Japanese made traditional paper mainly from three wood species: *Broussonetia kazinoki* termed *kōzo*, *Edgeworthia papyrifera* termed *mitsumata*, and *Wikstroemia sikokiana* termed *gampi*. Sometimes other materials such as bamboo or hemp were used. This traditional paper is called *washi*.

In Europe flax, hemp, cotton, and even, accidentally, wool were used for traditional paper production from the thirteenth to the nineteenth century. Sources of these fibers were textile wastes, both new cuttings and old rags. Living plants were another source for papermaking, but the process of deriving fiber from them started later. Pulping was usually accomplished using an alkaline process. The pulps were usually used as a blend of cut fibers. Before the nineteenth century, all paper was handmade. The traditional paper of East Asia that was made from bark fibers is commonly referred to as rag paper. Rag papers are generally very durable and can be preserved in favorable condition for hundreds of years.

The object of this study was to characterize the properties of paper used for artifacts in Korea, China, and Japan. Another topic was the comparison of Korean, Chinese, and Japanese raw materials and technology in order to differentiate the papers from East Asian countries. It was hoped that it would be possible to establish the origin of an artifact on the basis of the material and technology used in the process of papermaking. The first step for the analysis of historic materials is the identification of fiber composition. Fibers constitute the basic component of a paper sheet, and determination of fiber composition is essential to characterization of the paper.

Materials and Methods

Materials

The art objects examined are preserved in the collections of East Asian art in Chungbuk National University Museum in Cheongju, South Korea, and the Museum of Asia and the Pacific in Warsaw, Poland. Forty-five samples were taken from twenty-eight artifacts, such as a book leaf, paper cloth, and wallpaper. Their origins were known: eighteen from Korea, five from China, and five from Japan. All the artifacts were made of traditional, handmade paper. Because the dating of Chinese and Japanese objects was not confirmed, it is possible that twentieth-century materials may also be present. It should be mentioned that these twenty-eight artifacts may not fairly represent traditional methods of papermaking for these two countries, so Chinese and Japanese artifacts were used as reference material for the Korean samples.

Characteristics of Paper in the Samples

Generally the papers in the Korean samples had a bright color. The surface of the papers was semiglossy. The Japanese papers were very similar to the Korean papers, also with a semiglossy surface. The Chinese paper samples generally were darker in color and bore more visible traces of the mold on their surfaces than did the Korean and Japanese papers.

Preparation of Samples

Samples were examined by light microscopy, scanning electron microscopy (SEM), energy-dispersive x-ray analysis (EDX), and image analysis.

Light Microscopy
Each sample was divided into small pieces, placed in a small beaker, immersed in distilled water, and boiled in an autoclave. The water was decanted, and the samples were drained off. About 0.2 g of paper was dispersed on the glass slide. Fibers were treated with Herzberg stain. The stain was applied to the fiber field on the glass slide and covered with another piece of glass.

Scanning Electron Microscopy
A small piece of paper was placed in a horizontal position directly on wet graphite glue on the SEM stub. The specimens were dried in a drying oven at 100°C for about ten hours and coated with gold under vacuum using an Ion Coater Eiko IB-3.

Observations

Light Microscopy
The stained material was observed under an Olympus BX 50 microscope under transmitted light. The slide was examined for different fibers, according to the Polish Standards PN-72/P-04604 and PN-76/P-50125, with attention given to their morphological characteristics (the details of anatomical structure) and reaction to the Herzberg stain. Results were compared with reference materials. An Olympus camera attached to the microscope was used with transmitted light, and photographs were taken using color ISO100 Kodak film. The photographs were taken at magnifications of 40×, 100×, 200×, and 400×.

Scanning Electron Microscopy
The paper was observed with SEM (HITACHI S-2500C) in magnifications ranging from 150× to 2000×. The samples were examined for the characteristics of paper structure and fiber anatomy, as well as for information about fiber degradation. Photographic documentation of characteristic features was made using black-and-white Polaroid film.

Energy-Dispersive X-Ray Analysis
EDX (Kevex Sigma) connected with SEM analysis was used primarily to evaluate inorganic materials. This technique permitted simultaneous detection of many elements, together with some estimation of their relative amounts. Nonuniformity in distribution can be shown, and the presence of extraneous materials can be detected (Browning 1977). Two or more particles found on the fiber surfaces, chosen randomly from distinct parts of each paper sample, were analyzed.

Image Analysis
Paper was examined for the size of fibers using the Olympus Image Analysis Package. Each kind of fiber that took the stain on the glass slide was measured twenty times for length and width respectively. Minimum and maximum values were calculated, and results were compared with a reference key for the identification of fibers (Choi 1988).

Results and Discussion

Light Microscopy

Light microscopy is normally used to analyze samples for fiber composition (Nunome 1992; Rams and Jarminska 1999). The methods of fiber analysis make it possible to identify the kind of fibers present, but many bast fibers in highly beaten pulp are difficult to identify in this state. Fiber deformed by stamping and drying during the papermaking process can change its shape. It is much harder to identify fibers in historic papers. The main problem is the deterioration of the fiber with time. Because of this aging process, the morphology of the fiber can change. For example, changes in the dimensions, shape, and width of the lumen occur. In some cases of highly degraded pulp, it is impossible to recognize fibers from their morphology. In addition the presence of other ingredients in paper, such as glue, fillers, and sometimes fungi, can make these analyses difficult.

The analyzed material was generally a fiber-composed paper in good condition. Because of Asian papermaking technology, it did not contain many other ingredients, like glue. During observation, attention was given to the fibers' morphological characteristics and reaction to staining. In the identification of the fiber in pulp, attention was paid to the following features: the general shape of fibers, dimension of fibers (mainly width), cross-markings, type and dimension of lumens, shapes of the ends of the fibers, irregularities in fiber walls, the kind and size of vessel elements if any, and the size and shape of parenchyma cells if any (Florian et al. 1997; Helman-Ważny et al. 2000).

Of the thirty-five samples taken from Korean papers, thirty-four contained paper mulberry fibers, including two samples that additionally contained straw, mulberry, and unidentified fibers. The sample from the book cover of the certificate *Royal Writ of Appointment to Lee, Myung-joon, Lee, Jae-shim and Lee, Jae-Ki*, Chungbuk National University Museum, was established as composed of *Daphne* and grass; the exact type of grass could not be further specified. In comparison to these results, the Chinese and Japanese papers had various fiber components including paper mulberry, mulberry, hemp, sugarcane, rice straw, and Japanese *mitsumata* and *gampi*. Results were compared with reference material (Collings and Milner 1978, 1979; Ilvessalo-Pfäffli 1995, 307–51; Choi and Cho 2000) and in some cases with slides prepared directly from raw material (straight from the plant).

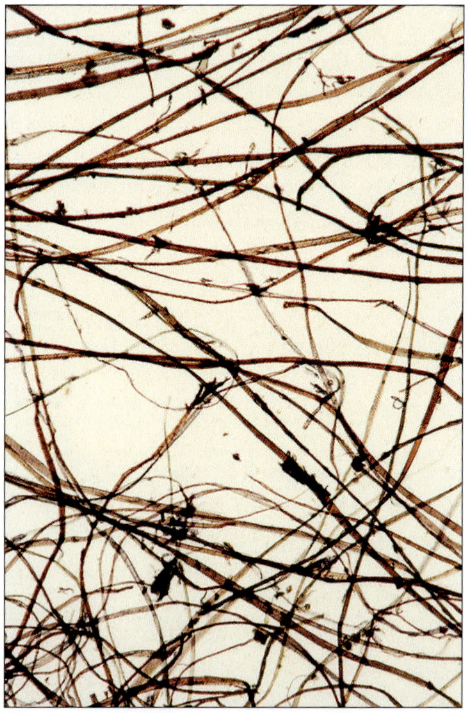

Figure 1*. Light microscope image of fibers from the colored woodcut print* Women in the Garden, *late ukiyo-e character, Japan, nineteenth century. Museum of Asia and the Pacific, Warsaw. Paper mulberry fibers stained with Herzberg stain, 40×*

Figure 1 shows an example of paper mulberry fibers in a Japanese historic paper. In some cases in the sample composed of paper mulberry, 20 percent of the fibers were stained black and differed in anatomical characteristics. In the past, Koreans also sometimes used additional fibers as a decoration. After staining with the Herzberg stain, the fibers of paper mulberry had a reddish brown to a pale purple color. Blunted fiber ends were found. The lumen was marked but interrupted at intervals. Cross-markings were clear but irregular. During observation under the microscope with transmitted light it was possible to find fibers and transparent membranes (stained by the Herzberg stain a violet to bluish purple) associated only with paper mulberry fibers. This transparent membrane in the raw material usually envelops the fiber. In historic papers, the membrane was torn on tested slides into small pieces or often separated from the fibers and appeared like a ribbon. In some slides, only a small quantity of the membrane was found. The condition of the membrane depends on the manufacturing technique used and the quality of the raw material.

Image analysis revealed that fiber length ranged from 2 to 17 mm. This information tells us rather more about the papermaking process than about fiber lengths. The widths of the fibers were found to be between 4.5 and 26 µm.

No morphological properties that could help identify country of origin of an artifact were observed.

Chinese samples varied in their content. The five samples were composed of (1) ramie with a very small amount of mulberry and *Daphne*, (2) bamboo, (3) *Daphne* and grass, (4) paper mulberry or mulberry, and (5) bamboo.

Three of the five samples of Japanese paper contained paper mulberry fibers (with the probable addition of hemp). One was composed of *mitsumata* (*Edgeworthia*); the final one contained sugarcane, jute, paper mulberry, *gampi* (*Wikstroemia*), and softwood.

Scanning Electron Microscopy

The scanning electron microscope can depict a paper's surface at higher magnifications and with more three-dimensional information than conventional light microscopy. Conventional light microscopy provides a maximum useful magnification of 1500–2000× and has a short depth of focus. SEM clearly reveals the three-dimensional structure of the fibers with a useful range of 10–100,000× magnification (Michaels and Boyd 1986). The SEM examination was undertaken to determine the condition of fiber components in historic papers from Korea, China, and Japan. This method also confirmed the fiber identification.

With SEM, it was possible to observe the topography of the paper in its finest details. Morphology in three dimensions, the extent of damage to the fibers, and mold growth were visible. Inorganic elements and glue were seen to surround the anatomical parts of the plant and fungi. Fibers were covered with a coating that consisted of many small particles. The coating was very difficult to identify clearly during SEM observation. Consequently, EDX was performed, as described below. The paper mulberry pulp was composed of good-quality long fibers, and the paper structure was in good condition. Some of the fibers looked degraded, but in general there were no significant differences between the samples from different artifacts. In a few samples, more glue and additional substances were used, illustrated in figure 2. The character of the surface depends not only on the influence of degradation but also on the papermaking process (fig. 3).

In magnifications ranging from 70× to 150×, the direction of the fibers on the paper surface was observed. On the surface there are no differences to indicate distinctions among Korean, Chinese, and Japanese samples. But under magnification, differences in fiber placement resulting from mold construction and sheet-forming technology were clearly visible. The mold for the traditional Korean paper does not have a surrounding frame or deckle; instead, a sheet of paper is formed from successive layers placed in alternate perpendicular and parallel directions. This traditional technique is called *oebal-cheoji*. In this method the sheet mold is dipped into the paper stock, which is then thrown in the forward direction. Then the paper stock is scooped again and rhythmically rocked from side to side; this procedure is repeated several times, causing the fibers to intertwine and paper layers to be formed. The water drains slowly through the bamboo screen (Choi et al. 2001). The result is the perpendicular arrangement of fibers in consecutive layers in the paper structure. SEM photographs of historic Korean papers showed that the fibers were alternately parallel and perpendicular to one another. That this characteristic was closely connected with paper forming was the conclusion of scientists from Chungbuk National University Museum in South Korea, based on their studies of historic papers (Choi et al. 2001). In Japanese paper, the paper pulp is taken into the mold and water is drained evenly through the sieve. The effect is that the long fibers are not placed in one direction. There were also differences related to the kind and condition of fibers. In the Korean papers a high glue content was visible only in samples taken from book covers and wallpapers.

Energy-Dispersive X-Ray Analysis

EDX was used to establish paper origin and to determine the presence of elements probably harmful to the paper, as well as to check the distribution of components in the paper. The high content of some elements should make it possible to predict the future condition of the papers. EDX can determine the elements on the surface of the fibers that are additional ingredients. Examples of the particles occurring on the fiber surface are shown in figure 4. During the process of papermaking particles get into the paper pulp in a number of ways. They may originate from the soil or in groundwater absorbed by the plant. More likely, they come from the water, fillers, and other chemicals added to the pulp during the process of papermaking. A study of the raw material should answer the question. Particles can also get into the paper during conservation and restoration treatments, as, for example, deacidification compounds and bleaching chemicals. Those elements originated from added compounds, and it is very difficult to recognize their origin. In the case of old European papers made of rags, we cannot expect the contents of trace elements to give any useful information. The waste rags, which bear accidental contamination, were treated with calcium hydroxide during processing. Also, for example, the hammers used for smashing and grinding fibers were made of different kinds of stone, and later iron or bronze. All are potential sources for trace elements. It is clear that trace elements are connected mainly with the papermaking technology, and tests for them are presently used to control the process of pulping in factories. But Asian papers produced in a traditional way for hundreds of years should be free of the chemicals used now. Also, the water used for papermaking should have been taken directly from the river and not from the tap. Potentially it may be possible to find features characteristic of specific factories or to determine differences among papermaking methods used in

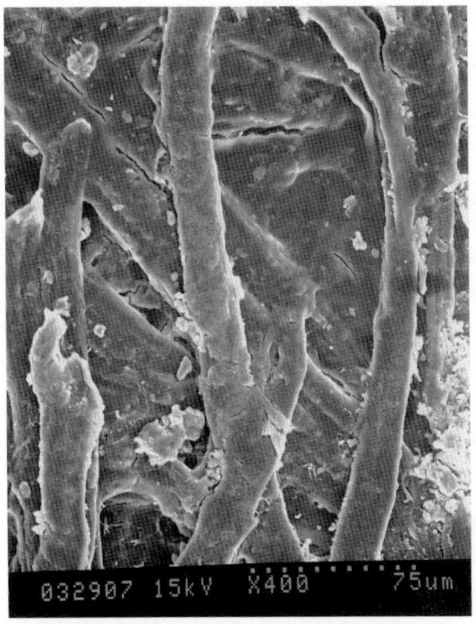

Figure 2. *Secondary electron image of fibers from the certificate* Royal Writ of Appointment to Lee Myung-joon, Lee Jae-shim and Lee Jae-Ki, *Korea, eighteenth century. Chungbuk National University Museum. There is a glue coating on the surface of the paper.*

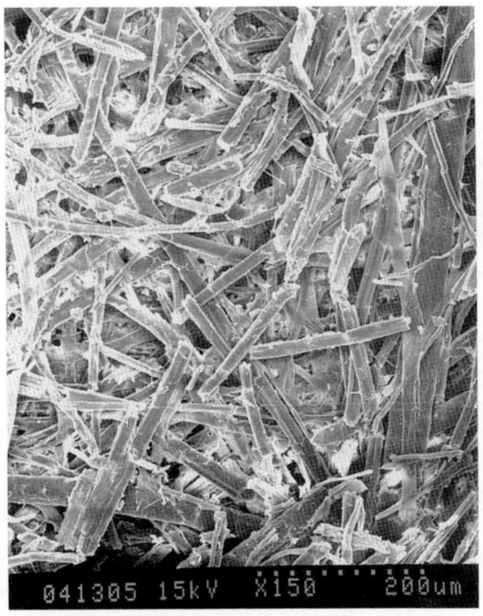

Figure 3. *Secondary electron image of fibers from* Letters of Mrs. Kim (Sunheon) to Her Daughter (Married), *Korea, sixteenth century. Chungbuk National University Museum. The paper is composed of unusually short paper mulberry fibers.*

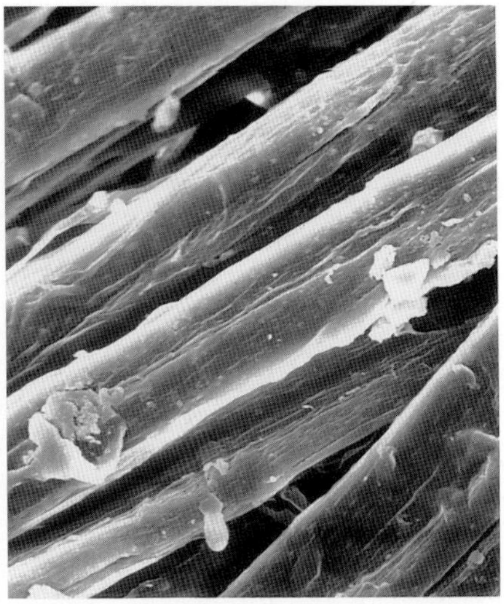

Figure 4. *Secondary electron image of fibers from the certificate* Royal Writ of Appointment to Mr. Lee Gyoe-Heung, *Korea, nineteenth century. Chungbuk National University Museum. Crystals are visible on the fiber surface.*

Korea, China, and Japan. Mabuchi used neutron activation analysis to detect differences among factories and to establish the origin of modern papers (Mabuchi et al. 1979). However, limited analysis created problems in interpretation and prevented the Mabuchi team from achieving conclusive findings. A similar difficulty occurs with EDX and is further aggravated by the historic nature of samples.

The content of trace elements occurring on the fibers was analyzed by EDX. EDX indicated that silica, potassium, aluminum, and calcium were the main components of particles in the paper pulp. Magnesium, titanium, chlorine, iron, sodium, chromium, phosphorus, and sulfur were also detected, but only in a small number of samples. The elemental content of the particles within the objects varied, but often a high content of silica and calcium, as well as aluminum and sulfur, was found repeatedly in different samples. Because for the most part two particles were analyzed from each sample, and only one sample was taken per object, results were not completely representative. Therefore, it is possible to make conclusions only about the amounts of trace elements in the different papers in one object.

One object is often composed of many kinds of paper, for example, books of letters or diaries. One such object is the certificate *Royal Writ of Appointment to Lee, Myung-joon, Lee, Jae-shim and Lee, Jae-Ki*, which consists of a book cover and four chapters dated to 1756, 1785, 1783, and 1814. Six samples were taken of the object, from each chapter and the book cover, because different kinds of paper were used. Within each sample, two particles were analyzed. The results are shown in figure 5.

It was possible to characterize the papers in the certificate by the content of trace elements. Generally, the content of calcium is high, as is the content of silica and sulfur. Graphs were made for the major and minor elements that occurred most often in the particles. The amount of one element is often similar, in spite of the fact that different kinds of papers were used in the object. It should be taken into consideration that the amount of trace elements in the paper is connected with the preparation of the object, because different kinds of paper were used and the date of each part is different. The amounts of elements in different parts of the book were discovered to be similar.

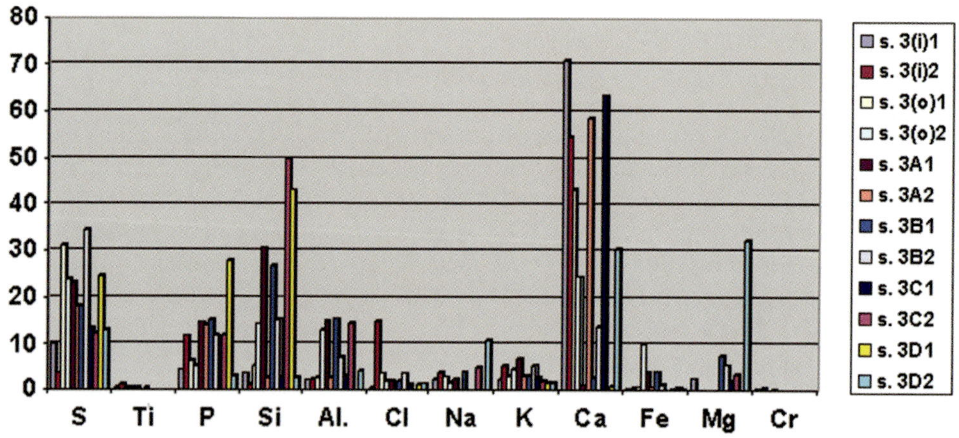

Figure 5. *Weight percent of major and minor elements found in the paper of the certificate* Royal Writ of Appointment to Lee Myung-joon, Lee Jae-shim and Lee Jae-Ki *(fig. 2). The variability of the elements within one object is due in part to the six kinds of paper used in the object. Two crystals were analyzed from each paper sample.*

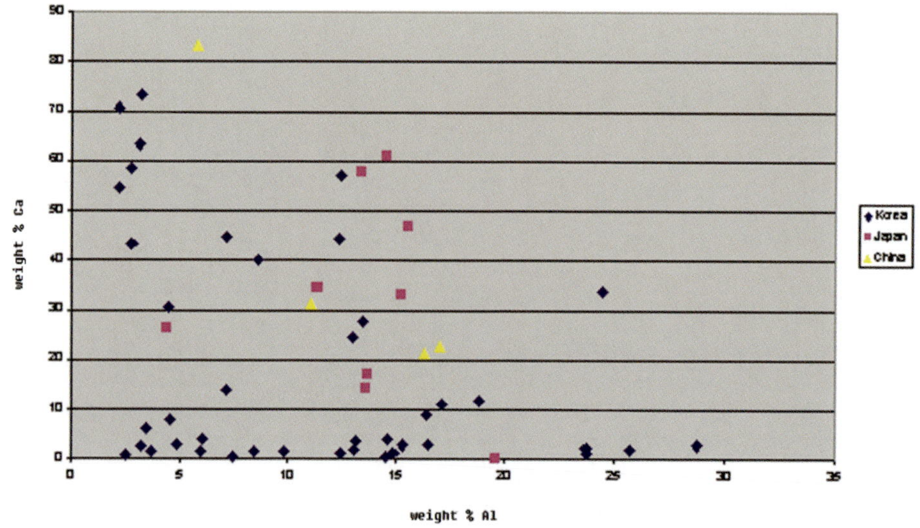

Figure 6. *Weight percent aluminum plotted versus weight percent calcium found in crystals attached to fibers from historic East Asian papers*

The correlation of calcium to aluminum, the two elements found most often in the particles, is shown in figure 6. The ratio of calcium to aluminum itself is unimportant. The content of aluminum in the Korean samples ranges from 2 to 30 weight percent in the particles. The content of calcium in the Korean samples is between 1 and 75 weight percent. The content of aluminum, as well as of calcium, is in the same range for samples from all three countries in the study, as illustrated in figure 6.

Unfortunately, it was not possible to attribute different papers to specific factories. The results of the investigation do not show a direct connection between trace elements and paper origin.

Assuming that manufacturing conditions, such as plant material or materials added, are fairly constant in one area or factory, EDX tests were made, but the results were not conclusive. Variations in the trace elements on the surface of fibers in the paper were not clearly present within the analyzed group of artifacts. No systematic studies have ever been made on the trace element content of historic papers, so it was possible only to report results, not to do comparative studies. To determine the difference between factories and areas, future research is needed, systematically focusing on historic papers in Japan and China to provide the reference data necessary for differentiating paper from different countries. It would be very helpful to know not only the basic distribution of the elements in the paper but also the content of compounds in the paper. Then interpretation would be easier.

Knowledge about the presence of potentially harmful elements in the paper, and their amounts, could be very useful for characterizing the material and can help in decision making before conservation treatment. The high content of some elements makes it possible to predict the future condition of the paper.

The papers made in East Asia are chemically durable. The pulping process involves cooking in solutions of alkalis, soaking, and beating. These procedures make the paper acid-free. The papers are also durable because of the characteristic long fibers of different kinds like paper mulberry or *gampi*, all of which have only a trace amount of lignin as compared to flax. And even if alum is added to the paper furnish, the papers are stable to acids (Winter 1985). The pH of newly made paper is neutral or basic. Quality depends also on the degree to which impurities were removed and to which the outer bark remained in the paper pulp.

Conclusions

Analyzing the components of paper helps characterize historic papers. Tests revealed that Korean historic paper from the sixteenth to the nineteenth century was made mainly of paper mulberry fibers called *dak* and that the objects were preserved in good condition. The condition of paper mulberry fibers observed under the light microscope and SEM confirmed the condition of the historic paper objects.

Analysis of Japanese samples showed that the papers were also composed of paper mulberry fibers called *kōzo* and also of *mitsumata* and *gampi*. Further research is needed into the raw material the Japanese used for papermaking in the past. Analysis of Chinese papers and previous work by the author revealed that they had a variety of fiber components, including paper mulberry, mulberry, grasses, and rice straw as well as bamboo and *Daphne*.

According to the EDX results, the papers examined differed in the content of silica, potassium, aluminum, and calcium. Unfortunately, analysis of the fiber content of one type—paper mulberry—did not permit differentiation of the art objects to establish the places of origin of the papers. These initial results indicate, however, that the trace elements in the papers may be characteristic of different factories and areas.

Acknowledgments

I am grateful to the following organizations for their support: Chungbuk National University Museum and Chungbuk National University in Cheongju, South Korea, as well as the Museum of Asia and the Pacific in Warsaw, Poland. Also, the following persons have provided generous information and advice: Professor Nam-Seok Cho, Professor Tae-Ho Choi, Professor Alicja Strzelczyk, and Professor Won-Kyu Park, as well as Iza and Wojtek Lubczynski, and Maya Slowinska. I also wish to express my gratitude to the Foundation for Polish Science for supporting my research through the Annual Stipend for Young Scientists in the year 2003 and 2004.

References

Browning, B. L. 1977. *Analysis of paper*. 2nd ed. New York: Marcel Dekker.

Choi, T.-H. 1988. On the qualities of bast fiber pulp from several woody plants. PhD diss, Department of Forest Products, Chungbuk University, Cheongju, South Korea.

Choi, T.-H., and N.-S. Cho. 2000. Identification of fiber in the Korean traditional paper (*hanji*). In *Identification and characteristics of archaeological woods and traditional papers*, ed. W.-K. Park, 18–27. Cheongju: Chungbuk National University.

Choi, T.-H., N.-S. Cho, I.-H. Choi, and T.-S. Jeong. 2001. Comparison of physical properties of *hanjis* made by different sheet forming processes. *Korea TAPPI* 33(4):21–27.

Collings, T., and D. Milner. 1978. The identification of oriental papermaking fibres. *Paper Conservator* 3:51–79.

Collings, T., and D. Milner. 1979. The identification of non-wood paper-making fibres, pt. 2. *Paper Conservator* 4:10–19.

Dąbrowski, J., and J. Siniarska-Czaplicka. 1991. *Ręzieło papiernicze*. Warsaw: Sigma Not.

Dąbrowski, J. 2001. Historia papieru: Papier zasadowy. In *Kwaśny papier: Zagrożenie zbiorów bibliotecznych i archiwalnych*, ed. T. Łojewski and Z. Pietrzyk, 12. Cracow: Biblioteka Jagiellońska.

Dam, V. Y. 1994. *Hanji*: The development and production of traditional Korean paper. In *Korean cultural heritage: Fine arts*, 154. Seoul: Korea Foundation.

Florian, M.-L. E., D. P. Kronkright, and R. E. Norton. 1997. *The conservation of artifacts made from plant materials*. Los Angeles: J. Paul Getty Trust.

Helman-Ważny, A., N.-S. Cho, T.-H. Choi, and W.-K. Park. 2000. Identification and characteristics of traditional papers as the first step of conservation treatment in Korea. In *Identification and characteristics of archaeological woods and traditional papers*, ed. W.-K. Park, 8–17. Cheongju: Chungbuk National University.

Hunter, D. 1978. *Papermaking: The history and technique of an ancient craft*. 3rd ed. New York: Dover Publications.

Ilvessalo-Pfäffli, M.-S. 1995. *Fiber atlas: Identification of papermaking fibers*. Series in Wood Science. Berlin and Heidelberg: Springer.

Jelisejev, D. D. 1988. K voprosu o rolji rukopisnoj knigi w kulturie Korei. In *Rukopisnaja Kniga w Kulturie Narodov Vostoka*, ed. O. F. Akimuškin et al., 294. Moscow: Izdatelstvo Nauka.

Kim, S.-K. 2001. Traditional papercrafts: Handicrafts of ordinary people. *Koreana: Korean Art and Culture* 15(1):73–77.

Kim, Y.-Y. 1983–84. Making paper by hand in Korea. *IPH Yearbook of Paper History* 4:253.

Koretsky, E. 1986. Papermaking today in Tibet and China. *Hand Papermaking* 1:2–6.

Mabuchi, H., T. Kadokura, and K. Notsu. 1979. Activation analysis of traditional Japanese paper. In *Cultural property and analytical chemistry*, 207–20. Tokyo: Tokyo National Research Institute of Cultural Properties.

Michaels, J., and J. Boyd. 1986. Damage and conservation: A book conservator borrows an SEM. *Paper Conservator* 10:73–94.

Nunome, J. 1992. *The archaeology of fiber before your eyes.* Kyoto: Senshoku To Seikatsusha.

Pan, J.-X. 1981. On the origin of papermaking in the light of newest archeological discoveries. *IPH Information* 15(2):38–39.

Rams, D., and D. Jarminska. 1999. Włókna papiernicze: Historia i identyfikacja. *Notes konserwatorski* 2:147–78.

Winter, J. 1985. Paints and supports in Far Eastern pictorial art. *Paper Conservator* 1:24–31.

An Evaluation of *Xuan* Paper Permanence and Discussion of Historical Chinese Paper Materials

M. Brigitte Yeh and Jesse Munn

ABSTRACT The motivation for studying the permanency of Chinese papers was the need to find suitable papers for the conservation of rare Chinese books in the collection of the Library of Congress, Washington, D.C. Fiber identification from an edition of a Ming dynasty printed pictorial, corroborated by bibliographic research into historical paper fibers, may point to a dating hierarchy for this rare book. Related to this study, a brief account of the history of the materials and techniques used in Chinese papermaking is given. These issues, which particularly concern artists and conservators, are addressed in a study that compares the archival quality of approximately twenty types of commonly used *xuan* papers collected from mainland China and Japan. In addition, samples of papers treated with a variety of adhesives and colorants used by the scroll mounting studio at the Freer Gallery of Art, Smithsonian Institution, were similarly evaluated to assess the aging properties of these preparations that are assumed to be stable. Tests for brightness, tensile strength, and pH were conducted on the papers following artificial aging. The testing procedures, conducted at the Research and Testing Division of the Library of Congress, and the reasons for their selection, are explained along with the results of this study and their interpretation.

Introduction

The connoisseur and collector Jan Tschichold offers a somewhat polar means for differentiating between Chinese and Japanese papers: Chinese *xuan* paper, he says, is "soft, yielding and feminine, thin and extremely sensitive [whereas] the Japanese is firm and masculine, tough, strong and durable." He continues, "*Xuanzhi* . . . is comparatively absorbent and has a smooth, soft surface suitable both for almost dry and for wet color" (Tschichold 1970, 41). The practice of painting and the early development of printing in China were predicated on the availability of good paper that was smooth yet durable. *Xuan zhi* 宣紙 (or *xuan* paper) has historically been the type of paper most frequently used by Chinese painters, calligraphers, and printmakers, and remains so today. It is still produced for use by artists and scroll mounters. Despite its long history of use in China, as well as its frequent utilization for *chine collé* prints in Europe and America, there has been scant analytical research on its permanence (Schenck 1999). The purpose of this paper is to help establish parameters and guidelines for researching the relative longevity of Chinese papers produced today in addition to creating an awareness of the gaps in knowledge that require further research.

This investigation into Chinese papers was prompted by two factors. One was a research trip made in 1997 to a handmade paper workshop in China, which revealed that pulp of inferior quality had begun to be substituted for traditional papermaking materials for economic reasons. Papermakers in China are notoriously reluctant to reveal the source of their pulp or how it is prepared. In a field in which secrecy is the norm, it was exceptional that this particular papermaker was so candid in disclosing his papermaking materials. Subsequently, the authors began to speculate whether other papermakers might likewise substitute less-archival pulps in their papers to reduce production expenses. As information on the materials used by handmade papermakers in China remains largely proprietary, the properties of these materials must be determined by independent analysis.

Coinciding with this trip was a proposed project to conserve a rare Chinese book from the collection at the Library of Congress: this was the second factor motivating the present study (fig. 1). The rare book is an edition of the printed pictorial *Shizhuzhai shu hua pu* 十竹斎書画譜 (The Ten Bamboo Studio Manual of Painting and Calligraphy), which was first published during the late Ming dynasty (1368–1644) in the city of Nanjing (Yeh 2002). Underlying the efforts to conserve the edition in the Library of Congress was the need to find an archival Chinese paper compatible for lining the embrittled pages of the manual and in-filling the lacunae. To identify which Chinese papers were archival, the authors evaluated their performance with artificial aging techniques. The accelerated aging and testing on the papers were performed according to standards set by the Technical Association of the Pulp and Paper Industry (TAPPI). With the cooperation of Chandru Shahani and his staff in the Research and Testing Division, the authors were able to utilize the accelerated aging equipment and expertise at the Library of Congress.[1]

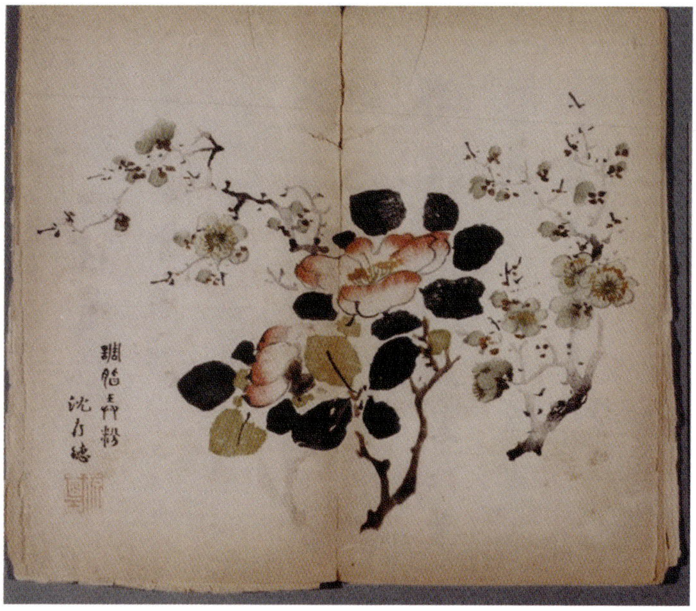

Figure 1. The Ten Bamboo Studio Manual of Painting and Calligraphy. *Library of Congress, Asian Division*

Additional experiments on the Chinese papers were done in collaboration with Xiangmei Gu, conservator of Chinese painting, and her staff at the Freer Gallery of Art, Smithsonian Institution, who were interested in ascertaining the effects of various colorants and coatings on the aging properties of Chinese papers. These colorants and coatings, primarily adhesives, were selected because of their frequent use in the conservation of Asian paintings and during the scroll-mounting process.

As a preface to this paper, one must first draw attention to a very thorough and timely article on the subject of Chinese paper conservation and its analysis by Dianne van der Reyden and Fei Wen Tsai (1997). They conducted similar accelerated aging studies on a variety of types of handmade Chinese papers, some of which overlap with this study. One significant difference was that in the laboratories at the Smithsonian Center for Materials Research and Education (SCMRE), they performed elemental analysis using scanning electron microscopy–energy-dispersive x-ray spectrometry (SEM-EDS), which provided valuable information but which was not available at the Library of Congress. At the time van der Reyden and Tsai published their research, the experiments for this study were already halfway completed. The results in this paper are not intended to duplicate their work.

Background of *The Ten Bamboo Studio Manual*

During the course of research, nine other exemplars of *The Ten Bamboo Studio Manual* were examined from collections around the world for comparison. The research confirmed that the Library of Congress edition is unique because it retains its original binding style, the *hudie zhuang*. This "butterfly binding" style is one in which the single folio images are folded with the printed text and images inside, whereas the unprinted surface remains on the outside of the fold. Each folio section is attached by adhesive applied along the outside edges of the fold.

Despite the rarity of its format, *The Ten Bamboo Studio Manual* in the Library of Congress has also suffered from this binding style: the paper itself and particularly the folds have become increasingly brittle and susceptible to breakage as the paper has aged. Thus it was considered necessary to line the book pages. The Library of Congress edition was printed on a paper that is considered a *xuan* type. The objective was to find an archival Chinese paper that was as similar to the original as possible, to preserve the volumes. Subsequently, the following questions needed to be addressed: What type of paper was used to print the original *Ten Bamboo Studio Manual*? What paper should be used for lining and fill repairs? At the time, the foremost concern was the archival quality of Chinese papers made in recent years.

The Raw Materials for Papermaking in China

The world's first paper has usually been credited to one man, Cai Lun, who was the director of the Imperial Workshops at Luoyang during the reign of Emperor Hedi of the Eastern Han dynasty (25–220 C.E.). In actual fact, he was responsible not for inventing paper but for improving upon existing techniques and extending the range of raw materials to produce good-quality paper. *Jin guan* 金关 paper from Juyan 居延 and *maquanwan* 马圈湾 paper from Dunhuang 敦煌 are some examples of early paper from the Western Han dynasty (206 B.C.E.–8 C.E.), which predates Cai Lun's era. It was not until 105 C.E. that Cai Lun is purported to have used a variety of materials such as tree bark, scraps of hemp, worn-out cloth, and fishnets to produce his high-quality paper. [2] The evolution of fibers used in early Chinese paper manufacture has been suggested by the respected paper historian Tsuen-hsuin Tsien (1973) and is presented in table 1. However, the information provided in this table is derived almost exclusively from historical bibliographic sources. A goal for future studies would be to incorporate archaeological and analytical evidence to document fiber usage in China.

Gradually, as papermaking technology spread, a greater variety of materials began to be utilized or substituted according to their availability. Suitable types of plant fibers include those that are cellulose rich, readily obtainable, and easy to treat: in other words, fibers that would be the most cost effective for use in papers. Plant fibers with higher yields of long cellulose but lower quantities of binding substances are generally the most desirable, since the latter must be removed by either chemical or mechanical means.

Xuan Paper

Xuan paper's history has been dated to the time of the Tang dynasty (618–907), when it was listed among the tribute items presented to the court. Originally produced in the city of Xuancheng in Anhui province, since the Ming dynasty it has been made principally of fibers from the inner bark of the *qingtan* 青檀 (*Pteroceltis tatarinowii*) plant, which is called blue sandalwood or wingceltis by its English common name.[3] During a time that predates the use of chemical bleaches, and when other plant fibers produced off-white or yellow-cast papers, *xuan* paper was prized for its whiteness. Once it was made available, the paper of choice was *xuan* paper. Later during the Qing dynasty (1644–1911), when bleaching was introduced, papermakers in other regions began to produce their own "imitation" *xuan* paper using alternative fibers, such as the ever-present bamboo, which could be whitened with bleaching agents.

Although high-quality *xuan* paper was originally well known as the paper produced around the Xuancheng region and made from *qingtan* pulp, the term *xuan* eventually came to describe any smooth, absorbent paper that was white and suitable for painting and calligraphy. Nowadays, true Anhui *xuan* paper composition is not 100 percent *qingtan* fiber but consists of a mixture of *qingtan* and rice straw. The mixtures will produce varying degrees of strength, smoothness, absorbency, and opacity. The typical compositions of the varying grades offered for sale in art supply shops may be seen in table 2. Other bast fibers that have been used in *xuan*-type papers include *Broussonetia papyrifera* (paper mulberry) and *Edgeworthia chrysantha*. Bamboo also came to be used in the late nineteenth century.

Papermakers in Asia are generally located in remote, mountainous regions near rivers that can provide the enormous quantities of water that the process demands. One of the authors (MBY) journeyed to some of these difficult-to-access locations in order to gain a better understanding of the Chinese handmade paper process firsthand. Zhejiang province is the location of one of the papermakers, where the local craftsmen have traditionally produced a *xuan* paper that is made purely of bamboo. The traditional process of preparing the bamboo pulp is extremely labor intensive and requires upward of six months if done properly. In light of this, one particular workshop owner has opted to use imported pulp in recent years, which he acknowledges is not archival.

Another paper workshop visited is located in Jingxian, western Anhui province (fig. 2). On the day the author visited, an average paper was being made from a pulp that the owner claimed was a mixture of paper mulberry and another fiber. The other half of the pulp mixture he declined to identify, but he stated that the workshop made the pulp. When asked, the owner admitted that chemical bleaching was being used to treat fibers for its average papers. However, special-ordered papers could be commissioned and made without chemical bleaches.

Paper Analysis

By using polarized light microscopy and scanning electron microscopy, the principal paper fibers in the copy of *The Ten Bamboo Studio Manual* at the Library of Congress were identified to be those of giant bamboo (*Bambusa arundinacea*) (fig. 3). Rice straw (*Oryza sativa*) and lesser additions of other, unidentified mixed fibers were also present. Two types of giant bamboo fibers appear in the sample: primary fibers that are long and pointy, and secondary fibers, which

Table 1. *A simplified history of plant fibers used in China (source: Tsien 1973)*

Fiber	Dynasty
Hemp	Western Han (206 B.C.E.–8 C.E.)
Paper mulberry	Eastern Han (25–220 C.E.)
Rattan	Jin (265–420 C.E.)
Bamboo	Tang (618–907)

Table 2. Xuan *paper compositions (source: Cao 2000)*

Type of *xuan* paper	*Qingtan* content (%)	Rice straw content (%)
Special *Jingpi*	70–80	20–30
Jingpi	55–65	35–45
Mianliao	30	70

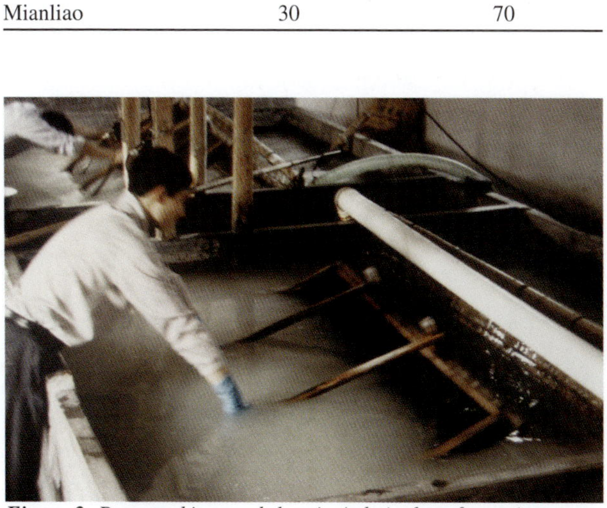

Figure 2. *Papermaking workshop in Anhui: sheet formation*

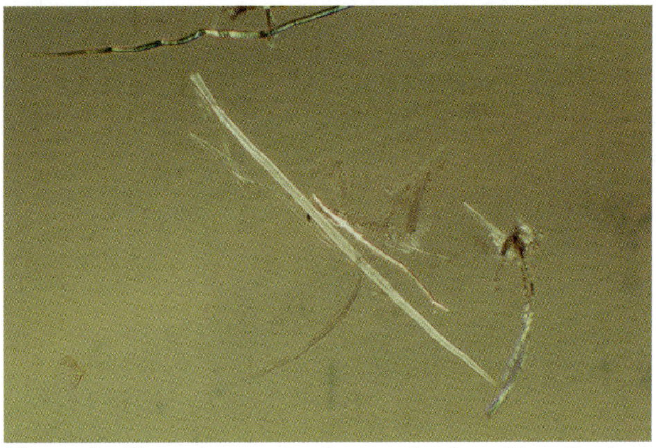

Figure 3. *Photomicrograph of giant bamboo fiber* Bambusa arundinacea

appear twisted. No *qingtan* or blue sandalwood fibers were detected.

That the paper from *The Ten Bamboo Studio Manual* is composed of bamboo is evidence for dating the edition. Bamboo fibers first began to be used in making a type of imitation *xuan* paper sometime in the middle of the Qing dynasty. Previously, in the Ming dynasty, during the time of *The Ten Bamboo Studio Manual*'s creation, pure blue sandalwood fibers were used for making *xuan zhi*. These facts suggest that the Library of Congress edition dates from the late eighteenth century or later—a welcome piece of information as the numerous extant editions of *The Ten Bamboo Studio Manual* have been notoriously difficult to sequence and date.

The use of bamboo fibers supports the likelihood that this particular edition was printed in southern China, since bamboo pulp papers have almost exclusively been manufactured in the south, even up until recent times. Zhejiang province is a southern region that is particularly noted for its papers made from bamboo fibers. This evidence corroborates with the fact that *The Ten Bamboo Studio Manual* was originally produced in Nanjing, a southern city adjacent to Zhejiang that has historically been a center for printing.

Accelerated Aging Experiments: Phase 1

Two phases of accelerated aging experiments were set up to answer the following questions: Which Chinese papers, particularly *xuan* papers, possess good aging properties? Does toning of the paper with pigments/dyes or treatment with various adhesives significantly alter the aging characteristics of *xuan* papers?

Phase 1 of the experiments was designed to answer the first question by comparing nineteen types of papers that had

Table 3. *Results of phase 1 testing*

No.	Paper type, origin	pH before	pH after	Brightness unaged	Brightness aged	Strength unaged	Strength aged
1	Toned *xuan* from Xiangmei Gu, Freer Gallery of Art	4.94	4.27	44.00	23.35	.00594	.00344
2	White *xuan* from Xiangmei Gu, Freer Gallery of Art	8.45	7.41	73.82	45.87	.01682	.0100
3	Anhui *xuan* purchased on Fujian Rd., Shanghai 安徽宣纸 福建路上海	8.41	7.73	66.28	40.7	.00446	.00230
4	*Fuchun Jiang xuan* paper mill, Zhejiang 富春江宣纸产 浙江省	7.25	3.78	77.41	18.12	.00680	.00012
5	*Dou An xuan* from Rong Bao Zhai 都安宣纸 荣宝斋	8.97	8.29	77.83	41.28	.00505	.00215
6	Two layer *xuan* from Rong Bao Zhai 四尺二层宣纸 荣宝斋	8.85	7.67	70.93	37.9	.01559	.00990
7	Heavy, single-layer *xuan* from Rong Bao Zhai 四尺重单 荣宝斋	9.01	7.83	79.18	47.14	.01216	.00608
8	*Jing pi*, single-layer from Rong Bao Zhai 四尺净皮单 荣宝斋	8.65	7.74	76.48	40.14	.01115	.00573
9	Special *jing pi*, single-layer from Rong Bao Zhai 四尺特净皮 荣宝斋	8.74	7.68	74.32	32.9	.01570	.00434
10	Anhui Special *jiqiu*, single-layer from Rong Bao Zhai 安徽鸡球特皮单 荣宝斋	8.48	7.05	77.49	33.63	.00712	.00157
11	*Jing xuan* single-layer *xuan* 精选单宣	9.07	8.03	72.8	38.15	.01021	.00387
12	Highest-quality single-layer *xuan* 超级单宣	9.21	8.23	73.95	50.05	.00646	.00334
13	Red Star single-layer *xuan* 红星 四尺单宣	8.94	6.64	77.55	38.17	.01372	.00610
14	Taiwan *mokuren,* Tokyo store 台湾木莲	8.28	6.95	74.71	54.04	.00648	.00295
15	Chinese Jade press stationery paper, Tokyo store 二双 中华玉版笺	8.59	7.52	77.50	54.23	.02311	.01021
16	Highest-quality net pattern *xuan,* Tokyo store 超级 四尺罗纹	8.71	*	77.18	*	.00617	*
17	*Hong Xing* single-layer *xuan,* Tokyo store 红星 四尺单宣	8.85	8.16	77.01	55.00	.01041	.00613
18	Single-layer *kyosen,* Tokyo store 四尺夹宣	8.88	8.02	72.95	50.04	.01678	.00640
19	Double-layer *kyosen,* Tokyo store 二层夹宣	8.72	7.94	71.76	49.90	.01529	.00895

* No data.

been gathered from various sources, mostly mainland China and Japan. They were tested to judge their aging properties and general archival suitability. The papers were artificially aged and evaluated for change in pH, brightness, and tensile strength (table 3).

Testing procedures followed standards set by TAPPI. The papers were artificially aged using two methods: the traditional free-hanging method in a humid oven and the newly developed method of sealing paper strips in individual incubation tubes maintained at 45% relative humidity and 90°C. The papers were also interleaved with sheets of purified wood pulp paper during the aging process to establish whether migration of the degradation products might occur. Many of the papers darkened after aging, and the discoloration had transferred to adjacent interleaving papers. The results from this initial survey indicated the need for further testing.

The paper strips were cut to a standard dimension of 1.5 cm × 17 cm. All of the samples were preconditioned in accordance with TAPPI method T-402 in a special, environmentally controlled chamber at 23°C and 50% RH for forty-eight hours prior to accelerated aging. Subsequent tensile-strength testing was also conducted within this chamber. The pH was measured by the cold water extraction technique with a slurried pulp. The samples were measured for brightness before and after aging using the Brightimeter Micro S-5 according to TAPPI T-452. The average measurement was taken for 20 data points for percent reflectance and the standard deviation computed. Tensile strength was evaluated using TAPPI test T-494 with an Instron machine that subjected samples to pneumatic pressure under even tension at 15 psi. The average tensile strength was recorded in the table as TEA or tensile energy absorption (in newtons per millimeter), which measures internal fiber-bond strength.

Figure 4. *Paper sample strips from phase 1 testing before (bottom) and after (top) aging*

Results showed that the papers all performed equally well with the anticipated exception of the sample from the handmade paper mill visited in Zhejiang province (fig. 4). On the bottom row of strips in figure 4 are the pre-aged paper samples. The top row shows the samples after thirty days of moist aging. The Zhejiang paper is the fourth sample from the left and is noticeably darker. It had been made of poorer quality, imported bamboo and proved to be significantly worse than other samples, whereas the other papers were of a more reasonable quality.

From these results, is there a quantitative way to correlate artificial aging to natural aging? Conversations with Shahani revealed that this relationship is still somewhat tenuous. For all types of artificial aging, researchers have become less reliant on calculations and Arrhenius plots in forecasting paper permanence but favor the use of a relative scale instead. Although conservators and scientists may desire to predict the longevity of a paper in a conclusive way, it is presently not possible to equate a certain number of artificially aged days with years of natural aging. Rather, one is encouraged to do relative comparisons of one paper's performance over another. In the case of handmade papers, which are so nonuniform, one sheet of paper should be compared to itself to reduce ambiguity in drawing conclusions.

The Development of a New Method for Accelerated Aging

As mentioned previously, samples were aged both as free-hanging sheets and as strips in incubation tubes. For several decades, scientists have been in universal agreement that a humid environment is essential for accelerated aging tests to approximate the natural aging process more closely than simply elevating the temperature alone. Some researchers determined that this method could not satisfactorily mimic the conditions of natural aging because it did not reproduce the autocatalytic reactions that were possibly occurring under museum and library storage conditions.

Around the time of this experiment, the Research and Testing Division at the Library of Congress had been involved in a ten-year project to develop a new method for accelerated aging to take an autocatalytic environment into account. Under the auspices of the American Standards for Testing Materials (ASTM), the Library of Congress was one of five laboratories from around the world commissioned to undertake research in developing a sound and practical accelerated aging methodology. At the time of this project, pilot-testing of the new accelerated aging method had commenced but had not yet been refined.[4] The new method for testing was proposed and described in the library's report to ASTM:

> Paper samples conditioned under specified temperature and relative humidity conditions are enclosed inside airtight glass tubes and aged in laboratory ovens

> that need provide only temperature control. Besides providing a better simulation of natural aging, this test offers several other advantages, most of which arise from the superior control of the aging environment within airtight tubes as compared with that provided by expensive and problem-prone aging chambers which must be employed currently to control relative humidity and temperature simultaneously. The new test is more precise and should therefore serve to improve reproducibility. (Arnold 2003, 2)

Prior to the development of this new test method, the acids formed by degradation during accelerated aging were volatilized and released during the free-hanging sheet tests, which took place in an open-environment chamber. Art and documents in collections, however, are often hung in sealed frames, stored in boxes and containers, or bound in books and stacked on shelves. In these instances, it is theorized that the acid by-products of aging present in a contained environment with the artwork or book may give rise to autocatalysis. Certainly this format is relevant to *The Ten Bamboo Studio Manual* since it is bound as a book. In the hopes of approximating these conditions, paper aging tests in sealed glass tubes were developed. The newer test method, once it is standardized, may provide a better simulation of real conditions in a typical library or art storage environment.

Accelerated Aging Experiments: Phase 2

In phase 2 papers were brushed with various preparations of animal glues, colorants, and wheat starch pastes, all of which are typically used at various times during the conservation treatment of Asian paintings or during the mounting process (fig. 5). The three types of *xuan* papers commonly used in the Freer's scroll-mounting studio were selected and coated with these preparations by Xiangmei Gu and her staff. The papers selected were Golden Bamboo (G), Red Star (R), and Special *jiqiu* 鶏球 (C).[5]

Comparisons were made among three animal glue combinations. One of the glue preparations was a studio recipe similar to the *dōsa* 礬砂 sizing used in Japan to which a small amount of alum is added to a warm animal skin glue solution. It is a liquid sizing used both to prevent ink bleeding and to discourage selective absorption of the glue binder from a pigment mixture. The other glue solutions were standard preparations of deer and sturgeon glue.

Colorants that are often used in toning papers to match the color of the aged artifact or painting were also studied for comparison. *Yasha* 矢車 is a Japanese dyestuff derived from boiling small pinecones of the deciduous alder tree *Alnus firma* Sieb. It is customarily used to dye papers and silks shades of yellow to brown. The same brownish tone can also be achieved using a Chinese pigment mixture of indigo blue, burnt sienna, rattan yellow, and black ink mixed with animal glue. The performances of these two coloring options were judged against one another.

Last, three types of paste were evaluated: a dry, powdered wheat starch paste purchased from the conservation supplier Talas, the same Talas wheat starch paste prepared by the Japanese method of presoaking before cooking, and Chinese flour paste (from which the gluten has presumably not been removed).

Results and Discussion

Alkalinity

In figure 6, one may readily see that all three brands of papers—Golden Bamboo (G), Red Star (R), and Special *jiqiu* (C)—were alkaline and that they demonstrated little change in pH after aging. The most likely reason for this result is the clay additions to the papers. Clay loading, common in Asian papermaking, serves as an opacifier and provides other desirable working properties. During the course of treatments with the various preparations, the pH of the papers remained in the alkaline range between 7 and 9 before and after aging (see fig. 6). No paper, treated or untreated, shifted more than 1 point in pH before and after aging. These results attest to the buffering capabilities of clay loading over an extended period of time.

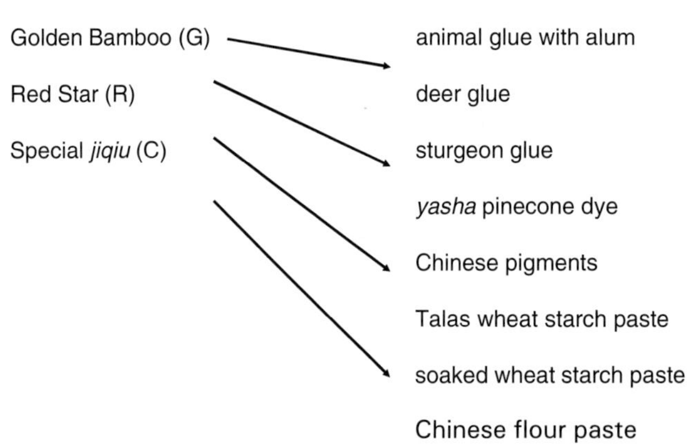

Figure 5. *Three brands of* xuan *paper treated with eight different preparations*

Tensile Strength

The tensile energy absorption or (TEA) reported is the average of ten to twenty runs per sample (fig. 7). The standard deviation of the results was somewhat high, but this level is not unusual for handmade papers, which may vary greatly in thickness within each sheet. The percent of TEA retained was also calculated and reported in figure 8. As a point of reference, a "permanent" paper has been roughly defined as a sheet that retains at least 50 percent of its strength after artificial aging for 30 days. This rule of thumb has been applied for Western machine-made papers. The target is often met and exceeded by high-quality, handmade papers that exhibit up to 80 percent retention in strength after aging (Library of Congress, Research and Testing Division, unpublished data).

For the three untreated Chinese handmade papers, the Golden Bamboo paper retained only 30 percent of its initial strength, making it suitable only for interim, in-process treatments, but not for long-term applications. The Red Star paper retained approximately 60 percent of its strength, and the Special *jiqiu* paper retained 50 percent. Therefore, according to the reference standard cited above, these two papers may be considered to be in the range of "permanent" paper quality in this category.

Among the three animal skin glue solutions applied to the papers, there was no significant difference. It is worth mentioning that the addition of alum in the *dōsa*-like glue preparation did not contribute appreciably to the deterioration of these papers, perhaps because of the low percentage that was used. The graph of the average tensile energy absorption for the aged and unaged samples shows that the

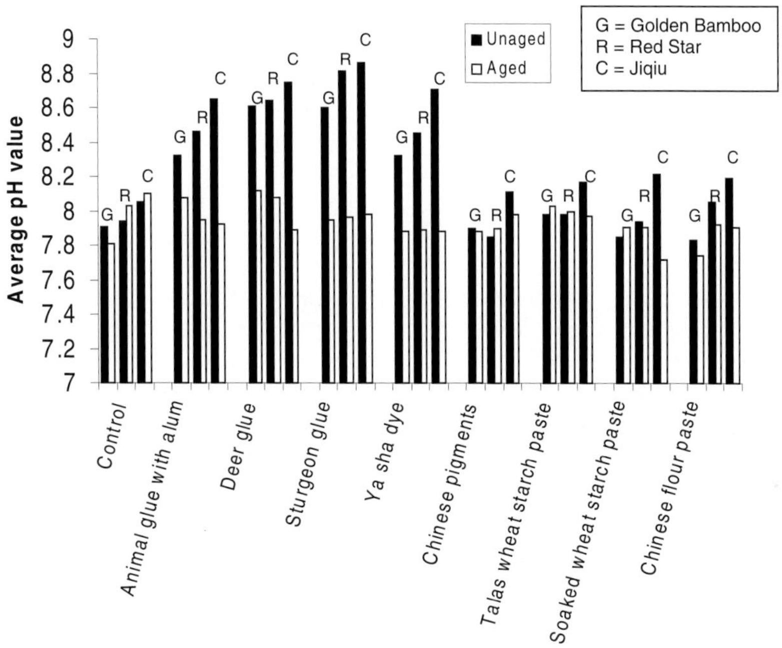

Figure 6. *pH levels of aged and unaged samples*

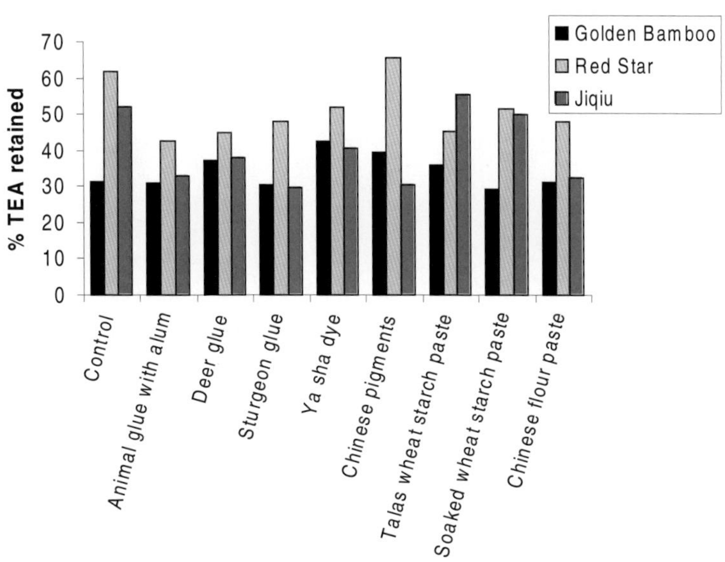

Figure 7. *Average tensile energy absorption of aged and unaged samples*

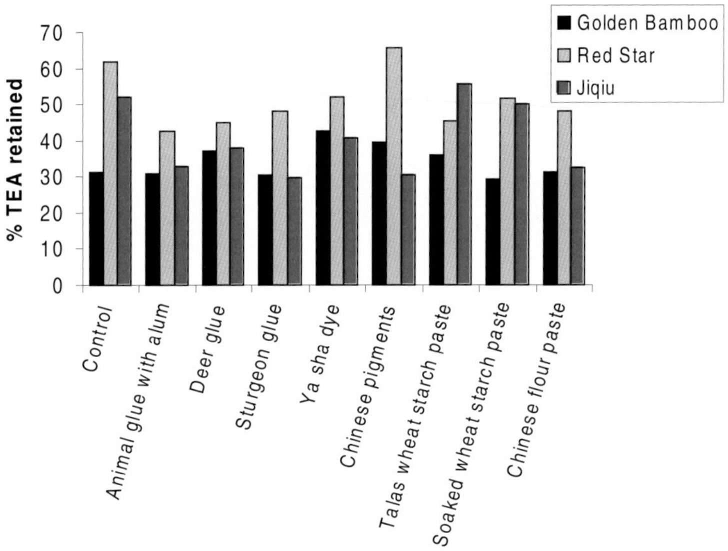

Figure 8. *Percent of tensile energy absorption retained after artificial aging*

dōsa-like mixture of alum and animal glue actually increased the strength of the papers to which they were applied by about 30 percent, whereas the sturgeon glue alone did not have a similar effect.

Upon first glance at the graph in figure 8 of the percent of TEA retained after artificial aging, it would appear that the glues have contributed to loss in strength. However, one should be careful to compare the animal glue TEA levels with the control group for the graph in figure 7 to see that the TEA measurements are approximately the same for the aged papers treated with animal glues and the untreated papers. This reading suggests that the initial gain in strength first caused by the glues is lost during the aging process; it does not appear that the glues contribute to additional degradation of the aged papers.

For the colorants tested, the *yasha* dye slightly increased the strength of the papers, whereas the Chinese pigments reduced the paper strength slightly. Once again, the graph shows that the aged control group papers have a tensile strength similar to that of the aged papers treated with the colorants. The papers treated with the Chinese pigments exhibit only slightly less tensile strength than the control group. The difference is not sufficient to suggest that one colorant is particularly favorable over the other.

Lastly, the papers in this series were applied with three paste solutions. As is normally done in the scroll-mounting studio, two sheets were pasted together for this experiment to create a double sheet thickness, which explains why their TEA measurements are appreciably higher than the other treated papers in figure 7. The percent of TEA retained (see fig. 8) after aging shows that the strength retained is similar to the animal glue and control samples. There is no significant change in strength to suggest that any one paste in particular will cause harm to the paper.

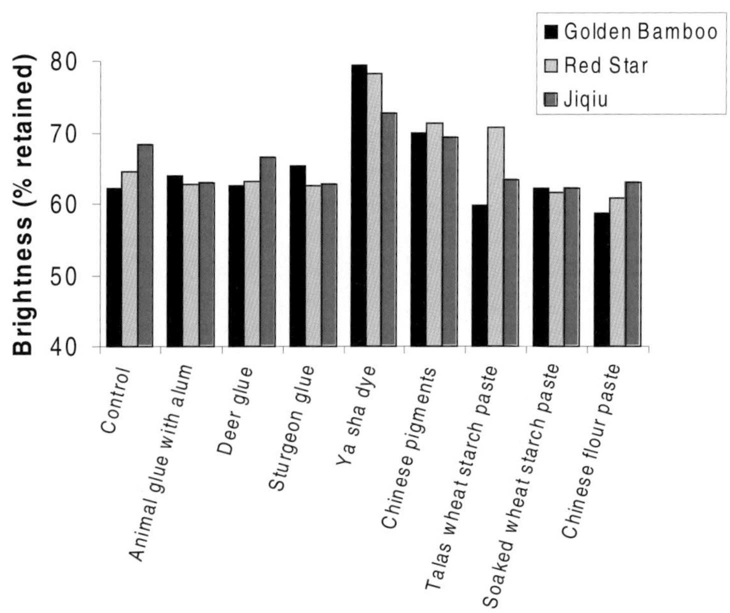

Figure 9. *Percent of brightness retained after artificial aging*

Brightness

All three untreated papers experienced significant loss in brightness despite the fact that these papers were aged in the absence of light (fig. 9). The darkening of the papers is not due to light but is due entirely to the chemical changes accelerated by the elevated temperature and humidity during artificial aging. As a guideline, the loss of 10 points after thirty days of aging is considered unacceptable for machine-made papers and is rather conspicuous to the unaided eye. In these experiments, the control for the Golden Bamboo paper changed 29 points, the Red Star changed 27 points, and the Special *jiqiu* paper changed 23 points.

Whether a paper will darken over time is a critical factor in selecting a repair or lining paper for conservation. As mentioned previously, these three Chinese papers have clay loading as opacifiers, which may contribute to their pronounced change in brightness under the conditions of accelerated aging. Previous research has revealed that some papers loaded with clay have darkened more than usual upon aging. The darkening effect may possibly be exacerbated by heat under the conditions of artificial aging. The migration of discolorants to adjacent papers is also a critical element in the choice of conservation-quality papers, yet is often overlooked.[6] The loss of brightness and potential to stain adjacent papers should not be considered trivial but ought to be the focus of future investigations.

Conclusions

Of the three paper types selected in phase 2, Red Star performed best in most categories. It has been one of the most widely used and distributed, quality *xuan* papers from China. Its manufacturers, based in Jingxian, Anhui province, keep their manufacturing information well guarded. However, none of the papers in this batch had satisfactory brightness ratings after aging, and for this reason they cannot be wholeheartedly endorsed until the aging of clay additives can be properly assessed. Conservators should be aware that the standards of accelerated aging tests established by the ASTM, although much improved, do not address the issue of paper's darkening upon aging. For conservators involved in the aesthetics of works of art on paper or documents, it remains essential to evaluate papers by their discoloration as they age.

After completing these tests, it was learned that the TEA test for tensile strength, though useful in measuring internal fiber-bond strength, is not necessarily an accurate measure of the paper's strength per se. This inaccuracy is due to an internal pulling mechanism that occurs when the fibers within the paper slide as it is being pulled apart. A better test was discovered to be the zero span tensile strength test, in which a sample strip is clamped to eliminate the gap between the pincers of the testing instrument, thereby precluding fiber-slide. This test could not be conducted at the Library of Congress because the instrumentation there is not calibrated to detect the sensitivity required for these very thin Chinese papers; however, the machine at the SCMRE laboratory is more finely adjusted for testing thin papers and was used in the study by van der Reyden and Tsai (1997).

In this study, the authors have surveyed Chinese papers that represent the range available to scroll mounters and artists and tested their permanence by subjecting them to typical studio treatments. During the current market changes in China, it is uncertain whether these papers will still be available or whether they will continue to be made with the same techniques and materials. Our ability to evaluate these papers is still impaired by the papermakers' reluctance to disclose the composition and processing techniques of their papers. An additional handicap is the possibility that, once a paper has been analyzed and evaluated through time-consuming laboratory testing, the formulation for the paper could change and undermine confidence in brands of paper. Further transparency of the papermaking industry in China is necessary if we are to be confident in the permanence of certain types of paper.

It is unlikely that we will ever be able to predict the precise lifetime of a given paper; however, reliable tests are being developed that can differentiate between longer-lasting papers and less-stable papers. Relative comparisons for accelerated aging remain the most reliable because relative rates are not susceptible to change. In the coming years the conservation community is likely to see more advances and become more assured in these methods.

Acknowledgments

The authors wish to acknowledge the following institutions and individuals for their assistance in this research: Dr. Chandru Shahani, Dr. Sang Lee, Dr. Ping Song, Carole Zimmerman, and Emily Olhoeft, Library of Congress, Washington, D.C.; Xiangmei Gu, Yuan-li Hou, and Valerie Lee, Freer Gallery of Art, Smithsonian Institution, Washington, D.C.; Dr. Tsuen-hsuin Tsien, professor emeritus at the University of Chicago; Dr. Weng-ott Yee; the Andrew W. Mellon Foundation; and the Fulbright Foundation.

Notes

1. From the time of the initial experiments, Jesse Munn has overseen the continuation of this project at the Library of Congress over the course of several years.
2. The use of cloth or textile fibers in the fabric of paper pulp is one explanation for the etymology of the Chinese character for "paper" or *zhi*, which is written with a "silk" or "thread" radical on the left.
3. One must be cautious when researching the plants used in making Chinese papers as well as their classification systems. Tsuen-hsuin Tsien (1973) refers to the same plants used for making *xuan* paper as *tan pi* 檀皮 and gives it the taxonomic equivalent of *Dalbergia hupeana*. In general, the authors found that the nomenclature in the literature on Chinese papermaking is highly inconsistent. Moreover, the translation of these botanical terms into their English equivalents sometimes adds to the confusion, and it is not always apparent which plant is being referenced. In these authors'

opinion, plant fiber identification remains problematic and will require coordination in the future among botanists, translators, and paper historians to sort out the precise taxonomy.

4. Regarding the experimental procedure: at the time of testing, thirty days had been the standard for the free-hanging, humid oven aging of papers, and therefore the duration for aging with the new incubation tube method was initially set at thirty days as well. Since that time, it has been determined that thirty days of aging in the tube is excessive. Depending on the temperature, the recommended aging time now has been refined to about two weeks.
5. The Chinese characters for *jiqiu* literally translate into "chicken ball" and were abbreviated as Chicken or "C" during the experiments due to the limitations of English-only character input for the analytical machines.
6. Migration of discoloration products was assessed in this study but not included in this report due to limitations of space.

References

Arnold, R. B., ed. 2003. *Accelerated aging of paper.* ASTM international paper aging research reports, RR no. D06-1004. West Conshoshocken, Pa.: ASTM International.

Cao, Tiansheng S. 2000. *Zhongguo xuan zhi. Di er ban.* Beijing: Zhongguo Qinggongye Chubanshe.

Schenck, K. 1999. The role of China paper in nineteenth-century French printmaking. Paper presented at Looking at Paper: Evidence and Interpretation Symposium, Toronto, 13–16 May.

Tschichold, J. 1970. *Chinese color prints from* The Ten Bamboo Studio. New York: McGraw-Hill.

Tsien, Tsuen-hsuin 1973. Raw materials for old papermaking in China. *Journal of the American Oriental Society* 93(4):510–19.

van der Reyden, D., and Tsai, Fei Wen. 1997. Analysis of modern Chinese paper and treatment of a Chinese woodblock print. *Paper Conservator* 21:48–62.

Yeh, M. B. 2002. Faithful to the brush: Chinese color printing technology and history in *The Ten Bamboo Studio Manual of Painting*. In *The broad spectrum: Studies in the materials, techniques, and conservation of color on paper*, 214–21. London: Archetype Publications.

Traditional Korean Papermaking

Hyejung Yum

ABSTRACT After Koreans developed papermaking skills, traditional handmade paper became an indispensable material of daily life in Korea. The aim of this research is to review the history, materials, and tools used in traditional papermaking, including scientific analysis of some old Korean paper samples using polarized light microscopy and scanning electron microscopy.

Introduction

Traditional handmade paper was an indispensable material of daily life in Korea. The diverse uses of paper can easily be seen in Korean culture throughout its history. Traditional handmade Korean paper was used for calligraphy, books, and envelopes; for doors, walls, and windows; for furniture, such as wardrobes, cabinets, and chests; for craft objects, such as writing brush holders, umbrellas, lanterns, boxes, baskets, fans, and kites; and for clothing and shoes. Koreans seem to be the only people who have been using paper for their floors.

Traditionally, the invention of paper has been attributed to Cai Lun in 105 C.E. in China (Laufer 1931). Papermaking skills were passed on to Korea and from there to Japan. Currently traditional Japanese papers are an essential material in paper conservation and, therefore, have been fairly well investigated. However, not many people know how Korea contributed to the development of papermaking or what kinds of paper Koreans have produced.

Old Chinese books state that Korean paper was regarded as one of the best by Chinese scholars from the tenth century on. Due to its excellent quality, Korean paper had long been one of the main tribute products levied by China. It is interesting that despite flourishing paper manufacturing in China, Korean papers were also exported to China for sale during and after the Koryeo dynasty (918–1392) (Pan 2002).

Since Korea was opened to the Western world, many traditions have declined, including traditional papermaking. There are still a few papermakers producing papers in the traditional way in Korea, but it has been difficult for them to earn a living because the traditional way of papermaking is so laborious and the market is limited. The aim of this research is to review the history, materials, and tools used in traditional papermaking. Some historic Korean paper samples have been analyzed using polarizing light microscopy (PLM) and scanning electron microscopy (SEM) to obtain close observations of each sample.

The History of Papermaking in Korea

The origin of papermaking in Korea is not known as no written records have been handed down. However, papermaking skill must have spread from China to neighboring countries at an early stage. Some domestic and foreign scholars hypothesize that papermaking might have existed in Korea as early as the third century or at least before the end of the sixth century. The belief that Korea started using paper before the fourth century is based on a piece of paper unearthed from the ancient tomb 116, Chechubchong, by the Choson Archaeological Site Research Group in 1931 (Cho 1996). As this tomb belonged to the Naknang period (108 B.C.E–313 C.E.), it is possible to conclude that at least around the beginning of the fourth century papermaking may have been carried out in Korea. In 610, Damjing, a Korean Buddhist monk, introduced papermaking skills to Japan along with ink sticks and coloring methods. Thus it seems likely that by the late sixth century papermaking skills in Korea were well developed.

The Era of the Three Kingdoms (First Century B.C.E.–Seventh Century C.E.)

Koguryeo (39 B.C.E.–668 C.E.) was a large Korean kingdom established during the first century B.C.E. It governed the northern part of the Korean peninsula and extended to Manchuria after the Han dynasty (206 B.C.E.–220 C.E.) in China had collapsed. Paekche (18 B.C.E.–660 C.E.) and Silla (57 B.C.E.–668 C.E.) were rival kingdoms that controlled the southern part of the Korean peninsula, dividing it east and west. It is believed that the kingdom of Koguryeo started to record its history from the beginning of its era and left the *Yugi* (One Hundred Volumes of Written History). The kingdom of Paekche compiled its history at the end of fourth century, and Silla did the same in the middle of the sixth century. There are no records of papermaking during this era, but it can be presumed that by that time traditional papermaking skills were well developed; without a large supply of paper, these volumes of history simply could not have been written. Although the kingdom of Silla was the last in terms of cultural development, eventually, in 668, it conquered the other two kingdoms and unified them. Several important historic artifacts have been found from this period, which is called Unified Silla (668–935). For example, the earliest known extant printed Buddhist text, *Dharani Sutra*, was found at the Pulguk-sa Temple in Kyeongju, the capital of Unified Silla, inside a stupa sealed in 751. During its conservation the paper was found to be made of paper mulberry, which Koreans term *dak*.

The Koryeo Kingdom (918–1392)

As the prosperity of Unified Silla declined because of civil war during the early tenth century, the last king of Unified Silla surrendered to a newly rising kingdom, Koryeo. During this period, the publishing of all kind of books flourished, including Buddhist texts, medical works, and history books. Paper money was also issued. The eleventh and twelfth centuries were the golden age of papermaking and publishing. As a result, the demand for paper increased, and, from 1145 on, the Koryeo government exhorted farmers to cultivate paper mulberry throughout the country and also encouraged people to make paper.

Koryo-ji, the paper made during the Koryeo dynasty, gained fame among scholars and emperors due to its excellent quality and was frequently mentioned in literature and history books in China (Pan 2002). It was said to be thick, strong, and white, with a lustrous surface acquired during the traditional finishing process, *jujibeop*.

Buddhism was the state religion, and the Koryeo government sponsored many Buddhist activities. One of its important achievements was the publication of the *Tripitaka*, the Buddhist sutras and commentaries. For just one set of sutras, 86,686 woodblocks were needed, and the entire edition required 162,890 sheets of paper (Cho 1996). It was also during this period that the first movable metal typeface was invented in Koryeo. Although the exact date of the invention is difficult to pin down, many scholars speculate that it was more than 150 years before Johannes Gutenberg used movable type in Europe in the fifteenth century. The earliest text printed was *Jikjisimcheyojeol* (1377), now housed in the National Library of France in Paris.

During the fourteenth century, the Koryeo kingdom was weakened by the intolerably heavy tribute duty the Mongols demanded, and as its power waned papermaking declined as well.

The Choseon Kingdom (1392–1910)

The Choseon kingdom, which replaced the Koryeo kingdom, was a neo-Confucian state. During its approximately 500 years, many literary and philosophy books were published. From the beginning, the government sought to revive papermaking skills by establishing a papermaking department. But as publishing increased, the supply of paper was inadequate. The Choseon government sent papermakers to China to learn their papermaking methods using different raw materials, and it also imported materials from Japan. In 1430 King Sejong sent an envoy to Daema-do Island to get fine Japanese mulberry for cultivation (Cho 1996). A wide range of different raw materials, such as pine tree bark, rice straw, and bamboo, came to be used for papermaking. Although this was not the first time that other materials had been mixed with paper mulberry, such mixtures became more common from this period on.

The peaceful era of the early Choseon culture ended in 1592 when the Japanese, newly unified under Hideyoshi Toyotomi, invaded. During the six years of war, Hideyoshi and his generals pillaged Choseon's cultural properties, and many craftsmen were taken as slaves and transported to Japan. The papermaking industry was greatly damaged by this invasion since it was based on the labor of farmers in the agricultural off-season. After this war the government pressured Buddhist monasteries to supply paper as monks were already producing papers for religious texts. At the same time, a heavy tribute of paper was demanded by China during the Yuan, Ming, and Qing dynasties (1279–1911). All these factors exhausted the papermaking industry, and the quality of paper produced declined. Near the end of this period, in 1884, a Korean government official brought Western paper equipment to Korea from Japan (Lee 2002).

The Korean Papermaking Process

There is not much written information on how the traditional Korean papermaking process has varied over time, but the principles employed seem to have stayed much the same. One-year old *dak* (paper mulberry, *Broussonetia kazinoki* or *Broussonetia papyrifera*) is usually harvested between November and February when the fibers of the inner bark are well formed and the moisture content of the outer bark is sufficient to make it easier to peel (fig. 1). When *dak* is cut, the diagonal cross-section is made to face toward the south, as it is believed this orientation will help *dak* sprout well the next year.

A steaming process is used to remove the outer bark of the harvested *dak*. Two connected pits are dug in a field in the shape of a gourd. In the bottom of the smaller pit, firewood is piled, and small stones are placed on top. The larger

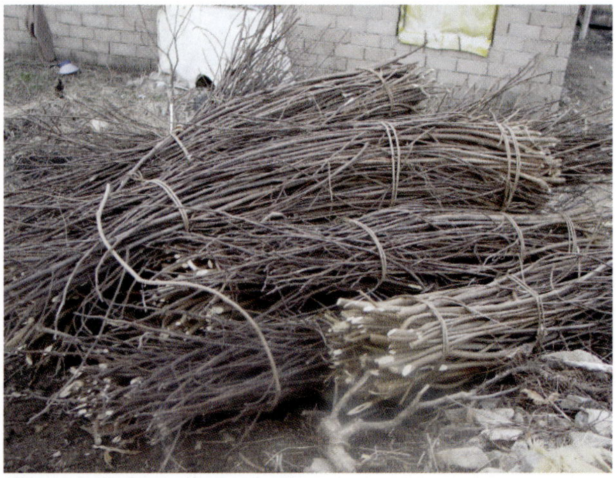

Figure 1. *Harvested paper mulberry*

pit is filled with the harvested *dak*. Branches are placed in layers and then covered with branches of pine. Then the pit is sealed with soil. Next the firewood in the smaller pit is lit. By the time the wood is almost burned up, the stones are very hot. Then this smaller pit is covered with soil, and the fire is extinguished with water. The heat of the stones turns the water to steam, which moves quickly through the connection to the larger pit. This steam cooks the *dak* and softens it.

Next the outer bark is removed. The outer bark consists of three layers—the brown outermost layer, the middle green layer, and the inner whitish layer. A worker grabs the bottom part of the branch with one hand and peels the softened outer bark off in strips with the other. The removed outer bark is called *jeopi* or *pidak*, and the outer bark with the top layer removed is called *nokpi* (*nok* means "green"). The lowest layer of the outer bark is called *baekpi*.

After the *jeopi* is soaked in a stream for about ten hours, the brown outermost layer and the green middle layer are peeled off with a knife to obtain the white inner bark. At this stage, workers are careful not to leave any dark spots, such as are formed near nodes or from damaged bark.

To obtain lye for cooking *baekpi*, the stems of plants such as soybeans and chili or the husks of buckwheat are burned. The ashes are placed in a big pot with holes in its bottom that are covered by a piece of fabric. Warm water is poured over the ashes. The fabric acts as a strainer. The solution that drains through is clear and alkaline, with an initial pH of approximately 10 to 12. Sometimes lime is also used along with plant ash. Today soda ash (sodium carbonate, Na_2CO_3) or caustic soda (sodium hydroxide, NaOH) replaces plant ash.

The white inner bark is soaked for another day and then roughly cut with a sickle to prevent its getting tangled during cooking. Next it is put into a big pot with the prepared alkali and boiled for four to six hours. Next the boiled *baekpi* is put into a stream for half a day to remove the excess lye and substances dissolved in the lye. During rinsing the *baekpi* is also bleached by sunlight and turned regularly so the bleaching is even. This process takes about five sunny days. Afterward dark particles (believed to be inclusions of soil) are removed by hand, as shown in figure 2. Then the bleached *baekpi* is removed from the water and squeezed. It is put on a flat stone and pounded with a wooden club until it is well spread out and softened. The beaten *baekpi* is called *dakjuk*.

The prepared *dakjuk* is put into a rectangular wooden container filled with water and stirred with a bamboo stick. After it is mixed thoroughly to break up the flocs of pulp, *dakpul*, a mucilage extracted from the root of *Hibiscus manihot* L. (fig. 3), is added to help the fibers disperse evenly and prevent them from sinking in the water during the sheet formation process. Too much *dakpul* makes the stock thick, delays drainage through the screen, and results in less fiber deposition prior to discharge, resulting in the formation of a thin sheet. Too little *dakpul* will speed drainage, and the sheet will be thick.

The traditional method of sheet formation, called *oebaldugi*, is one of the features that differentiates the Korean papermaking technique from other Asian methods. The papermaking screen rests upon the mold frame. During the process the craftsman holds the screen and the mold together. The far side of the mold is attached to a string hung from above. The craftsman scoops the vat stock from his side first and then lets it flow toward the far side by tilting the mold. This first scoop forms the ground of the sheet. The second scoop is taken laterally from either side, and as the mold is tilted it flows to the opposite side. This process is repeated until the desired thickness is achieved. Unlike a Japanese mold, in a Korean mold there is no deckle on top of the screen to contain the stock. Thus after dipping there is only enough time to allow a change of direction before the stock discharges off the opposite side. Consequently the sheet formation process must be done quickly and requires great skill. The paper on the craftsman's side is usually thicker and gets gradually thinner toward the far side.

After a sheet is formed, the screen is removed from the mold frame, turned over, and put on a prepared flat place. A round wooden log is lightly rolled over the screen to squeeze excess water from the sheet. Then the screen is carefully lifted in a rolling motion, leaving the newly formed sheet of paper smooth and unwrinkled. The paper formed in this way is called *ilhapji*, meaning one-ply paper. As one side of the paper is slightly thicker than the other, the papers are piled in alternate directions during couching, so that an even thickness may be achieved. Two sheets make one piece of paper called *ihapji*, meaning two-ply paper. These sheets of

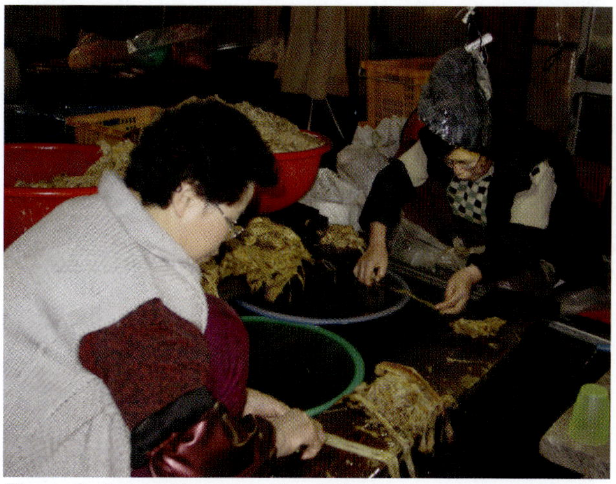

Figure 2. *Papermakers removing dark spots*

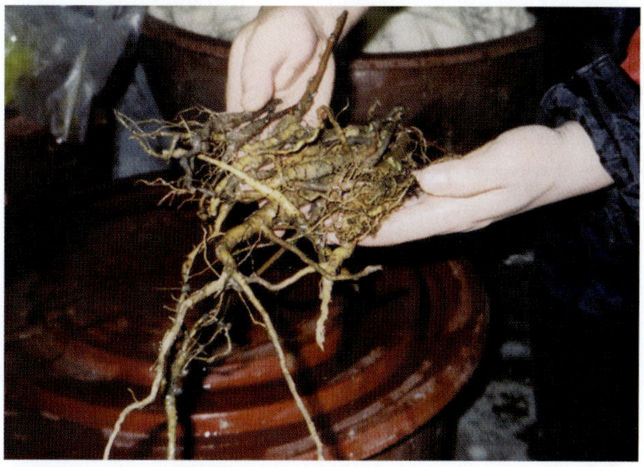

Figure 3. *Roots of* Hibiscus manihot *L.*

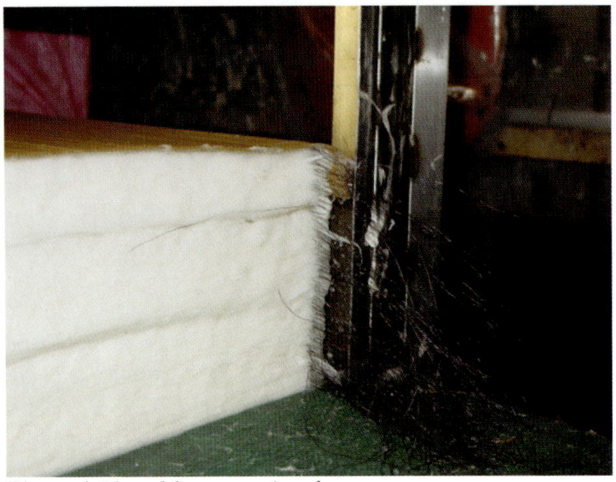

Figure 4. *Thread for separating sheets*

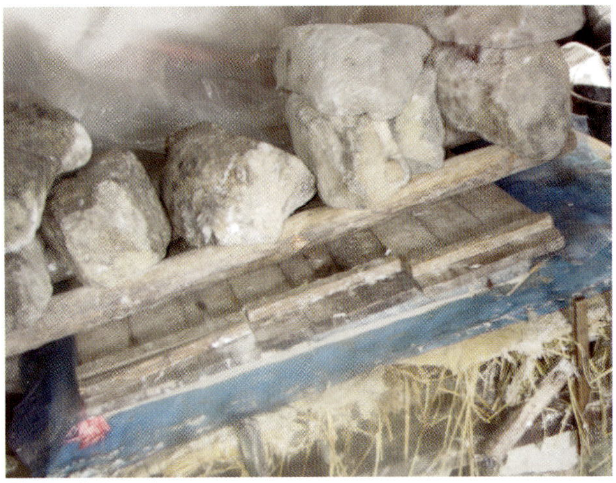

Figure 5. *Pressing*

paper are stacked in a pile, but each completed two-ply sheet is marked with a thread to make it easy to separate them later (fig. 4).

For drainage, the stack of sheets is covered with a thick board and weighted with stones overnight. Later the sheets are separated and put on wooden boards to dry in the sun. Today steam-heated stainless steel dryers are generally used for drying.

Finally, ten dried sheets are piled up and overlapped with one damp sheet. This stacking continues until about one hundred sheets are piled together. The stack is covered with a thick wooden board, weighted with stones (fig. 5), and left for a day to allow the moisture to spread evenly throughout the whole stack. The stack is uncovered and beaten with a wooden club about two to three hundred times. Through this process some of the sheets become dried. The dried sheets are then alternately piled with damp sheets and beaten again the same way. After that the whole stack is left in a shady place for half a day and the beating process is repeated again if it is too wet. This process is repeated until the stack is dry. The craftsman decides how many times the stack of sheets should be beaten. Through this process the surface of each sheet becomes smooth and the fibers closely packed together. Sometimes diluted rice starch paste is applied to the surface of the paper as a sizing material.

Materials

There is little information on materials used in traditional papermaking in Korea. However, paper mulberry has clearly been the main material. There are two main kinds of paper mulberry in Korea: *Broussonetia kazinoki* Siebold and *Broussonetia papyrifera* (L.) Vent. Both belong to Moraceae family. *Broussonetia papyrifera* is native to Korea, Japan, and China (Ilvessalo-Pfäffli 1995). Over time, some varieties have crossbred, and it is rather difficult to characterize some of them. Paper mulberry grows naturally throughout the country to a height of about three meters. *Broussonetia kazinoki* is called *chamdak* in Korean and *Broussonetia papyrifera* is called *kkujinamu*.

From the early days Koreans have used additives to facilitate their unique sheet formation technique. The general term for additives is *dakpul*. *Hibiscus manihot* L. has been the common additive since the middle of the nineteenth century, but before that elm tree bark along with a plant that belongs to the mallow family (but is not *Hibiscus manihot* L.) was used (Cho 1996). There are two or three species of mallow used in papermaking. All are planted in May and harvested in October or November (Lee 2002). The viscous liquid from the roots is indispensable for traditional Korean papermaking, as it prevents mulberry fibers from tangling and at the same time makes them disperse evenly in water during sheet formation.

Papermaking Screens

It is not clear when bamboo screens were first used for papermaking in Korea, but it seems as if bamboo and stems of native grasses have been used as the main materials for papermaking screens for a long time. Today, however, bamboo is the main material used. For making a screen only the outer layer of bamboo, taken from between the nodes, is used because it is stronger and lasts longer than inner layers. The split bamboo is carefully trimmed to make fine splints with an even thickness. Then the splints are bound together using fine thread on a loom (fig. 6). Traditionally silk threads

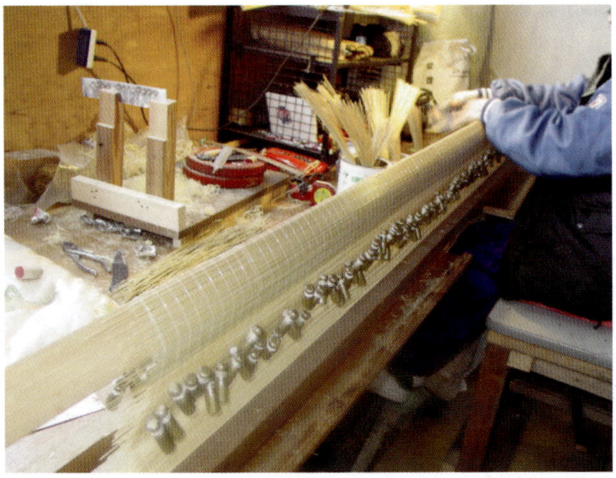

Figure 6. *Making a bamboo screen*

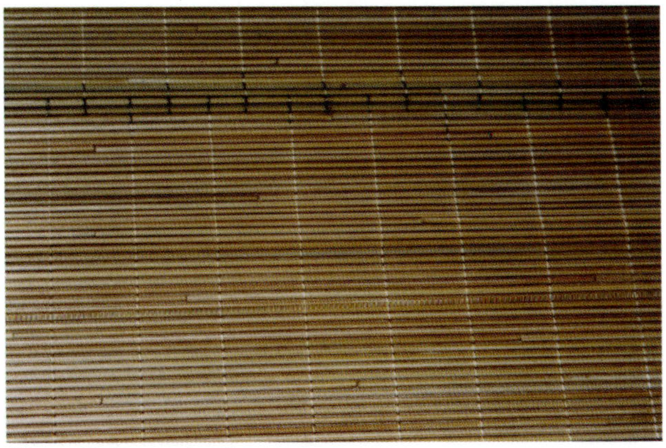

Figure 7. *Korean bamboo screen*

were twined and coated with persimmon juice, which makes silk threads strong and durable in water. Twined horsehair was also used for this purpose, but today nylon threads are used. To provide even strength while making a screen, small weights are tied to the end of each thread. The edges of a completed screen are wrapped with fabric.

There are three distinctive features of traditional Korean papermaking screens. First, their length is longer than their width. Second, the silk threads, which usually leave imprints as chain lines on the paper, run lengthwise, changing position in the middle of the screen by moving over half a space (fig. 7). By changing the position of the silk threads, the craftsman could avoid forming imprints on the paper that could weaken parts of a sheet. The third feature of traditional Korean papermaking screens is the absence of paired chain lines. Normally, Japanese and Chinese bamboo screens show some paired chain lines, which are more narrowly spaced than others. The length of bamboo splints is limited as the parts of bamboo between the nodes are used. Therefore, to make the desired width of screen, often splints would be connected using silk threads. In Japanese and Chinese screens the joined ends of the splints appear alternately, and as a result both ends of the splints are bound with threads to secure the joined areas (Barrett 1983; Pan 2002). A traditional Korean screen, however, does not show these paired chain lines, as the joined parts of two splints are usually scattered throughout the screen.

Scientific Analysis

Traditional Korean papers have some distinctive characteristics due to the unique sheet formation and finishing process that allows the paper to have not only multiple grain directions and layers but also a smooth, compact surface (Yum 2001). Fifteen Korean paper samples dated to between the fourteenth and seventeenth century and five modern papers produced between 2000 and 2002 were selected for close observation of their paper characteristics. The historic papers appear to have similar characteristics: their surface is smooth and lustrous. They are a warm, off-white color. Two papers of a brown color were probably colored with natural dye, possibly with *Phellodendron amurense* Ruprecht, although no analysis was done. Apart from their color, they are similar in appearance to the other historic papers. The contemporary papers are porous and do not have the same kind of compactness or smoothness of the surface as the historic papers. One of the contemporary papers went through *dochim*, the traditional finishing process, to make its surface smooth, but its scanning electron microscopy (SEM) image does not show the same degree of flatness as the historic papers do. Each sample was examined using polarized light microscopy (PLM) and SEM.

A polarizing microscope enables clear observation of the morphological characteristics of the fibers. Analyzing fibers can be one method for obtaining important information on the materials used for an artifact, and this information can also be used as a reference to deduce the date when the paper was produced. The surface texture and cross-section of each sample were examined with the aid of SEM.

Observation

A polarizing microscope does not generally make it possible to define the structural features of bast fibers, as many features, such as the shape of the fiber ends and the fiber width, may vary greatly even within a single fiber preparation. Moreover some immature cells may not show typical diagnostic features. The fiber analysis did not reveal much information in terms of materials used except where paper mulberry was used as the main material. The most distinctive characteristics of paper mulberry are that its fibers are enveloped in a transparent membrane. Moreover, cross-markings occur frequently, but they are finer than those of hemp (fig. 8). The fiber in figure 9 shows well defined and

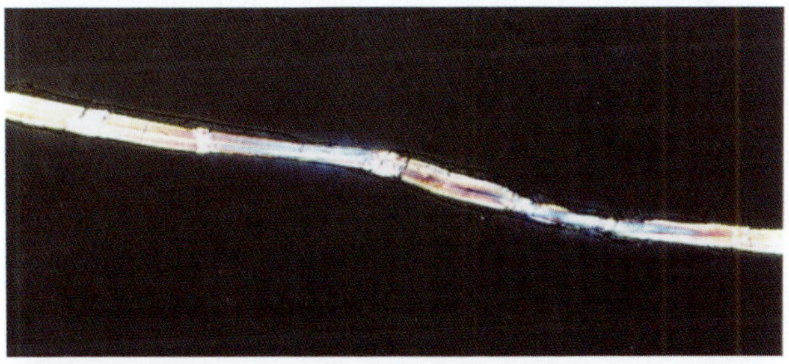

Figure 8. *PLM image of a paper mulberry fiber*

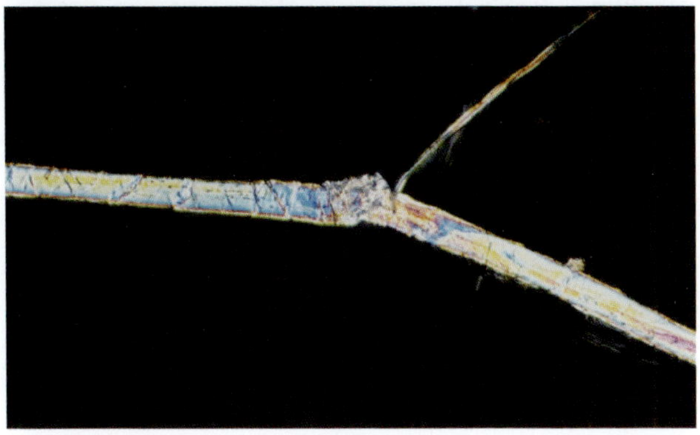

Figure 9. *PLM image of a hemp fiber*

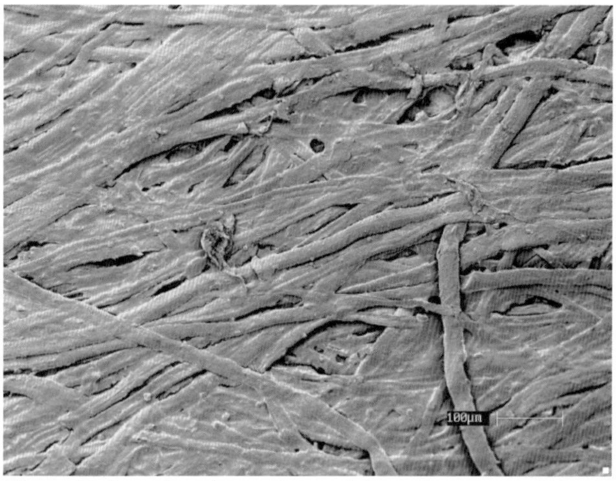

Figure 10. *Secondary electron image of the surface texture of historic Korean paper*

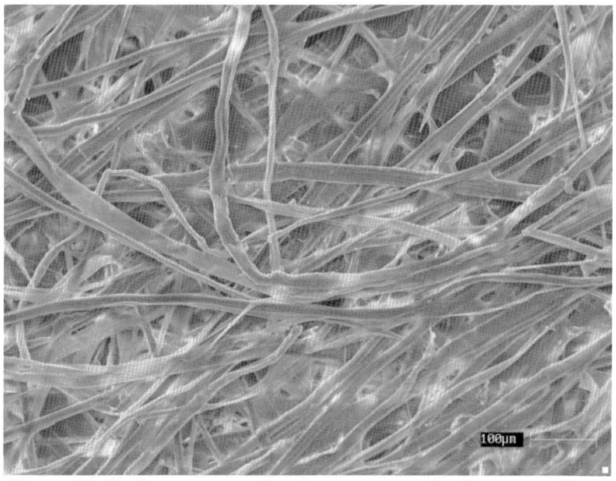

Figure 11. *Surface texture of modern paper made of paper mulberry*

smooth cell walls with partial longitudinal splitting, which is one of the characteristics of hemp.

SEM images show the distinctive features of traditional handmade Korean paper. The surface textures of ten historic paper samples (fig. 10) present a densely packed and quite flattened surface. Some of them have ill-defined fibers that appear to be covered with an extremely thin layer of material, suggesting that sizing material was also applied during the traditional finishing process. This feature is clearer when they are compared with images of modern paper samples (fig. 11). One traditional papermaker in Korea attested that sometimes during the final process the paper was sprayed with extremely diluted rice paste, although another papermaker said such a paste was not used. Some cross-sections also revealed several layers, as if several thin sheets had been couched and pressed together to achieve a desired thickness.

Conclusions

Once Koreans developed papermaking skills, they produced many different types of handmade paper. In ancient Korea handmade paper was not only used for writing and drawing but it was also an essential material of daily life. Strong but porous mulberry paper was ideal for windows, doors, and clothing. Traditional Korean paper can be defined as being thick, strong, and a warm white color. These characteristics were due to several factors. First, the main material used for paper production, paper mulberry, has long fibers. Second, during the papermaking process, maceration was mainly done by cooking in natural alkali and beating, which cause less damage to the fibers than cooking in caustic soda (sodium hydroxide) and grinding (which shortens fibers). Third, traditional sheet formation enables paper to have multiple fiber directions. Fourth, the traditional finishing process makes the paper surface smooth and compact. This might be the reason Korean papers were favored by the Chinese, who produced fine and delicate paper that was not as strong as Korean paper. The Chinese also valued the smoother surface of Korean paper for use in calligraphy.

Korea played a vital role in the development of papermaking and in the transition of papermaking skills to Japan. Knowledge of the techniques and materials used in traditional papermaking in Korea will help in understanding the history of papermaking in East Asia.

Acknowledgments

This work was carried out during the Getty Postgraduate internship at Cornell University. I would like to thank John Dean, director of the Preservation Department, for his generous support. I warmly thank Dolina Millar for her valuable advice throughout writing this paper. I am indebted to Lee Kyu-Shik for providing me with the historic paper samples.

References

Barrett, T. 1983. *Japanese papermaking: Traditions, tools, and techniques*. New York: Weatherhill.

Cho Hung-Kyun. 1996. *A research report on traditional Korean papermaking*. Seoul: Korean Science Foundation.

Ilvessalo-Pfäffli, M.-S. 1995. *Fiber atlas: Identification of papermaking fibers*. New York: Springer.

Laufer, B. 1931. *Paper and printing in ancient China*. Chicago: Caxton Club.

Lee Sung-Cheul. 2002. *Urihanji* [Traditional Korean paper]. Seoul: Hyeonamsa.

Pan Chi-Hsing. 2002. *Chungguk jejigisulsa* [History of Chinese papermaking technology]. Trans. B. Cho. from Chinese into Korean. Seoul: Sungchang.

Yum Hyejung. 2001. Conservation of four Meiji Japanese prints by Gekko and fibre identification of Korean paper. Master's thesis, School of Arts and Social Sciences, Northumbria University.

An Investigation of Alum in the Gelatin Sizing of Far Eastern Paintings on Silk

Ekaterina Pasnak, Season Tse, and Alison Murray

ABSTRACT The traditional Far Eastern practice of painting on silk involves sizing the support with collagen glue mixed with alum. The purpose of the research was to determine whether alum promotes silk deterioration and chalking of the paint film. It was found that silk sized with alum at 5% and 15% of the weight of dry glue concentrations performs much better than unsized silk or even silk sized with pure gelatin. The highest tensile strength, the least color change, and the smallest drop in pH were shown by the 15% samples. In addition, it was determined that the degradation of silk molecules was not affected by alum concentration up to 15%. Slight changes in color of the paint layer were not caused by the sizing, regardless of alum presence. Further work must be undertaken using longer aging periods in order to be able to extrapolate the degradation profile with certainty.

Introduction

In collections of Far Eastern art are many paintings executed on silk with a thick application of mineral pigments mixed with collagen glue, a technique called, in Western practice, distemper. In most of the Chinese ancestral portraits of the Ming (1368–1644) and Qing (1644–1911) dynasties in the collection of the Royal Ontario Museum (fig. 1), the silk is dark and very brittle. The purpose of the present research was to study the influence of alum in the sizing of silk, to determine whether the acidity of alum causes degradation in silk in spite of its own inherent acidity, and to monitor the deterioration of gelatin as a sizing and binding medium. This experiment was designed to follow Chinese practice as closely as possible.

The Kwungtung silk used in the experiment is similar to that traditionally used in paintings: a very fine, lightweight, loosely woven ivory fabric with a double untwisted thread in the warp direction and single in the weft direction (warp, 44 ends per cm; weft, 33 ends per cm; fig. 2). Silk is composed of fibroin protein fibers secreted by the cultivated silk worm, *Bombyx mori*. Much has been written about the structure of silk (e.g., Becker and Tuross 1994; Kaplan et al. 1994; Kerr 1992; Iizuka 1985; Rose and Von Endt 1984). In microscopic cross-section, the silk from which the Kwungtung fabric is made appears to be degummed, with brins separated and not paired.

While ox and hide glue is recommended for sizing, as it is "clear, thin and transparent" (Silbergeld and McNair 1988, 17), lime-processed mammalian hide glues have similar chemical composition, and therefore purified, light-colored rabbit skin glue was chosen as an appropriate substitute. The concentration of the glue in a sizing solution is rather thin, approximately 1–2%. The proportions of alum to dry weight glue mentioned in the seventeenth-century *Mustard Seed Garden Manual* are 3:11 (approximately

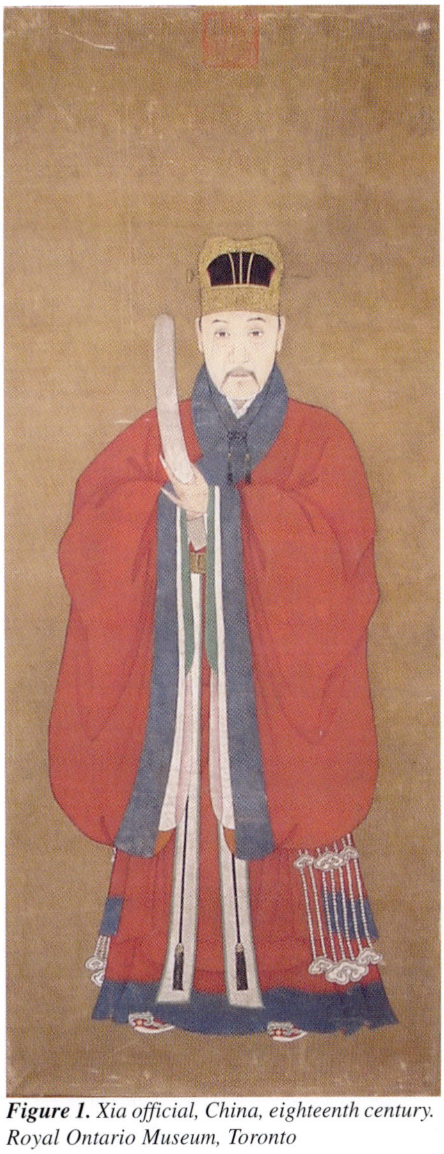

Figure 1. *Xia official, China, eighteenth century. Royal Ontario Museum, Toronto*

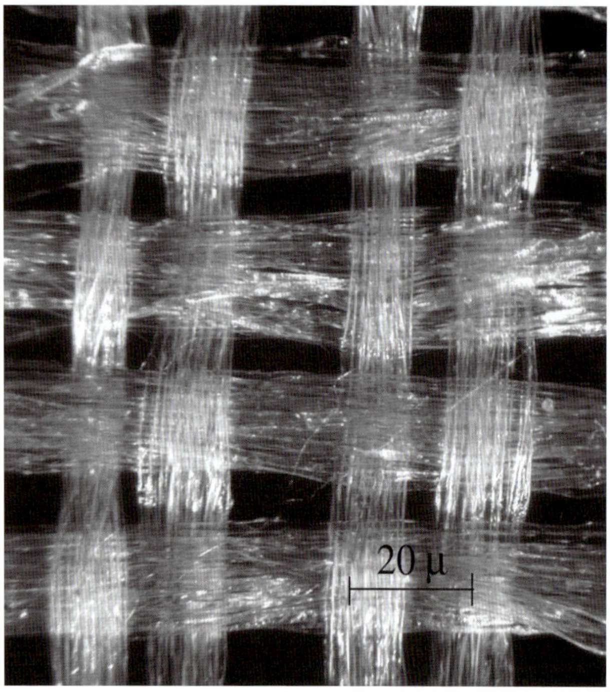

Figure 2. *Unsized Kwungtung silk*

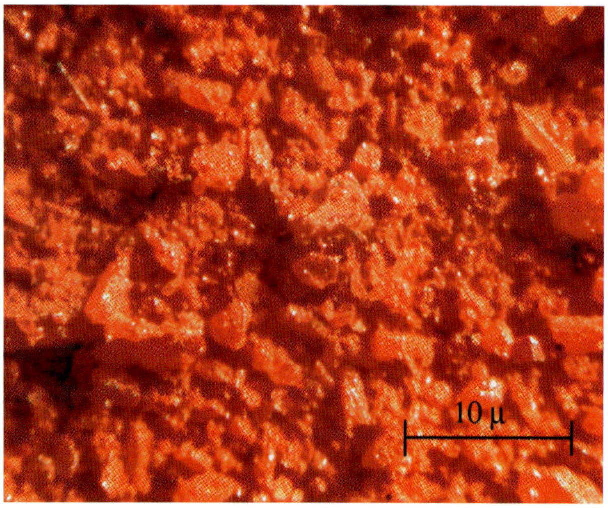

Figure 3. *Cinnabar paint layer*

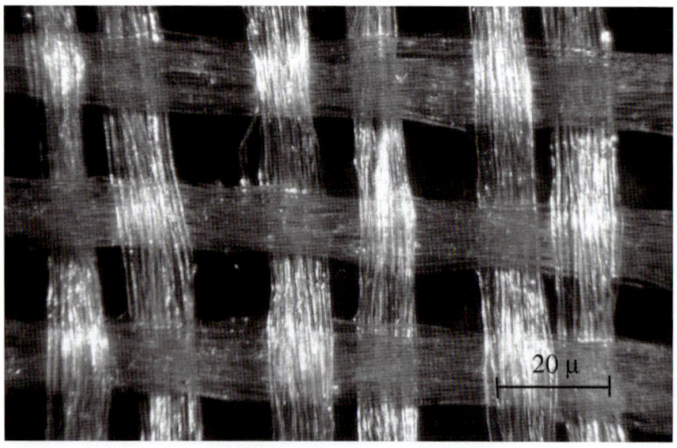

Figure 4. *Kwungtung silk sized with 1% rabbit skin glue and 15% alum based on dry weight of glue*

30%) in winter and 3:7 (approximately 45%) in summer (Sze 1977).

The term "alum" refers to double sulfates of trivalent metal ions, usually aluminum. X-ray diffraction (XRD) analysis confirmed that alum purchased in a Beijing art supply store[1] and Japanese alum (*myōban*) are identical and have the same crystal structure as the laboratory grade of aluminum potassium sulfate; however, personal communication with vendors of Chinese paper in Toronto[2] revealed the use of Chinese pharmaceutical alum to size papers. XRD analysis showed this alum is a mixture of aluminum potassium sulfate and aluminum ammonium sulfate. This product does not dissolve as readily as pure aluminum potassium sulfate, and it precipitates easily; it could be the cause of efflorescence found on some scrolls. For the present experiment the laboratory grade of aluminum potassium sulfate was used.

Alum plays a significant role in Far Eastern painting by hardening the glue, which reduces the dimensional change of the sized support while the painting is being executed and when the finished painting is exposed to varying humidity. As it makes the glue partially insoluble, alum protects the painting from water action during mounting. Alum also increases the vibrancy of colors, keeping them on the surface of the painting. Finally, aluminum salts minimize biological attack on the glue.

The main problem associated with alum is its high acidity. The concentration corresponding to 30% alum sizing used in the present experiment (0.3% alum in water) has pH of 3.82. The detrimental effect of alum has often been discussed by Chinese and Japanese artists, mounters, and conservators. If too much alum is used, the paint may exfoliate or white streaks or salt efflorescence may occur (Belozerova 1995), and "the colors will lose their brilliance" (Sze 1977, 47). Darkening and embrittlement of the silk are also attributed to alum, and mounters advise using it moderately or not at all when a consolidant is required during relining (Koyano 1979). In modern practice, Japanese conservators use either traditional *nikawa* (bovine skin glue) without alum[3] or synthetic adhesives.[4]

Colors are prepared by mixing the pigment with 3–5% of the same kind of rabbit skin glue that was used for the sizing. For this experiment natural cinnabar from Hunan province was chosen as a pigment, as it was assumed that mercury sulfide would not introduce additional reactions. It was decided to use roughly ground cinnabar from a Chinese pharmacy because it most closely resembles the pigment encountered in actual paintings (fig. 3). The intent was to see if gelatin degradation under the action of alum might result in exfoliation of the paint layer.

Since silk and gelatin are both protein molecules, having a similar chemical makeup, a homogeneous material is formed when sizing is thinly applied onto silk. White silk becomes semitranslucent; the loose, flexible weave turns into a rigid network with more compact threads; and the interstices of the weave become more pronounced (fig. 4).

Experimental

Sample Preparation

For the experiment, three identical sets of eighteen samples each were prepared. In each set of eighteen, six served as unaged controls, six were aged for ten days, and six were aged for thirty days. In each group of six samples, one was unsized and unpainted; the second was sized with 1% rabbit skin glue (RSG)[5] with no alum added; the third was sized with 1% RSG and 5% alum solution (i.e., 5% of the weight of dry glue); the fourth was sized with 1% RSG and 15% alum; the fifth was sized with 1% RSG and 30% alum; the sixth was sized with 1% RSG and 45% alum.

The silk was washed with Orvus paste (a nonionic laurate sulfate detergent used in textile conservation) and stretched wet on a custom-made window screen frame with the warp parallel to the long axis of the frame (fig. 5). Then the appropriate size solution was applied.

The border around each sample was adhered with wheat starch paste to two-ply Chinese paper,[6] in imitation of the traditional format of hanging scrolls, so that only the front of the silk was directly exposed to environmental conditions during aging. For testing, the sealed margins were cut off and discarded, the paper was removed, and sized silk uncontaminated with starch was analyzed.

Strips of silk were used for tensile testing. Afterward they were tested by cold pH extraction to study the increasing acidity of the sized silk. After cleaning the silk from the experimental sizing, they were next analyzed by high performance size exclusion chromatography (HPSEC) to determine the decrease in molecular weight of fibroin. A small area painted with hand-ground cinnabar[7] mixed with 4% RSG and an area not covered with paint were used for morphological analysis and to observe colorimetric changes.

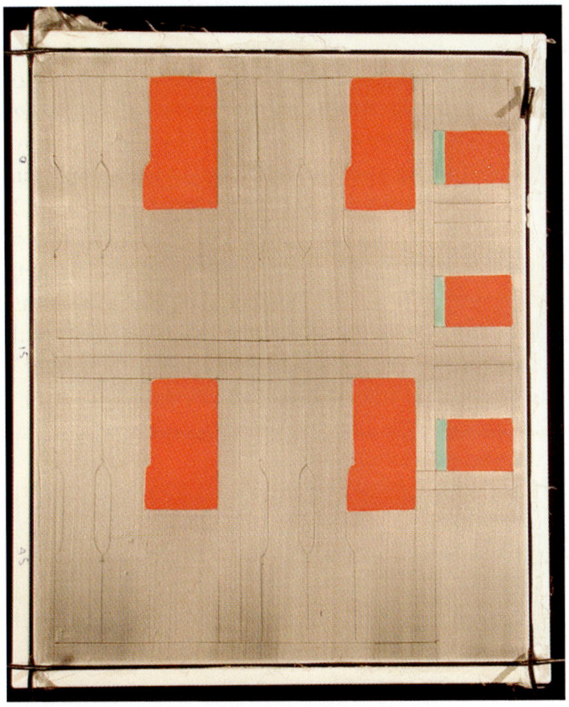

Figure 5. *Prepared and painted silk on a frame*

To monitor possible changes in the gelatin due to the action of alum and avoid contamination with soluble components of silk, the same sizing solutions were applied to samples of the inert nonwoven polyester fabric Reemay, which were aged together with the silk samples. This size was then extracted for pH and infrared analysis.

Artificial Aging

To promote hydrolytic cleavage by the action of alum, samples were aged in the dark at 80°C, 50% RH for periods of ten and thirty days. Higher humidity would have accelerated silk degradation (Hansen and Sobel 1992), but gelatin would have reached its glass transition temperature (Tg) with increased humidity and temperature (McCormick-Goodhart 1995). A shorter aging time was intentionally chosen to capture the first deterioration processes that might occur in silk and gelatin.

pH Measurements

Gelatin solutions prepared for sizing and alum-gelatin solutions extracted from aged Reemay strips were measured with an Omega pHH-25 Microprocessor pH meter. Later three strips of sized silk from each set previously used in tensile testing were put into 8 ml of cold Aquafina (commercially available deionized water), and a pH measurement of the whole system (silk and size) was taken.

Colorimetric Measurements

To determine the extent of yellowing, the unpainted test area was measured in three adjacent areas with a Minolta ChromaMeter CR300, using the Hunter L*a*b* system for color measurements. The changes in a* (red-green axis) and b* (yellow-blue axis) as well as ΔE were calculated for sized silk and painted areas.

FTIR Analysis

Dupont (2002) suggested that when paper is sized with gelatin with a low concentration of alum, the gelatin is degraded first in preference to the cellulose. Since the same sequence might be occurring with the sized silk, the Nicolet Avatar 320 Fourier transform infrared spectrometer (FTIR) was used to monitor changes that might occur in the gelatin. In the infrared spectrum of gelatin, the intensities of the two amide peaks at 1,540 cm^{-1} and at 1,655 cm^{-1} decrease as the gelatin is hydrolyzed. Ratioing these areas against that of the S-O peak from the alum in the sample, which should remain unchanged by the aging process, gives a semiquantitative analysis (fig. 6). For the analyses of unaged size, a few ml of size solution were air-dried. For the aged size, the gelatin-bearing Reemay was soaked in 30 ml demineralized water and left in a warm water bath until the gelatin was dissolved. The pH was measured, and the solution was then evaporated to obtain the dry

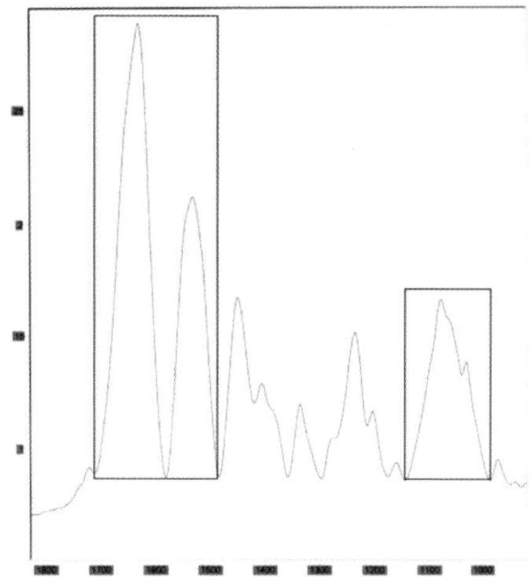

Figure 6. FTIR absorption spectrum of the 15% alum size before aging. The analysis calculations are made by dividing the area corresponding to protein peak (1750–1479 wavenumber) by area corresponding to the alum peak (1180–945 wavenumber).

gelatin. The spectra were recorded using the Golden Gate attenuated total reflection (ATR) accessory on the Nicolet Avatar FTIR and the EZ E.Z.P.52aOMNIC software program. Peak areas were calculated using the peak fitting tool in GRAMS 32A1 program. The areas corresponding to the protein frequencies between 1750 and 1479 cm^{-1} were obtained and divided by the area of the S-O absorption bands of alum corresponding to 1180 and 945 cm^{-1}. The peak areas do not provide absolute values and are not exact; thus the ratios allow only a semiquantitative analysis.

Tensile Strength

To determine the degree of strength loss, tensile testing was performed according to ASTM standard D5035-95 using a constant rate of elongation testing machine in the Dupont Canada Testing Laboratory, Kingston, Ontario, where the environment is kept constant at 21°C and 50–55% RH. The machine used was an Instron 1127 model with a D 1000 lb cell, set at 150 mm/min elongation. Calculations were plotted in nominal stress (load/initial area) versus nominal strain (extension/original length), using Mtest software.

High Performance Size Exclusion Chromatography

High performance size exclusion chromatography was performed to determine the average molecular weight of the silk fibers before and after aging and to see if the molecular length, calculated in daltons, had been reduced.

Sample Description

Tests were performed on aged strips used in tensile testing (six sets from the first aging period and six from the second) and three strips of unaged silk, which served as a control. Silk strips were washed repeatedly with warm water to ensure all the size was removed, and then air-dried.

Sample Preparation for SEC Analyses

Silk was cut into 0.5 mm pieces, and approximately 4 mg were weighed into a 5 ml beaker. Saturated lithium thiocyanate (0.5 ml LiSCN; 1.7 g LiSCN/ml of water) was added to the silk and allowed to stir overnight (~ sixteen hours) on a magnetic stirrer. After stirring, 0.75 ml of mobile phase (0.5M tris-urea buffer; pH7) was added to the silk solution. This solution was centrifuged at 3,000 rpm for five minutes and filtered using a 0.4 μm syringe filter prior to injection.

SEC Analyses

SEC analyses were carried out at the Canadian Conservation Institute by Season Tse using a Waters HPLC system equipped with an autosampler, column heater, and a photodiode array detector. Separation was carried out using two Zorbax columns GF-250 and GF-450 (9.4 mm × 250 mm) connected in series (separation range of 10,000 to 900,000 daltons). The column temperature was maintained at 30°C, with a flow rate of 1 ml/min and column pressure of 840 psi. A 0.5M tris-urea buffer of pH 7 was used as the mobile phase. Prior to separation, the buffer was vacuum-degassed using a Viscotek VE7510 in-line degasser. Elution profiles were monitored at 275 nm using a photo diode array detector. Millennium32 GPC software was used for data handling and calculation of molecular weight averages based on the elution times of protein molecular weight standards.

Results and Discussion

pH Measurements

For the conservator, pH measurements are one of the easily available methods of tracking changes of the materials. The pH of the glue solution from the Reemay was usually much lower than that of the corresponding size extracted from silk. Figure 7 demonstrates that the higher the concentration of alum, the lower the pH. The pH of the pure glue solution did not change with aging, but the pH of the alum-gelatin size increased with aging, attributable to the buffering effect of gelatin, since it was previously observed by Dupont (2003) that the pH of paper sized with gelatin and alum does not exhibit the same drop in pH as paper sized only with alum.

Figure 8, illustrating the pH of the sized silk, demonstrates a different trend. The pH range between the samples is quite narrow (6.85 for 0% and 6.01 for 45%), and upon aging the drop is rather slight, so that after thirty days of aging all of the samples fall within the neutral range (5.9–6.3). In the end, the highest pH value among sized samples was demonstrated by 15% alum sizing, which was effectively the same as 0% alum.

Colorimetric Measurements

It was shown that heat aging causes silk to undergo significant color change regardless of whether it is unsized, sized with alum, or sized without alum. Initially samples changed approximately to the same extent, with the exception that

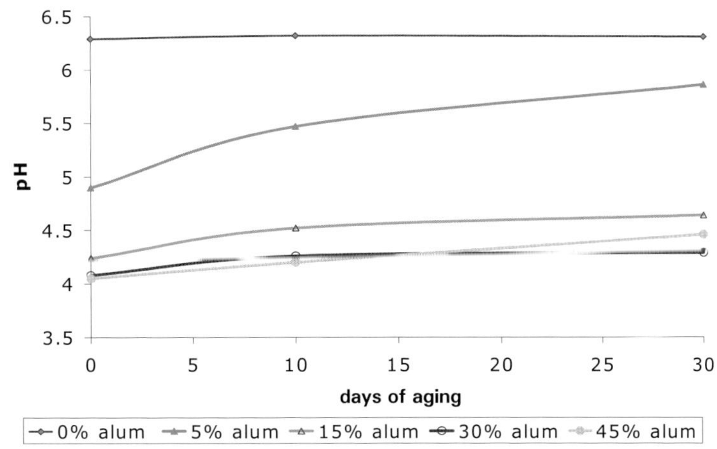

Figure 7. *pH of the gelatin-alum solution*

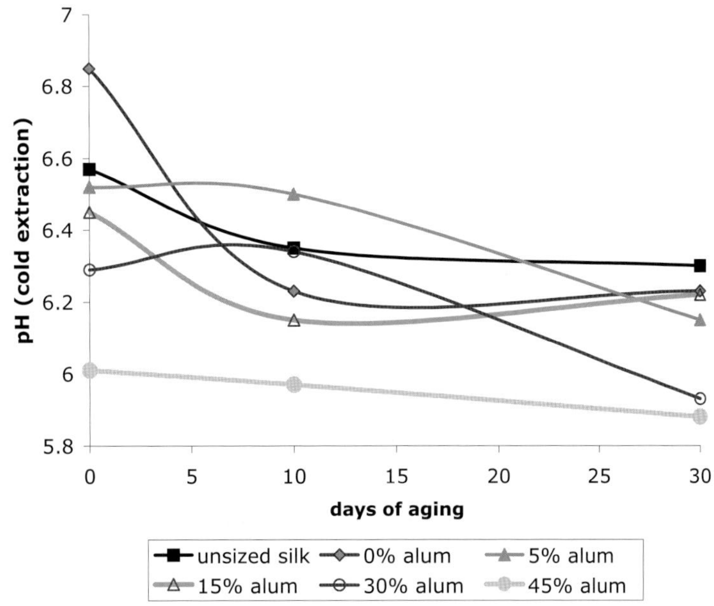

Figure 8. *pH of the sized silk*

samples sized with pure gelatin yellowed the most. After a longer aging period the presence of alum promoted greater discoloration. The 30% and 45% samples yellowed the most, followed by the 5% and 0% samples. Again the 15% alum samples were an exception, yellowing the least.

Morphological Analysis and Colorimetric Measurements of the Paint Layer

To approximate the Far Eastern scroll-mounting practice, the samples were aged with paper attached to the back. To apply this backing, the prepared samples were wetted, the warp and weft were aligned, and the paper was adhered using wheat starch paste. It was discovered that alum concentration above 5% sufficiently hardened the gelatin and made the fibers rigid enough that they maintained their alignment while wet; the higher the concentration of alum, the easier the procedure was. Cinnabar was chosen because it is a mobile pigment, and the wetting of the silk samples was an opportunity to test whether the sizing would in any way protect the paint layer, but in fact on the recto this pigment exhibited an extreme tendency to run no matter what kind of sizing was applied to the silk. The cinnabar also migrated to the verso of the silk support when it was painted onto 0% and 5% alum-sized silk. When wetting the back of these two types of silk, the paint struck through to the verso. Thus a higher alum concentration has advantageous properties for a painter or mounter. This finding explains why a concentration of 25–30% is most often encountered in Japanese and Chinese painting manuals.

Examination of the paint layer revealed no changes, neither after ten days nor after thirty days of aging. Neither flaking nor cracking was observed. With the chosen concentrations and aging conditions, the paint layer seemed to be well bound to the substrate.

Colorimetric measurements were taken of the paint layer to see whether the color changes occurring to cinnabar could be in any way related to alum sizing. The overall change (ΔE) measured after thirty days of aging was rather small

(ΔE ≤ 2.22) and barely noticeable to the unaided eye. After the first ten days of aging the color change was the least for samples with a high concentration of alum (30% and 45%) and the greatest for 0% alum. However, upon further aging, all the differences reached almost the same level. These color changes may suggest that in the beginning a harder size protects the paint surface better from the verso than unhardened sizes but that at a later stage the size has no effect on paint layer conditions.

FTIR Analysis

The spectrum of the dried film extracted from Reemay fabric was taken. In figure 9 the ratio of the protein area over the sulfate area is plotted. High numbers on the y-scale signify a greater contribution of carbonyl and amino groups of the protein in comparison to the sulfate group of the alum. The two lines on the plot corresponding to gelatin before aging and after ten days of aging are roughly parallel, with the aged curve beneath the unaged. This result indicates a decrease in the protein region that could be caused by its degradation and cleavage. The third line on the graph, representing results after the second aging, appears anomalous, since 5% shows a large decrease in protein content, whereas 15%, 30%, and 45% rose to figures larger than the initial ones.

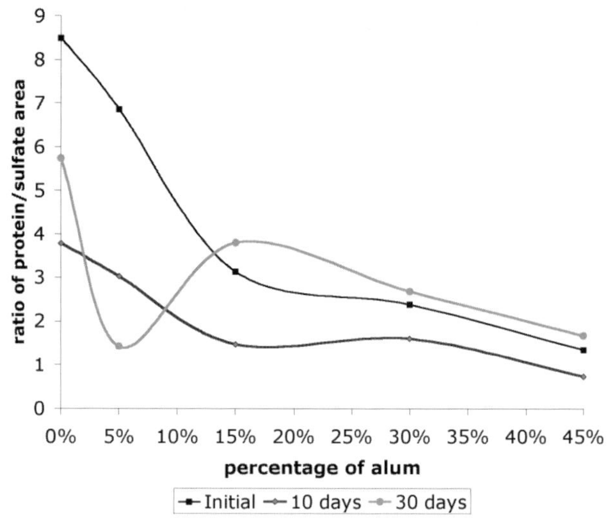

Figure 9. The ratio of the protein area over the sulfate peak is plotted before aging, after ten days, and after thirty days of aging.

From her experiments, Dupont (2003) concluded that two processes are occurring during the aging of gelatin mixed with alum: cross-linking caused by alum forming bridges, and hydrolysis accelerated by the acidity of the alum. Two simultaneous processes occurring at different rates for samples with different alum concentrations might explain the seeming abnormality of the FTIR graph. The primary decrease of protein content on the second aging curve might represent the effect of gelatin hydrolysis. The increase of the protein signal in 15%, 30%, and 45% might be explained by the onset of gelatin cross-linking. The sharp decrease of protein in the 5% samples could suggest that hydrolytic reactions are preferred at this concentration and at this aging period. In the future, other, more sensitive techniques such as viscometry or chromatography must be used to determine more accurately the processes occurring in the gelatin.

Tensile Testing

Tensile testing is one of the primary ways to determine the properties of the fibers and examine whether alum improves cohesion between the size and the silk, makes the support stronger, or accelerates its deterioration. Tensile strength is calculated by a ratio of the peak load at break over the thickness of the tested strip.

Unsized silk and silk sized without alum degrade similarly (fig. 10). The addition of alum contributes to an increase of tensile strength, but the increase becomes more

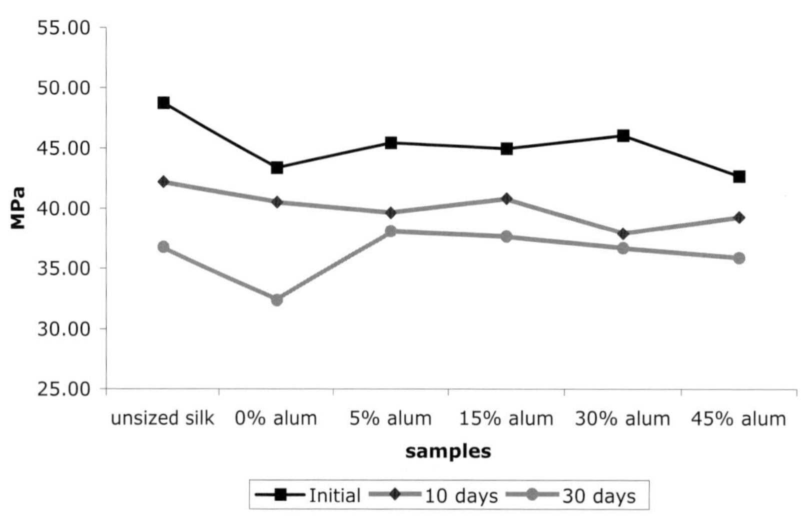

Figure 10. The tensile strength of the six samples before, after ten days, and after thirty days of aging

apparent with aging. Initially the difference between the samples with alum and without alum is slight, but after thirty days of aging, samples with 30% and 45% alum performed slightly better than simple gelatin sizing. The best performance was observed with 5%, 15%, and 45% alum sizing; these samples retained the largest proportion of their initial strength.

Elongation at Break

The properties of a textile are greatly influenced by its elongation at break; the shorter the elongation, the more brittle the material. In the beginning all sized samples broke around 18%. After thirty days aging, 0%, 5%, and 15% alum-sized silk strips showed greater breaking extension in comparison with other samples.

In engineering terms, the toughness of a material is described by the area under the stress/strain curve; the bigger the area, the tougher the material. Stress/strain curves of the aged unsized silk and silk sized only with gelatin (fig. 11) fall below the original curve, a finding which shows that both materials weaken. Figure 12 illustrates the stress/strain curves of the 45% samples. After only ten days, the profile rises steeply upward and the extension at break dramatically shortens, indicative of embrittlement. For samples sized with 5%, 15%, and 30%, all three curves, before and after two periods of aging, follow the same profile, neither rising nor dropping (fig. 13). All three were toughened from aging, and only the elongation at break of 30% was much shorter than the rest. Thus in the tensile testing, after a short, one month aging period, 5% and 15% alum-sized silk perform the best.

HPSEC Results

The molecular weight of a polymer is one indicator of its properties. An increase in molecular weight results in an increase in tensile strength, elongation, toughness, hardness, abrasion resistance, and chemical resistance. The distribution

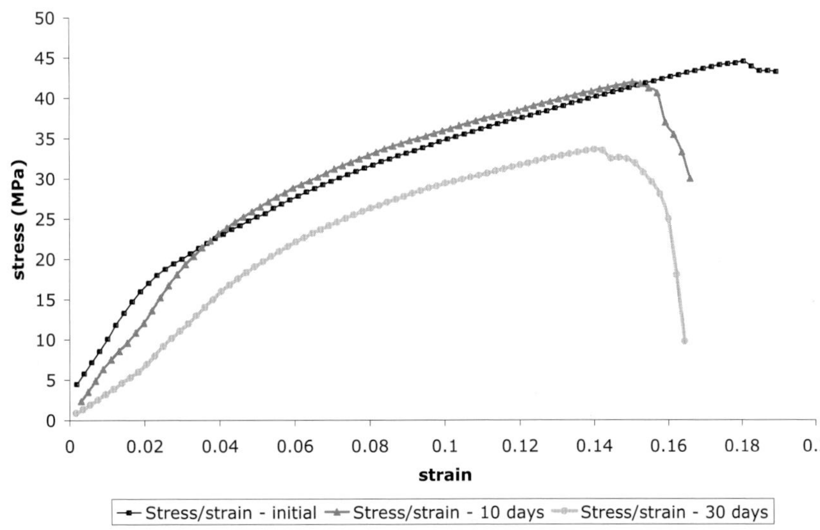

Figure 11. *Engineering stress, silk sized without alum, before aging, after ten days, and after thirty days of aging*

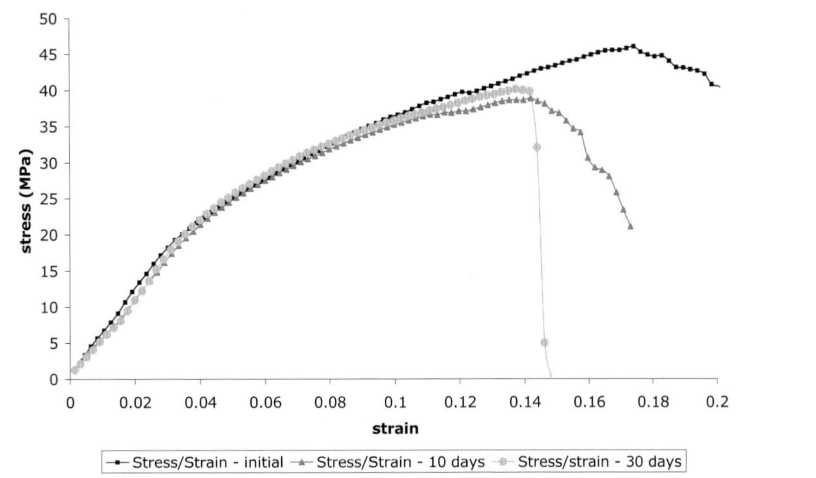

Figure 12. *Engineering stress, 45% alum-sized silk, before aging and after ten and thirty days of aging*

of molecular weight also affects polymer properties. A narrower distribution often means an increase in tensile strength.

The usefulness of HPSEC to determine silk deterioration has been reported in various sources (Garside and Wyeth 2002; Tse and Dupont 2000; Howell 1992). With calibration using standards of known molecular weights, the raw data can be converted to a molecular weight distribution curve (fig. 14).

The calculated molecular weight values are not actual but apparent molecular weights of silk protein. The three calculated apparent molecular weight averages—Mn, Mw, and Mp—of all the treated and aged silk samples showed similar trends, but the Mp values showed the greatest change and are the most sensitive to degradation.[8]

During aging the heat caused a decrease in molecular weight of all the silk samples, regardless of the size. The effect of alum concentration became more obvious after thirty days of aging, evident especially at 30% and 45% alum. After the second aging period, the curves of unsized, 0%, 5%, and 15% samples overlap each other, while 30% and 45% are eluted later (see fig. 14). Thus, the results showed that addition of up to 15% alum did not significantly change the rate of aging of silk; the addition of 30% and 45% showed an increased rate of deterioration.

Comparison to Tensile Strength Measurements

Tensile strength tests and HPSEC provide complementary results. While the molecular weight measurements of HPSEC are much more sensitive to changes of silk due to deterioration, tensile strength provides information about the properties of the whole system. In spite of the high tensile strength contributed by higher percentage of gelatin-alum size (fig. 15), the HPSEC results showed (fig. 16) that,

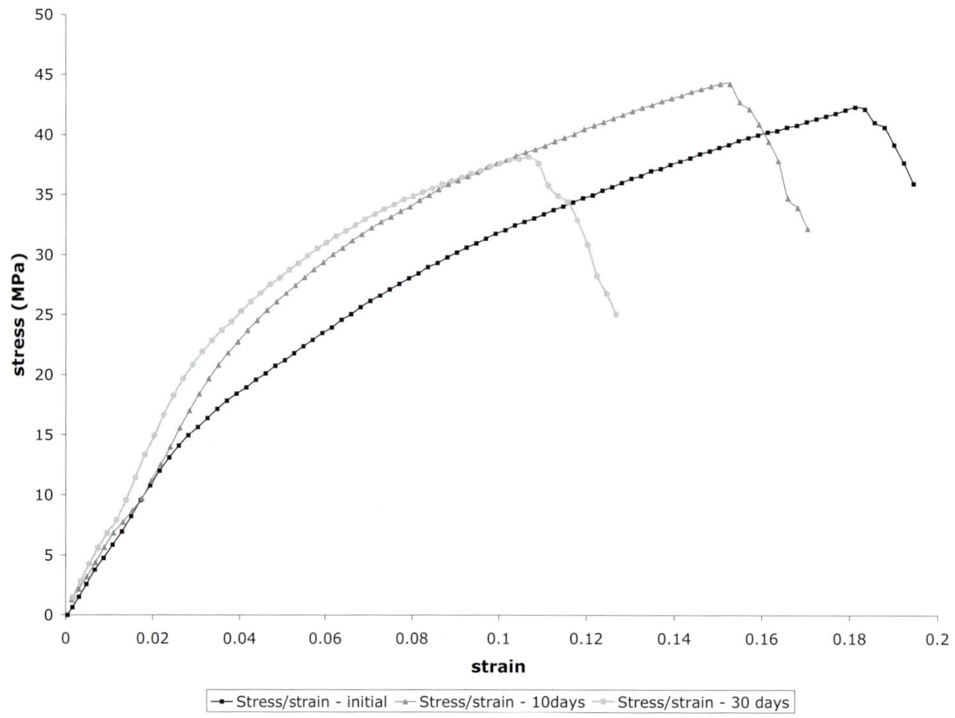

Figure 13. *Engineering stress, 5% alum-sized silk, before aging, after ten days, and after thirty days of aging*

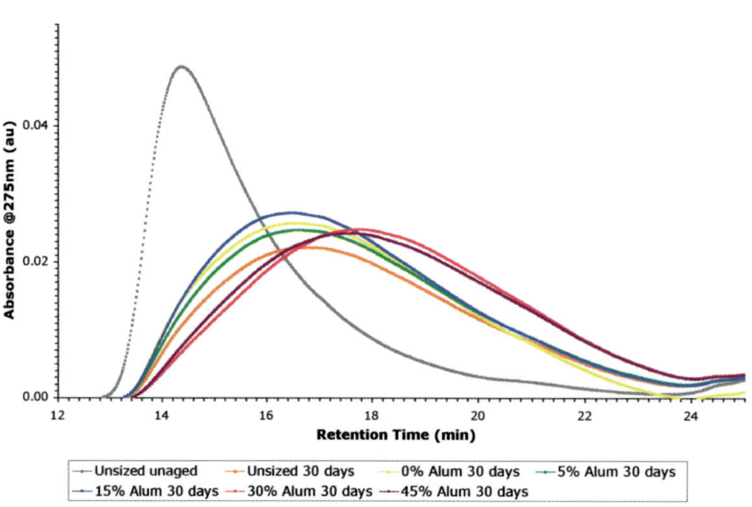

Figure 14. *HPSEC curves of Kwungtung silk aged at 80°C, 50% RH for thirty days*

after thirty days of heat aging, at 30% and 45% alum silk deteriorates at a greater rate than lower alum concentrations. It is suggested that high tensile strength of 30% and 45% samples is due largely to the tanning effect of aluminum salts. With longer aging, it is expected that silk with 30% and 45% alum will degrade at an increasing rate, and the strength of these systems will eventually become less than those with lower concentrations.

The Tanning Effect of Alum

It is suggested that aluminum potassium sulfate acts as an agent similar to the tannins used in the leather industry. When aluminum sulfate dissolves in water, it dissociates into a cationic aluminum complex and an anionic sulfate:

$$Al_2(H_2O)_{12}(SO_4)_3 \leftrightarrow 2H^+ + 3SO_4^{2-} + [Al_2(OH)_2(H_2O)_{10}]^{4+}$$

These large cations interact with polyelectrolytes represented by the sizing dispersions and silk proteins, undergo hydrogen bonding, and form a "quasipolymeric" alumina structure (Strazdins 1986). In figure 17, based on the models given by Chambard (1977) and Covington (2001), the cross-linking mechanism between protein molecules and alum is proposed. The cross-linking is not permanent, however, and can be easily reversed by the action of water. Nevertheless, with aging the hydrolysis of the bonds becomes more difficult.

Conclusions

The present research showed that alum does not significantly contribute to the deterioration of the silk supports at a concentration of 15% or lower. The SEC results demon-

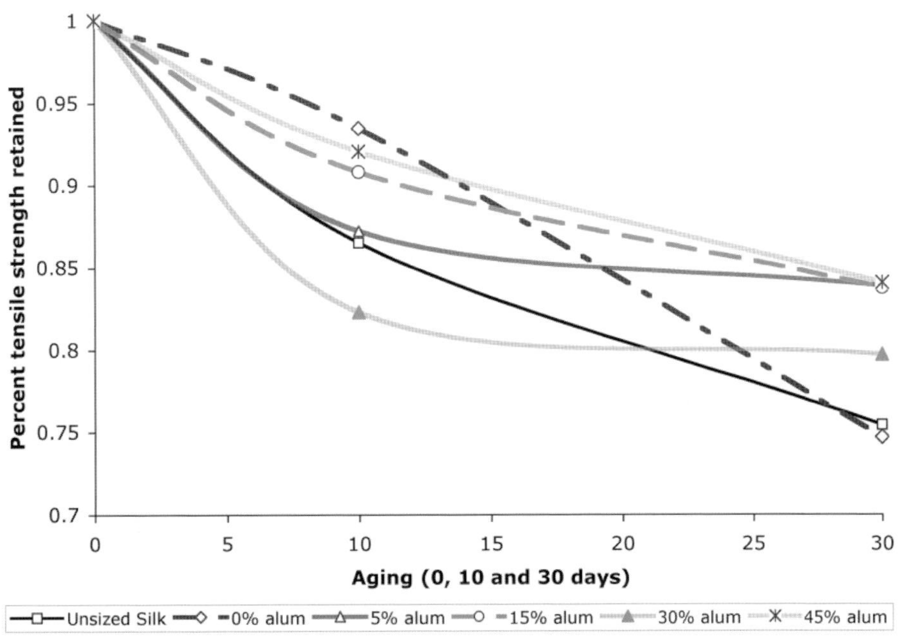

Figure 15. *Percent loss of the tensile strength of the unsized and sized silk after ten days and after thirty days of aging*

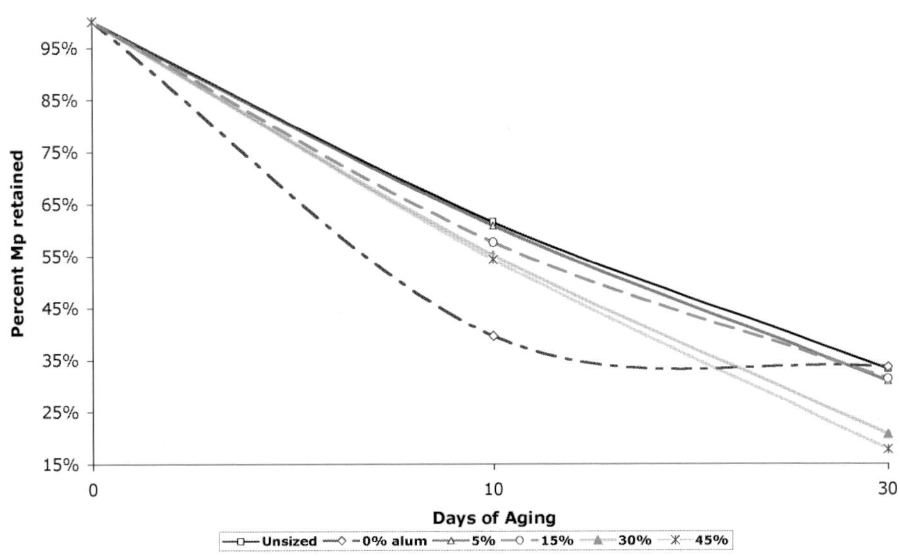

Figure 16. *Percentage of the peak molecular weight retained after two periods of aging: ten days and thirty days*

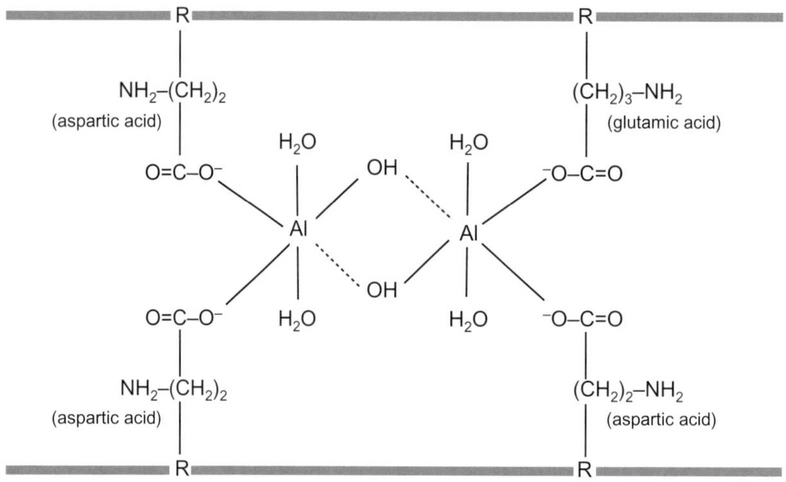

Figure 17. *Model of the tanning effect of aluminum sulfate on protein molecules*

strated that after ten days and thirty days of aging at 80°C and 50% RH, alum accelerates the degradation of the silk molecules only at concentration of 30% and 45%. In tensile testing 30% alum exhibited good initial properties, but upon aging both 30% and 45% alum showed greater undesirable changes than other alum-sized samples. The best performance in tensile testing, ability to retain original pH, and show least color change was demonstrated by 15% alum-sized samples. Results almost as good were obtained from 5% alum. Thus aluminum potassium sulfate at concentrations of 5–15% can be considered beneficial. The preference for 5% or 15% cannot be estimated from the present research; samples would have to be aged for a period of sixty or ninety days in order to extrapolate with certainty the degradation profile. It is possible that a small concentration of alum, for example 5%, might be the best as far as long-term stability is concerned, but time could be required to allow bridges to form between gelatin and silk so the gelatin would harden sufficiently to act as a size for painting supports.

Results from the FTIR analysis suggest the interaction between alum and gelatin is complex and alum causes gelatin to undergo hydrolysis and cross-linking, either sequentially or simultaneously. The better performance of 15% compared to 5% alum-sized samples might indicate that some unknown stoichiometric laws dictate certain processes to occur within the silk-alum-gelatin network upon aging. In the future, size exclusion chromatography or viscometry could be employed to estimate the rates of hydrolytic scission and cross-linking.

It was discovered that pure gelatin without the addition of alum degrades in a way similar to unsized silk and, in humid heat aging, does not provide a protective barrier. Both sized and unsized silk exhibit significant color changes upon aging, and at large alum concentrations, such as 30% and 45% alum, discoloration was the most pronounced.

Unsized silk and silk sized with pure gelatin and various alum concentrations did not exhibit a large difference in their pH values, neither before nor after thirty days of aging, and showed acceptable neutral values between pH 6.9 and 6.0. The behavior of the paint layer was not influenced by different sizes and exhibited minimal degradation.

The results suggest the use of a gelatin-alum solution at 5–15% alum concentration per dry weight of gelatin as a size for silk intended for in-fills or as a consolidant for distempera paintings on undegraded silk; the higher concentration of 25%, 30%, or even 45% encountered in manuals on Far Eastern paintings do cause degradation of the silk proteins, embrittlement, and yellowing. The best performance of 15% alum-gelatin sizing in all the experiments undertaken contradicts the long-held assumptions of the detrimental effect of alum, but further tests with longer aging periods and perhaps different aging conditions are certainly required.

Acknowledgments

The work described was performed as part of the science project for a master's degree in art conservation at Queen's University, Kingston, Ontario (EP). The following are thanked for technical assistance: Herbert Shuvell (FTIR spectroscopy); John Lawrence, Peter Apsley, and Dupont Canada (tensile testing at the Dupont Technical Research Laboratory); Alan Grant, Geological Department, Queen's University (XRD analysis); Charles Cooney, Metallurgical Department, Queen's University (tensile testing). Janet Cowan, Royal Ontario Museum, is thanked for providing study materials.

Notes

1. Alum was purchased in an art supply store in Beijing called Rong Bao Zhai Studio.
2. Shi Tai Arts and Crafts, Scarborough, Ontario.
3. Philip Meredith, personal correspondence.
4. Among synthetic adhesives are acrylic emulsions (Higuchi 1979), polyvinyl alcohols (Koyano 1979), and water-soluble ethyl hydroxy ethyl cellulose (Grantham and Cummings 2002).
5. Granular rabbit skin glue was purchased at Lee Valley Tools, Ottawa, Ontario.
6. Shi Tai Arts and Crafts, Scarborough, Ontario.

7. Cinnabar was purchased at the Yan Yan Chinese Herbs Centre, Scarborough, where it was lightly ground by an electric grinder; it was further worked by hand with a muller to eliminate large granules.
8. Molecular weight averages: Mn, number-average molecular weight, is a convenient way of measuring the "averaged" chain length in a polymer mixture and can be characterized directly by laboratory measurement of colligative properties. Mn is defined as the total mass of the sample (Wi) divided by the total number of chains (N).

   Mw, weight averaged molecular weight, is a weighted average of a polymer mixture, taking into account that heavier molecules contribute more to the total weight of the mixture. It can be experimentally determined by light scattering and ultracentrifugation. It is a more representative value than Mn and more influenced by the higher molecular weight fractions. Mp, peak molecular weight, is the molecular weight of the fraction with the highest population. This value is very close to Mw in this series of silk samples.

References

Becker, M. A., and N. C. Tuross. 1994. Initial degradative changes found in *Bombyx mori* silk fibroin. In *Silk polymers: Material science and biotechnology*, ed. D. Kaplan, W. W. Adams, B. Farmer, and C. Viney, 252–69. American Chemical Society Series 544. Washington, D.C.: American Chemical Society.

Belozerova, V. G. 1995. *Kitayskiy svitok* [Chinese scroll]. Moscow: GosNIIR, State Institute of Scientific Research for Conservation.

Chambard, P. 1977. Miscellaneous mineral tannages. Chap. 25, pt. A, Alum tanning. In *The chemistry and technology of leather*, 2:349–64. Malabar, Fla: Krieger Publishing Co.

Covington, A. 2001. Chrome tanning: Exploding the perceived myths, preconceptions and received wisdom. *Journal of the American Leather Chemists Association* 96:467–79.

Dupont, A.-L. 2002. The role of gelatine/alum sizing in the degradation of paper: A study by size exclusion chromatography in lithium chloride/N,N-dimethylacetamide using multiangle light scattering detection. In *Works of art on paper: Books, documents and photographs—Techniques and conservation*, ed. V. Daniels, A. Donnithorne, and P. Smith, 59–64. Contributions to the IIC Baltimore Congress, 2–6 September 2002. London: International Institute for Conservation.

Dupont, A.-L. 2003. Gelatine sizing on paper and its impact on the degradation of cellulose during aging: A study using size-exclusion chromatography. PhD diss., Amsterdam University, Amsterdam.

Garside, P., and P. Wyeth. 2002. Characterization of silk deterioration. In *Conference preprints—Strengthening the bond: Science and textiles*, ed. V. J. Whelan, 55–60. North American Textile Conservation Conference (NATCC) 2002, Philadelphia, 5–6 April 2002. Philadelphia: North American Textile Conservation Conference.

Grantham, S., and A. Cummings. 2002. Painted Japanese paper screens: The consolidation of paint layers on a paper substrate. In *The broad spectrum: Studies in the materials, techniques, and conservation of color on paper*, ed. H. Stratis and B. Salvesen, 198–206. London: Archetype Publications.

Hansen, E. F., and H. Sobel. 1992. Effect of the environment on degradation of silk: A review. In *Postprints of the American Institute for Conservation 20th Annual Meeting*, 14–30. Buffalo, N. Y. Washington, D.C.: Textile Specialty Group, American Institute for Conservation.

Higuchi, S. 1979. Treatment on painting of sliding screens and wall panels to prevent exfoliation in Japan. In *Conservation of Far Eastern art objects*, 69–78. International Symposium on the Conservation and Restoration of Cultural Property, Tokyo, 26–29 November 1979. Tokyo: Tokyo National Research Institute of Cultural Properties.

Howell, D. 1992. Silk degradation studies. In *Silk: The Harpers Ferry Regional Textile Group*, 11–12. 11th Symposium, Washington, D.C., 12–13 November 1992. Washington, D.C.: National Museum of American History.

Iizuka, E. 1985. Silk—An overview: U.S.-Japan Seminar on Polymer Liquid Crystals, 1983, ed. J. L. White and S. Onogi. *Journal of Applied Polymer Science* 41:163–73.

Kaplan, D., W. W. Adams, B. Farmer, and C. Viney. 1994. Silk: Biology, structure, properties, and genetics. In *Silk polymers: Material science and biotechnology*, ed. D. Kaplan, W. W. Adams, B. Farmer, and C. Viney, 2–16. American Chemical Society Series 544. Washington, D.C.: American Chemical Society.

Kerr, N. 1992. Understanding the behavior of silk: Structure-property relationships. In *Silk: The Harpers Ferry Regional Textile Group*, 5–9. 11th Symposium, Washington, D.C., 12–13 November 1992. Washington, D.C.: National Museum of American History.

Koyano, M. 1979. *Japanese scroll paintings: A handbook of mounting techniques*. Washington, D.C.: Foundation of the American Institute for Conservation.

McCormick-Goodhart, M. 1995. Moisture-content isolines of gelatin and the implications for accelerated aging tests and long-term storage of photographic materials. *Journal of Imaging Science and Technology* 39:157–62.

Rose, C. L., and D. W. Von Endt. 1984. *Protein chemistry for conservators*. Los Angeles: Objects Specialty Group, American Institute for Conservation.

Silbergeld, J., and A. McNair, trans. 1988. *Chinese painting colors: Studies of their preparation and application in traditional and modern times*, translation of Yu Feian, *Chung-Kuo hua yen-se ti yen-chiu*. Hong Kong, Seattle, and London: Hong Kong University Press and University of Washington Press.

Strazdins, E. 1986. The chemistry of alum in paper making. *TAPPI Journal* April:111–14.

Sze, M.-M., trans. 1977. *The Mustard Seed Garden Manual of Painting*. Bollingen Series. Princeton, N.J.: Princeton University Press.

Tse, S., and A.-L. Dupont. 2000. Measuring silk deterioration by high-performance size-exclusion chromatography, viscometry, and electrophoresis. In *Historic textiles, papers and polymers in museums*, ed. J. Cardamone and M. T. Baker, 98–113. American Chemical Society Symposium Series 779. Washington, D.C.: American Chemical Society.

Foxing on the Backs of Chinese Paintings

Xie Yulin and Chen Yuansheng

ABSTRACT In this study foxing on the backs of three Chinese paintings was investigated with analytical techniques such as ultraviolet light examination, x-ray fluorescence spectroscopy, and scanning electron microscopy with energy-dispersive x-ray spectroscopy. The results indicate that foxing is related to the growth of fungi and may involve both fungal development and a series of chemical processes. No evidence was found to prove that the foxing was caused by inorganic factors. Foxing causes considerable oxidation of the cellulose chain. The pH of the foxed area of paper is more acidic than that of the unaffected area. Paper affected by foxing becomes weaker and friable. It can be considered a widespread danger to Chinese painting and therefore must be studied.

Introduction

Foxing is found on drawings, engravings, manuscripts, books, archival documents, and cotton fabrics (figs. 1 and 2). It frequently occurs on Chinese paintings on *xuan* paper, a special paper handmade by traditional Chinese methods. Usually foxing spots appear on paintings or books only after a long period of time. Previous researchers reported that they had not seen rusty red foxing spots appear in less than twenty-three years (Meynell and Newsam 1978). Their report indicates that it takes many years for the oxidation of cellulose to result in colored breakdown products. In our case, however, foxing spots appeared on the backs of three Chinese paintings after just twenty days. We were surprised by the large number of spots, the speed of their formation, and the size of the affected area. Thus we were obliged to investigate and analyze their formation.

Experiment

Samples were cut from the edges of the three Chinese paintings. The paintings consist of a layered structure with five sheets of paper and a sheet of silk, on which the painting is executed. Each sample has two parts: an unaffected area and a foxed area. Several analytical techniques were used to analyze the samples.

The fluorescence of both the unaffected areas and the foxed areas was determined using ultraviolet (UV) light of 365 nm, which shows the foxing better than light of 254 nm. Metallic elements on the unaffected and foxed areas were analyzed by x-ray fluorescence spectroscopy (XRF) (TN Spectrace QuanX). The pH of both the unaffected and foxed areas was determined by flat electrodes (WTW330I with Sen Tix Sur electrode). Samples were directly coated with platinum, then examined using scanning electron microscopy (SEM) (JEOL JSM-6360LV). Samples of foxing without a platinum coating were also examined using an environmental SEM (PHILIPS XL30E). Samples were also analyzed by infrared microscopy (Bio-Rad Win-IR). Energy-dispersive x-ray spectroscopy (EDS) and x-ray diffraction (XRD) were used to identify crystals found in the foxed area of sample 2.

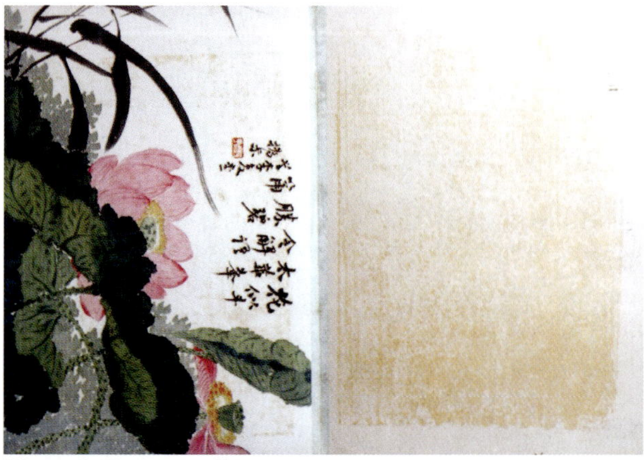

Figure 1. *Foxing on Chinese paintings*

Figure 2. *Foxing on Chinese paintings*

Results

UV Light

The fluorescence analysis method is simple and sensitive in differentiating some compounds that appear similar under visible light. The color of the foxing on sample 1 and sample 2 is the same; both appear brown under daylight. But the color of the foxing on sample 3 appears yellow. The fluorescence of sample 1 and sample 2 appears as a bull's-eye pattern—a dark brown bull's-eye with a blue-white halo. The fluorescence of sample 3 appears as a snowflake pattern in light blue. Sample 2 and sample 3 under sunlight and UV light of 365 nm are shown in figures 3 and 4.

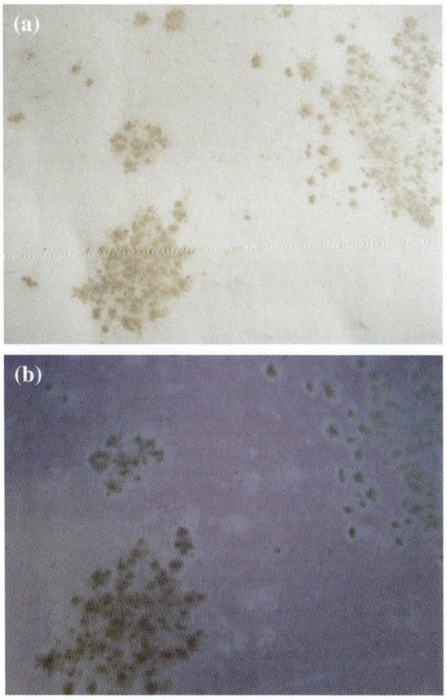

***Figure 3** a. Sample 2 under daylight (10 × 8 cm). **b.** Sample 2 under UV light, 365 nm*

XRF

Foxing in a bull's-eye pattern is usually considered to be due to the presence of a metal impurity, but the metallic elements of this foxing were found to be the same as those of the foxing in the pattern of snowflakes. We also found no difference in the amount of iron and other metallic elements between affected and unaffected areas of the samples (Tang 1978; Cain 1993; Daniels and Meeks 1993). Such elements are inherent in *xuan* mounting paper. The two kinds of foxing on the backs of Chinese paintings are not related to iron. The XRF results of sample 2's foxed and unaffected areas are shown in figure 5. The peak heights and shapes of the metallic elements are almost superimposed. The same is true for sample 3, which is shown in figure 6.

pH Measurements

While pH of *xuan* paper is usually higher than 7, the pH of samples 1–3 is lower than 7. The pH of the foxed areas is lower than that of unaffected areas. See table 1 for a summary of pH measurements.

SEM

A difference is visible in SEM micrographs between the surface of foxed and unaffected areas of sample 1 (figs. 7 and 8). The foxed area looks very dirty and has many fiber fragments, while the unaffected area looks clean, without any fragments. Under lower magnification, the foxed area of sample 1 looks the same as that of sample 3 (fig. 9). Only at

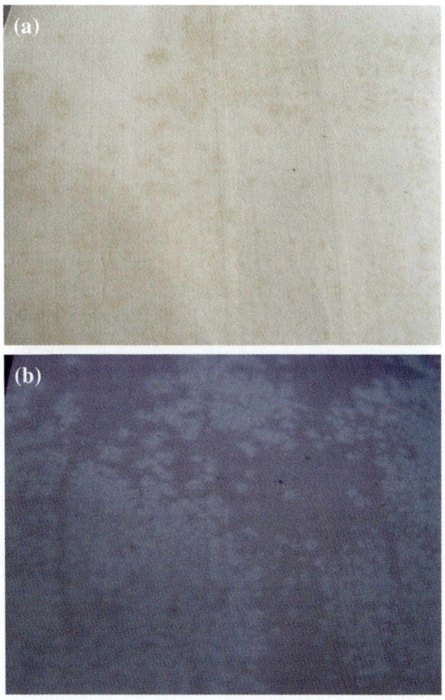

***Figure 4** a. Sample 3 under daylight (10 × 8 cm). **b.** Sample 3 under UV light, 365 nm*

***Table 1.** pH for foxed and unaffected areas of the samples*

Name		pH	Average
Sample 1	Unaffected area	6.35, 6.24, 6.42, 6.52, 6.62, 6.18	6.39
	Foxed area	4.83, 5.26, 4.73, 4.86, 4.56, 5.02	4.88
Sample 2	Unaffected area	6.24, 6.52, 6.42, 6.30, 6.48, 6.54	6.42
	Foxed area	4.43, 4.51, 4.53, 4.52, 4.82, 4.54	4.56
Sample 2 b (back surface)	Unaffected area	6.05, 5.93, 5.97, 6.02, 5.97, 5.74	5.95
	Foxed area	4.43, 4.68, 4.78, 4.82, 4.32, 4.86	4.65
Sample 3	Unaffected area	6.66, 6.84, 6.74, 6.87, 7.02, 6.91	6.84
	Foxed area	5.34, 5.17, 5.10, 5.15, 4.60, 5.32	5.11

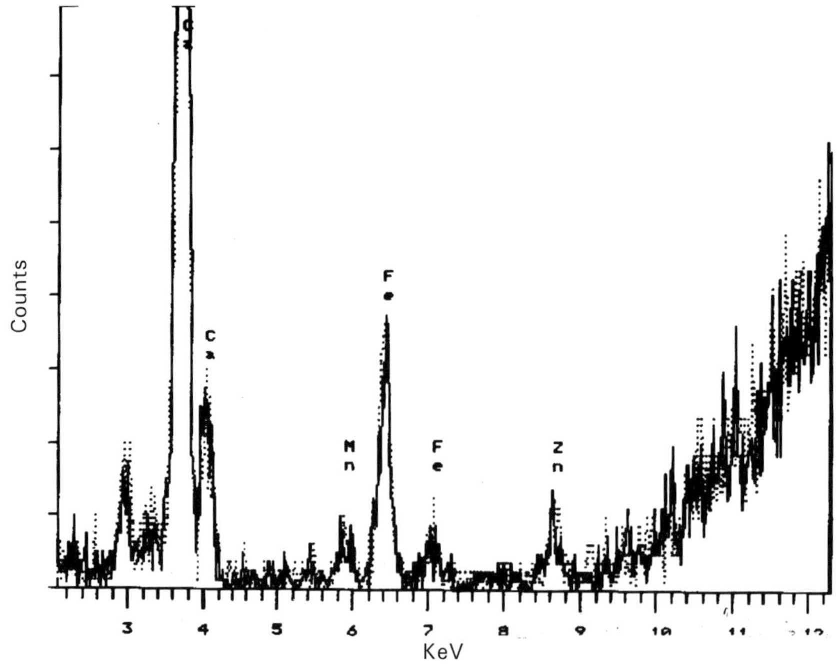

Figure 5. *XRF spectra of sample 2, unaffected area shown in line, foxed area shown in dots*

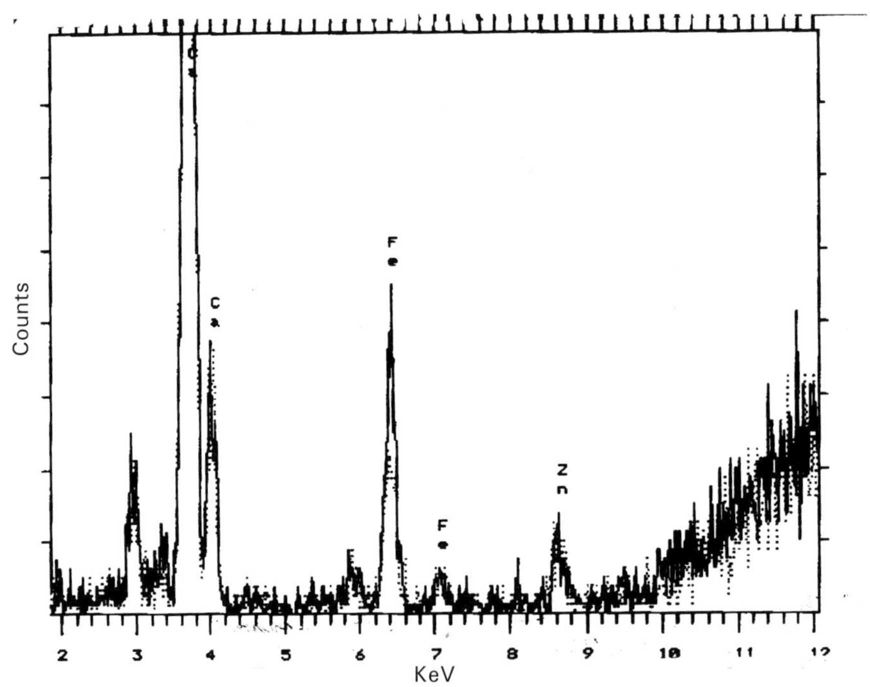

Figure 6. *XRF spectra of sample 3, unaffected area shown in dots, foxed area shown in line*

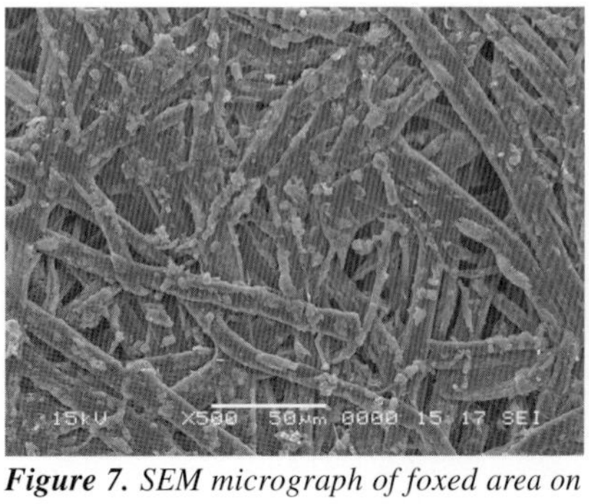

Figure 7. *SEM micrograph of foxed area on sample 1*

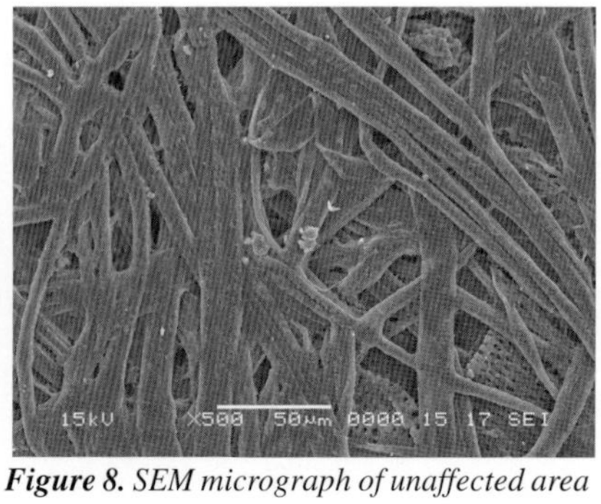

Figure 8. *SEM micrograph of unaffected area on sample 1*

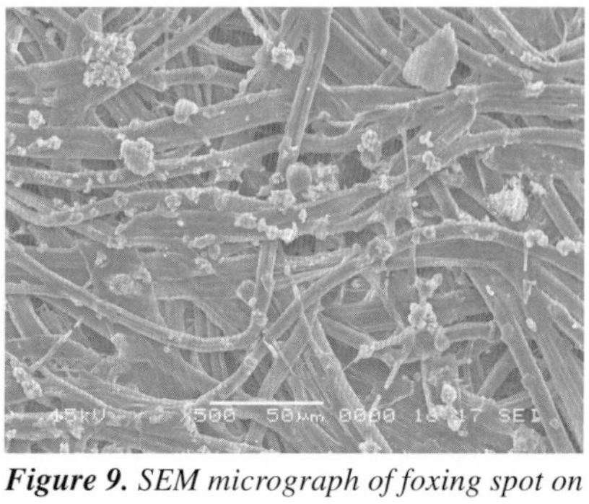

Figure 9. *SEM micrograph of foxing spot on sample 3*

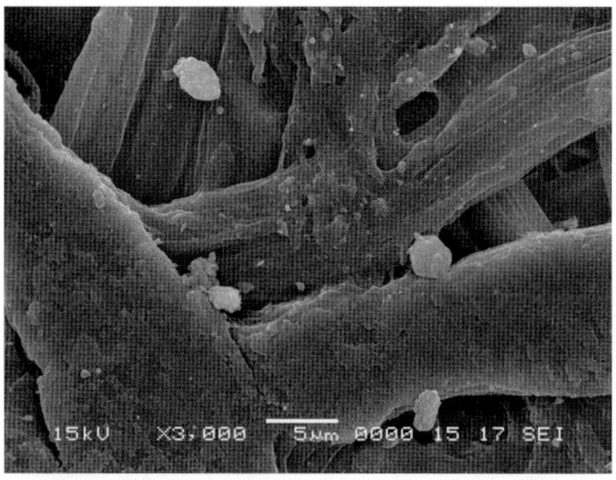

Figure 10. *SEM micrograph of the detail of foxing spots on sample 1*

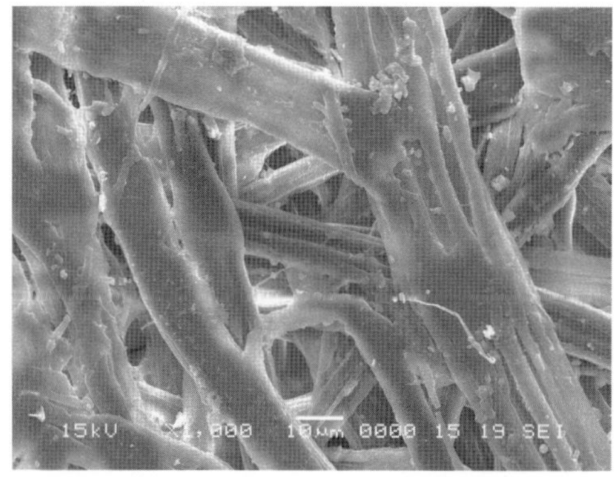

Figure 11. *SEM micrograph of the detail of foxing spots on sample 2*

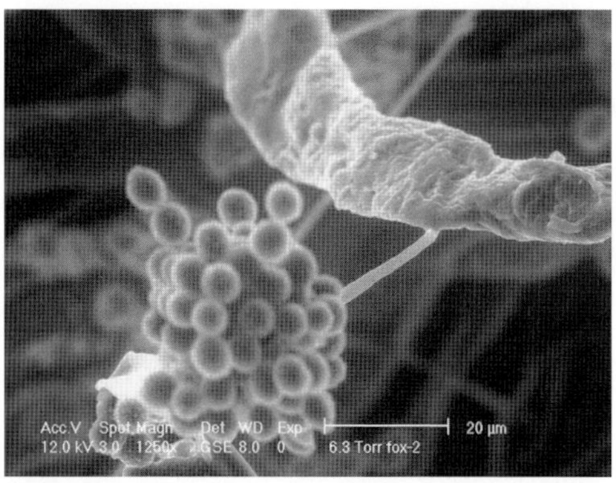

Figure 12. *SEM micrograph of foxing spot on sample 3, after incubation*

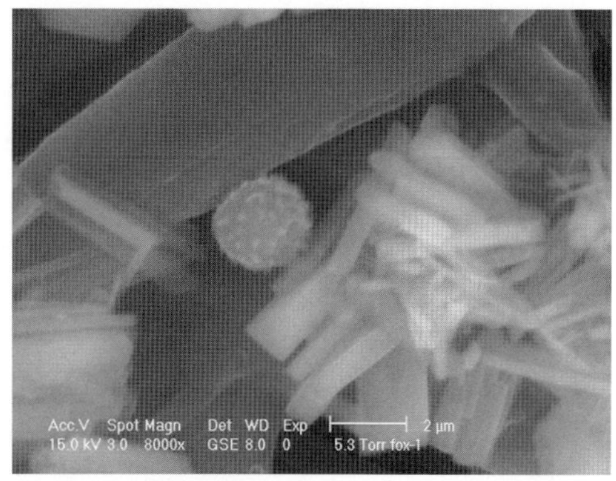

Figure 13. *SEM micrograph of a conidium and crystals on sample 2, after incubation*

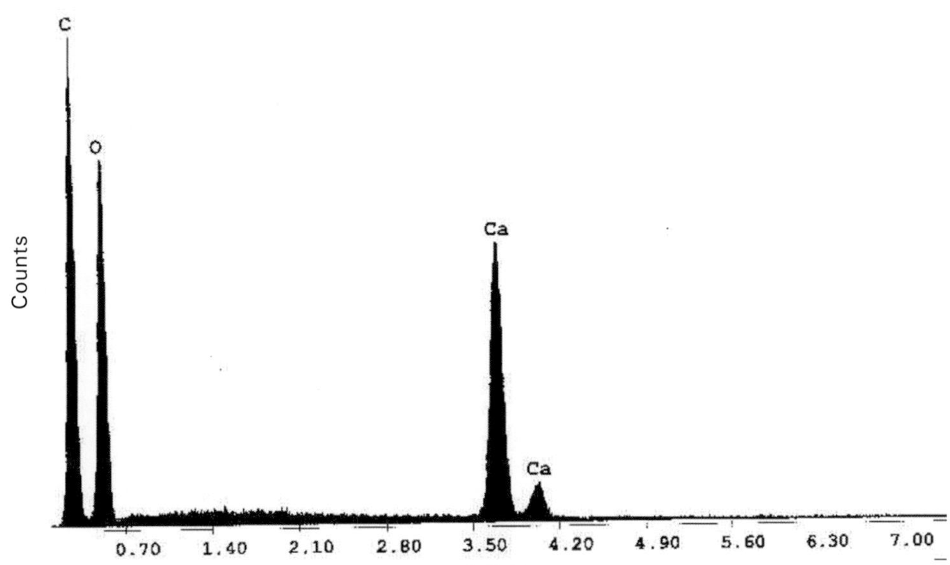

Figure 14. *The EDS spectrum of a particle on incubated foxing spot of sample 2*

higher magnification were many fibers on the foxed areas of samples 1 and 2 seen to be covered with a thick layer of material (figs. 10 and 11). A few fibers in figures 10 and 11 look torn apart.

There were many conidia and hyphae in sample 3 (fig. 12), but few conidia and hyphae were observed in sample 2 (fig. 13).

Surprisingly, crystals appeared in sample 2. The crystals were analyzed by EDX and XRD. The results are shown in figure 14 and table 2. From the analysis, the crystals were identified as calcium oxalate.

Table 2. *XRD results for crystalline particle found in affected area of sample 2 after incubation*

	Phase ID	File ID	I%
No. 1	C-Carbolite	PDF#50-0926	104.1
No. 2	Calcium oxalate (CaC_2O_4)	PDF#20-0838	21.5

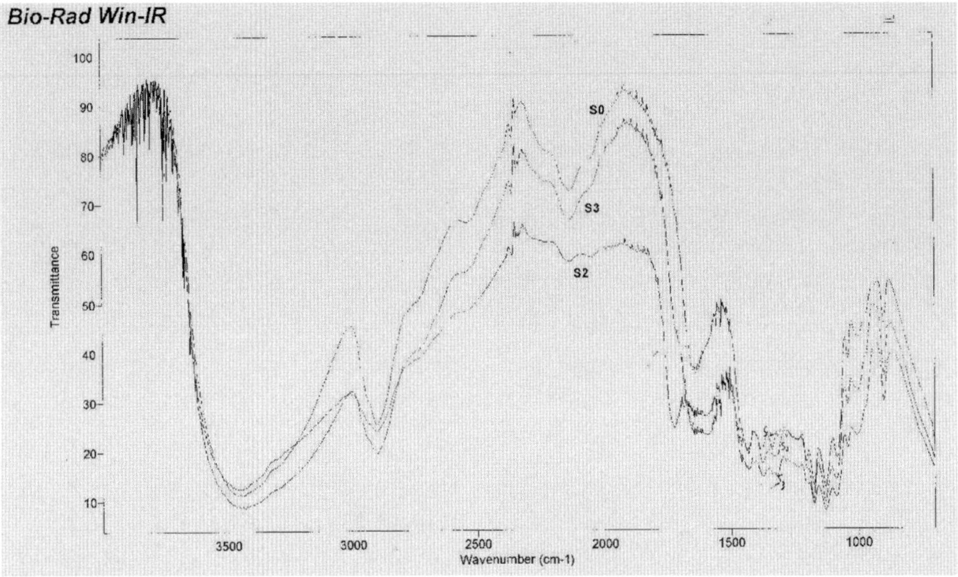

Figure 15. Infrared spectra of unaffected area of sample 2 (S0) and foxed areas (S2 and S3)

IR Spectroscopy

IR spectroscopy of sample 2 resulted in a spectrum with a peak at 1,726 cm^{-1} in the foxed area that was absent in the spectrum of the unaffected area (fig. 15). The peak represents the typical absorption for the carbonyl bond ($R_1R_2C=O$).

Discussion

As is well known, chemical reactivity and fungal growth occur on paper in a humid environment (Peters 2000). First, the moisture level for the three Chinese paintings affected by foxing was considered and determined. Even though the case was kept at 20°C and 60% RH, the humidity of the microenvironment between the wooden boards and the back of the painting was very high. Because of condensation on the back of the boards, the moisture content there was found, using a moisture-meter, to be 16–18%. As a consequence of the warm temperature (20°C), an excessive volume of water vapor was adsorbed by the porous *xuan* paper, adding to the localized capillary condensation. Because the paper is porous, water is adsorbed and desorbed quickly at the interface, and the moisture runs over the back of the painting. Condensed water in the paper is believed to accelerate chemical reactivity and the degradation of cellulose and other materials, and to induce fungal growth.

The environment of the case was also considered and investigated. Nitrogen dioxide (NO_2) is an active air pollutant that can participate in photolysis under the action of 405 nm visible light (Zhang 1999). It can also decompose and accelerate the oxidation of cellulose. Air pollutants such as NO_2 are possibly formed by condensation reactions that take place when cellulose oxidation interacts with nitrogen-containing compounds (Bicchieri et al. 2001; Choisy et al. 1997). The products of cellulose oxidation impart a peculiar color to the stains, brown in the case of those containing nitrogen. The concentration of nitrogen dioxide was deter-

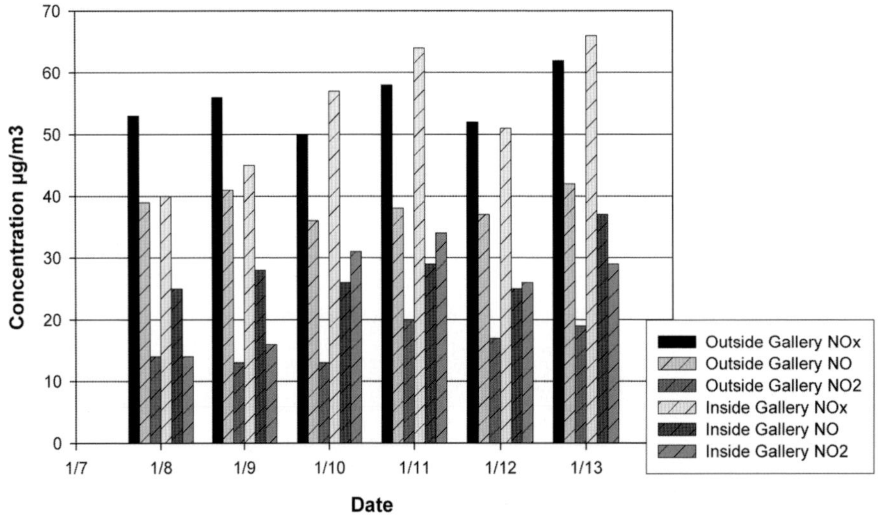

Figure 16. The concentration of nitrogen oxides in the storage environment

mined for the storage environment. The data are shown in figure 16.

It was found that there was a higher concentration of nitrogen dioxide and nitrogen oxide in the period during which the foxing spots formed than was seen during the entire year. The concentration of nitrogen dioxide inside the case was higher than outside. The three paintings were exhibited in that environment. But the typical absorption peak for -NO_2 is not evident in the IR spectrum.

Cellulose is a macromolecular material with polar groupings. Its 1,4-glycoside bond is very sensitive to acid. Under the right conditions of acidity and temperature, and with sufficient time, the glycosidic bond is broken and the degree of polymerization begins to decrease. The hydrolysis of cellulose is induced by acids produced by fungal activity. Analysis of foxed areas by infrared spectroscopy has shown absorption due to the presence of carbonyl bonds, which can be formed as the result of cellulose degradation. Acids are not consumed but rather catalyze bond scission during the chemical reaction, and the rate of cellulose degradation is increased with acid concentration in a linear manner (Strlic and Kolar 2002). At the same time, the mobility of water probably redistributes acidity in the paper. The pH of the front of the painting, where the moisture has passed through five sheets of *xuan* paper and a sheet of silk, is almost the same pH as that of the back, around 4.6. Not only is the foxed area on the back of the painting acidic, but the opposite side is acidic also.

On the basis of SEM micrographs, the formation of foxing spots on sample 3 can be related to the growth of fungi (Arai 1987a; Florian 1996). Many conidia and hyphae can be seen on the foxed area of sample 3, including a mass of conidia. As they are very small, only 0.1–1 µm, we suspect that they may be the initial conidia. The conidia are nearly spherical with aculeate ornamentation (sharp warty extensions) and have an average size of 5 µm according to SEM micrographs. They strongly resemble those in the literature in morphology and size. The fungus belongs to the genus *Aspergillus* (Florian and Manning 1999, 2000; Nol et al. 1983; Arai 1984, 1987b). A close-up view shows the yellow-brown drops of water that are produced by fungi on the foxed area (fig. 17). The role of fungal infestation is clear, but we have not succeeded in producing brown hyphae and yellow conidia in each foxing spot through incubation. Moreover, few conidia and hyphae on sample 2 can be seen using SEM. Many paper fibers covered with a thick layer of material can be observed on the foxed areas in samples 1 and 2 (see figs. 10 and 11). The surface of the foxed area (very dirty) is seen to be very different from that of the unaffected (clean) area in the SEM micrographs. The cause is believed to be fungal self-dissolution because of the large amount of acidic secretion produced by their fast growth. After incubating, we wondered why only a few conidia and hyphae could be found and were surprised to find crystalline calcium oxalate produced on a foxed area of sample 2.

On the basis of experiments, Arai and co-workers concluded that foxing occurs due to an amino-carbonyl reaction known as the Maillard reaction (Arai et al. 1989, 1990). We think the reaction may occur on paper. We have identified cellulose oligosaccharides, organic acids, and amino acids on foxed areas (Chen and Xie 2002). The kinds of acids and their amounts in foxed and unaffected areas of paper are so different that the Maillard reaction is probable. But in Arai's opinion, the location of microcolonies of xerophilic fungi could be the areas where foxing forms. We have observed that foxed areas do not always have fungal growth. Foxed areas may not grow fungi; instead, fungal secretion products could be transferred here as sediment by water transport and through the amino-carbonyl reaction form foxing spots. Alternatively, the foxing spots may bloom to grow fungi; then the amount of fungal acidic secretions causes self-dissolution, and aminophenol forms from the original fungal protein to take part in the reaction.

In addition to the action of fungi, there is another theory for the formation of foxing spots. According to Rebrikova's free radical mechanism, foxing is initiated by the presence of hydrogen peroxide formed at the wet and dry interface (Rebrikova and Manturovskaya 2000). Oxidation of cellulose by molecular oxygen results in the formation of hydrogen peroxide and peroxide radicals. Because of the presence of significant trace metals, hydrogen peroxide catalyzes and forms cellulose oxidation by free radicals. The products of cellulose oxidation can enter an amino-carbonyl reaction with nitrogen-containing compounds. The results of the reaction are the formation of melanoidin-type colored compounds. All of the conditions that can cause free radical cellulose oxidation exist, but this mechanism cannot be confirmed.

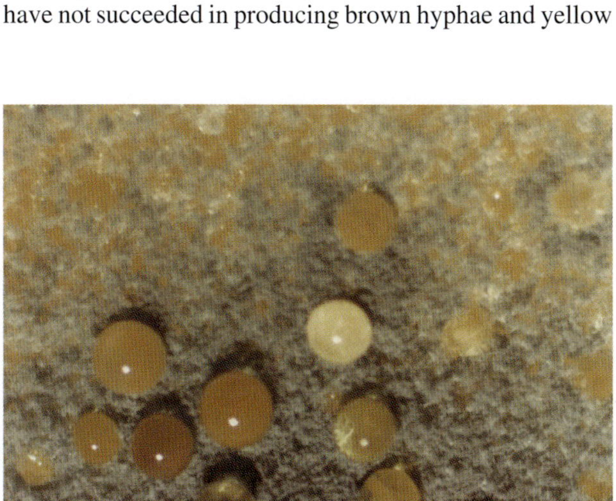

Figure 17. *The yellow-brown drop of water produced by fungi on foxing*

Conclusions

On the basis of the XRF spectra, the type and amount of the metallic elements are similar not only for the foxed areas and the unaffected areas, but also for both the bull's-eye and snowflake foxing spots. For this reason the cause of the rapid formation of foxing spots on the back of Chinese paintings is not related to inorganic factors.

According to the SEM micrographs, in which many conidia and hyphae can be seen, the rapid formation of foxing is related to the growth of fungi. Further proof is given by the incubation experiment in which yellow-brown drops of water were produced by fungi in the location of the foxing spots.

The rapid onset of foxing in the case of the three paintings is also related to chemical reactions. Foxing is a complex process and results from a series of reactions caused by moisture, air pollutants, and fungal growth. Which of these plays the most important role should be taken into consideration, and more research is necessary to reach a complete understanding. It is very important for the conservation of Chinese paintings to monitor and control the environment; it is especially important to control moisture.

References

Arai, Hideo 1984. Microbiological studies on the conservation of paper and related cultural properties. Pt. 1, Isolation of fungi from the foxing on paper. *Science for Conservation* 23: 33–39.

Arai, Hideo 1987a. On the foxing-causing fungi. In *Preprints: ICOM Committee for Conservation*, ed. K. Grimstad, 3:1165–67. 8th Triennial Meeting, Sydney. Marina del Rey, Calif.: Getty Conservation Institute.

Arai, Hideo 1987b. Microbiological studies on the conservation of paper and related cultural properties. Pt. 5, Biological and morphological characteristics of fungi isolated from foxing and the formation and prevention of foxing. *Science for Conservation* 26:43–52.

Arai, Hideo, Chichiro Nemoto, Norie Matsui, Noritaka Matsumura, and Miroyuki Murakita. 1989. Microbiological studies on the conservation of paper and related cultural properties. Pt. 8, On the components found in foxing. *Science for Conservation* 28:7–16.

Arai, Hiedo, Noritaka Matsumura, and Hiroyuki Murakira. 1990. Microbiological studies on the conservation of paper and related cultural properties. Pt. 9, Induction of artificial foxing. *Science for Conservation* 29:25–34.

Bicchieri, M., G. Pappalardo, F. P. Romano, F. M. Sementilli, and R. De Acutis. 2001. Characterization of foxing stains by chemical and spectrometric methods. *Restaurator* 22(1):1–19.

Cain, C. E. 1993. Foxing caused by inorganic factors. *Proceedings of the Second International Conference on Biodeterioration of Cultural Property*, 278–91. Japan: ICBCP-2 Organizing Committee.

Chen Yuansheng and Xie Yulin. 2002. The causes of foxing on the Chinese painting. *Sciences of Conservation and Archaeology* 14(suppl.):63–76.

Choisy, P., A. De La Chapelle, D. Thomas, and M. D. Legoy. 1997. Noninvasive techniques for the investigation of foxing stains on graphic art material. *Restaurator* 18(3):131–52.

Daniels, V., and N. Meeks. 1993. An investigation into foxing phenomena with particular attention to the inorganic components. *Proceedings of the Second International Conference on Biodeterioration of Cultural Property*, 292–304. Japan: ICBCP-2 Organizing Committee.

Florian, M.-L. 1996. The role of the conidia of fungi in fox spots. *Studies in Conservation* 41(2):65–75.

Florian, M.-L., and L. Manning. 1999. The ecology of the fungal spot in a book published in 1854. *Restaurator* 20(3):137–50.

Florian, M.-L., and L. Manning. 2000. SEM analysis of irregular fungal fox spots in an 1854 book: Population dynamics and species identification. *International Biodeterioration and Biodegradation* 46:205–20.

Meynell, G. G., and R. J. Newsam. 1978. Foxing: A fungal infection of paper. *Nature* 274(5670):466–68.

Nol, L., Y. Henis, and R. G. Kenneth. 1983. Biological factors of foxing in postage stamp paper. *International Biodeterioration Bulletin* 19(1):19–25.

Peters, D. 2000. An alternative to foxing? Oxidative degradation as a cause of cellulosic discolouration. *Papier Restaurierung: Mitteilungen der IADA* 1:55–60.

Rebrikova, N. L., and N. V. Manturovskaya. 2000. Foxing: A new approach to an old problem. *Restaurator* 21(2):85–100.

Strlic, M., and J. Kolar. 2002. Evaluating and enhancing paper stability: The needs and recent trends. *Proceedings of the Fifth EC Conference on Cultural Heritage Research: A Pan-European challenge*, ed. R. Kozlowski, 79–86. Cracow: European Commission.

Tang, L. C. 1978. Determination of iron and copper in 18th and 19th century books by flameless atomic absorption spectroscopy. *Journal of the American Institute for Conservation* 17(2):19–32.

Zhang, C. 1999. *Tu shu, dang an, bo wu guan cang pin bao cang xue yuan li* [Theory on the conservation of books, archival materials and museum collections]. Beijing: Beijing Ke Xue Ji Shu Chu Ban She.

Chinese Painting Colors: History and Reality

Jennifer Giaccai and John Winter

ABSTRACT Seven Chinese paintings from the collection of the Freer Gallery of Art and Arthur M. Sackler Gallery, Smithsonian Institution, Washington, D.C., were examined to determine the pigments used in their making. The paintings examined were a Yuan dynasty handscroll dated to 1347, a Ming dynasty handscroll, and five late Qing dynasty ancestor portraits. Examination of the paintings showed a complex use of pigments to produce a wide variety of colors and shades of color. In contrast to Japanese paintings, the study and identification of pigments on Chinese paintings has been much more limited. This study indicates that Chinese painters typically used lead white rather than the calcium carbonate white pigments favored by Japanese painters. Also, the five Qing dynasty portraits studied were done using ultramarine, emerald green, and a variety of arsenic sulfides, pigments rarely found in Japanese paintings. All of the paintings showed an intimate knowledge of the working properties of the pigments used, and a variety of pigment mixtures was employed to create the desired effects. This paper also compares the pigments identified on the seven paintings examined with the information recorded in contemporary literature and modern research results on pigment usage in China.

Introduction

The history of the use of pigments in East Asian painting owes a great deal to work in the area of Japanese painting. There are studies on paintings themselves, the long continuing investigations of Yamasaki Kazuo (e.g., Yamasaki and Emoto 1979) being followed and supplemented by work at the Freer Gallery of Art (e.g., FitzHugh 1979) and occasionally in other locations. It is notable that the laboratory work was kept firmly in a historical context, with art historians such as Akiyama Terukazu cooperating from the beginning in Japan, while the Freer work was promoted initially by former director Harold P. Stern, with later assistance from both curatorial staff and from visiting scholars.

In China, things have taken a different course. There is, in some respects, a strong historical record, but investigations of artifacts themselves have heavily focused on wall paintings (especially in caves and temples), painted sculpture, and to some extent archaeological material. The reasons for this focus lie partly in the accidents of how such studies develop and partly in the ways that the Chinese have traditionally viewed painting.

The earliest Chinese historical records appear to have few accounts concerned with painting, but from around the fourth century we do start to see such writings. Three essays have been attributed to the well-known fourth-century painter, Gu Kaizhi (ca. 345–ca. 406). Later, in 847, Zhang Yanyuan (ca. 815–after 875) finished an extensive manual, *Litai ming huaji* (Record of Famous Painters of All the Dynasties), which contained sections on painting materials and on mounting (Acker 1954 [1979]). It mentions the use of cinnabar, azurite, malachite, red lead, lead white, orpiment, and what may have been ultramarine. It also refers to the "ant-ore of Nan-hai," generally taken to be a reference to red insect dye or *yan zhi*. Over the succeeding centuries a variety of books dealing with various aspect of painting appeared (Franke 1950; Acker 1954 [1979], 1974; Soper 1958; Shen 1959; Maeda 1970; Bush and Shih 1985), including the important seventeenth-century *Chieh tzu yüan hua chuan* (Mustard Seed Garden Manual) (Sze 1977). Most of these accounts have relatively little to say, however, about the materials used by the painter, whether pigments, silk and paper support materials, or binders and sizes. Rather, the emphasis tends to be on the assessment and ranking of known painters, the skill needed in the depiction of this or that type of feature, and so forth: what we refer to broadly as connoisseurship. The *Mustard Seed Garden Manual*, and to some extent, the Zhang Yanyuan manual, are unusual in the detail that they give to the materials of painting. But even the *Mustard Seed Garden Manual*, which has been much cited in this context, devotes less than twenty pages (in translation) to pigments out of almost six hundred and perhaps as much again to other painting materials. It is significant, moreover, that the work tended initially to be ignored in Chinese scholarly circles on the grounds that it was a book for beginners in painting.

This point brings us to the attitude of the more influential of Chinese painters and critics themselves. It has been observed many times that these have played down the importance of color in artistic expression by the painter. In their introduction to a translation of a twentieth-century painting manual, Silbergeld and McNair (1988, ix) refer to the cultivation of "an unspoken prejudice against any prominent use of, or even discussion of, color in painting." For many centuries, the mainstream of Chinese painting as a self-conscious art form has been strongly under the influence of, if not controlled by, the literati, a loosely defined class of classically educated scholars who emphasized the close relationship of painting and calligraphy as well as the expressiveness possible with brush and ink, and

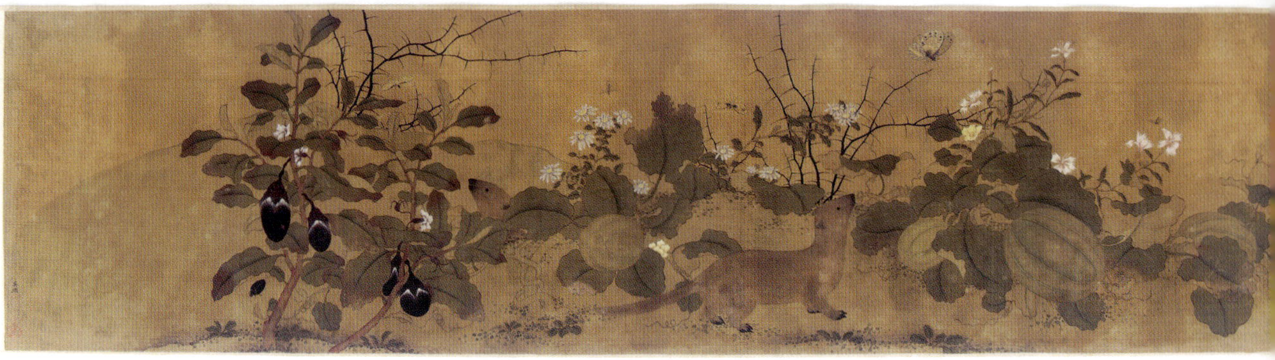

deemphasized the role played by color, certainly by bright or intrusive colors. Indeed, even Zhang Yanyuan precedes his list of pigments with the warning,

> For grasses and trees spread forth their glory without depending upon cinnabar and jasper; clouds of snow swirl and float aloft and are white without waiting for ceruse. Mountains are turquoise without needing [the color] "sky blue" and the phoenix is iridescent without the aid of the five colors. For this reason one may be said to have fulfilled one's aim if one can furnish [a painting] with all the five colors by the management of ink [alone]. (Acker 1954 [1979], 185)

This attitude was communicated to Western historians of Chinese painting, and only recently have we started to see any serious questioning of it. It is certainly true that on a good many Chinese paintings from all periods, colors other than the brownish or bluish nuances of Chinese ink are, in hard fact, obviously there. Nevertheless, the lack of emphasis on their importance or significance by the historians and connoisseurs may be one reason why relatively little has been done to study them.

The Chinese historical record does contain a great deal of historical information on pertinent materials on what might be called the technological level. A well-known example is the seventeenth-century *Tian gong kai wu* (Chinese Technology in the Seventeenth Century) by Song Yingxing (b. 1587) (Sung 1966), which gives accounts of the manufacture of cinnabar, paper, and ink, among other things. The Chinese scientific and technological literature has been studied extensively by Joseph Needham and his collaborators for his volumes in the Science and Civilisation in China series. These give us valuable information on the history or origin and manufacture of many known or probable painting materials. Much of this information comes from the Daoist literature. To a large extent the Daoists were responsible for the development of Chinese alchemy, a context that fostered curiosity about materials and their properties and possible applications, whether as some kind of medical elixir, for the production of precious metals, or other applications deemed useful.

Where there has been significant progress in China on pigment identifications in paintings and other examples of use, it has been made in studies of murals (primarily in caves and temples), painted sculpture, and, to a lesser extent, archaeological materials. The best-known corpus of results here has been from the Mogao caves, a large complex of caves or grottoes situated a few kilometers from Dunhuang in northwest China and extensively decorated with Buddhist painting dating from as early as the third or fourth centuries up to perhaps the fourteenth century. A great deal of work on the Mogao cave and other murals has been done at the Dunhuang Research Academy and other places and has been published (e.g., Hsu et al. 1983; Kuchitsu 1997). A little is known also about periods earlier than Dunhuang from such archaeological finds as painted textiles from the tomb of Ma Wang Dui (Wang 1975). We might also mention at this point the handful of studies done on Chinese antiquities in museum collections. In the 1930s cinnabar was identified in the incisions of Chinese oracle bones (Benedetti-Pichler 1937). A notable piece of work in this class was the demonstration of the existence of two barium copper silicates, a blue and a purple version, on painted objects and tiles of about the Han dynasty (206 B.C.E.–220 C.E.) (FitzHugh and Zycherman 1983, 1992). These two pigments, named Han blue and Han purple, evidently fell out of use, and so far no reference to them in the historical literature has been noted.

We have mentioned already the significance of the Chinese literati scholar in Chinese painting history. Such scholars regarded themselves as, in some sense, "amateurs." Whether or not this is a literally valid designation, there is a distinction to be drawn between them and professionally trained painters, if only because the latter are more likely to have been schooled in the use of colored paints, particularly those applied in thick, opaque paint layers. The relation between these two classes is not going to be a simple one, or necessarily the same at all periods. Still, the professional painter, producing works to order for particular requirements, was there, doing his painting, whether or not the "amateur" overlapped with his activities. We mention this situation mostly because of our own ventures into looking at the materials of portraiture from more recent periods in China, surely a quintessentially professional area and one that certainly displays an abundant use of pigments, both alone and in mixtures, as will presently be seen.

We find ourselves, then, with a rather rich general historical background, but with little of it referring specifically to painting materials. On the experimental front, the last two or three decades have seen steady and continuing publication of identifications (and associated observations on deterioration mechanisms) of materials of mural paintings and

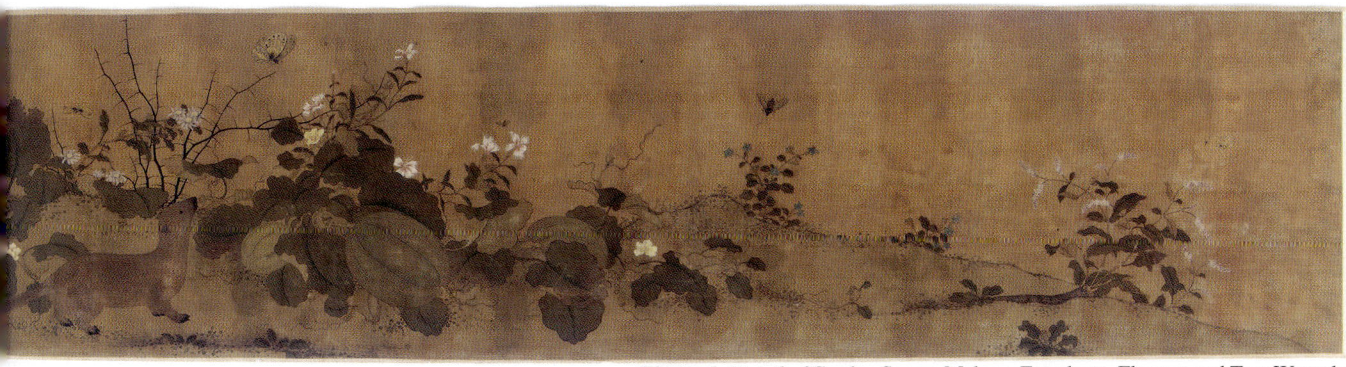

Figure 1. *Detail of* Garden Scene: Melons, Eggplants, Flowers and Two Weasels, *Ming dynasty (?), handscroll on silk. Freer Gallery of Art, F1911.222*

painted sculpture, especially religious works, many of them historically early. This research has tended to be regarded as of lesser artistic importance by the Chinese scholar, and there has been little or nothing on the more highly regarded studio painting. The picture does not show a total lack of information and results, but rather a heavily skewed situation, with notable blank areas.

This context can serve as a starting point for our concern with Chinese painting materials; certainly the collections of the Freer Gallery of Art and Arthur M. Sackler Gallery, Smithsonian Institution, Washington, D.C., are well able to supply the works needed to begin correcting this imbalance. The existing evidence suggests a number of specific areas of potential interest for the study of pigments used in the studio paintings of China. The work on the murals of the Mogao caves indicated the frequent use of natural ultramarine at certain periods: it would be of interest whether this use extended into other parts of China. The history of the introduction of Prussian blue into China has not been explored and links up with our own preceding work on the appearance of this pigment in Japanese paintings (Leona and Winter 2003). Another link to our own work lies in the question of organic colorants. It appeared (from unpublished observations here and some evidence in the literature) that organic red colors have been used in Chinese as well as Japanese works, probably over a fairly wide period. Moreover, historical studies (Laufer 1919; Schafer 1957, 1963) have suggested the importation of a number of such possible colorants. Considerations such as these suggested some points of focus for our own move into the further study of Chinese painting materials.

Paintings Studied

Seven Chinese paintings from the collection of the Freer Gallery of Art and Arthur M. Sackler Gallery were examined. They include a Yuan dynasty (1279–1368) handscroll dated to 1347, a Ming dynasty (1368–1644) handscroll, and five late Qing dynasty (1644–1911) ancestor portraits. All are listed below. Paintings with accession numbers beginning "F" are in the collection of the Freer Gallery of Art; those with accession numbers beginning "S" are in the collection of the Arthur M. Sackler Gallery. With exception of S1991.133, they are described further in Stuart and Rawski (2001), and the locations in that book are noted here in parenthesis.

1. F1911.222. Unknown artist, *Garden Scene: Melons, Eggplants, Flowers and Two Weasels*, Ming dynasty (?), handscroll on silk. (fig. 1)
2. F1945.32. Zhao Yong, *Horse and Groom, after Li Gonglin*, dated 1347, handscroll on paper. (fig. 2)

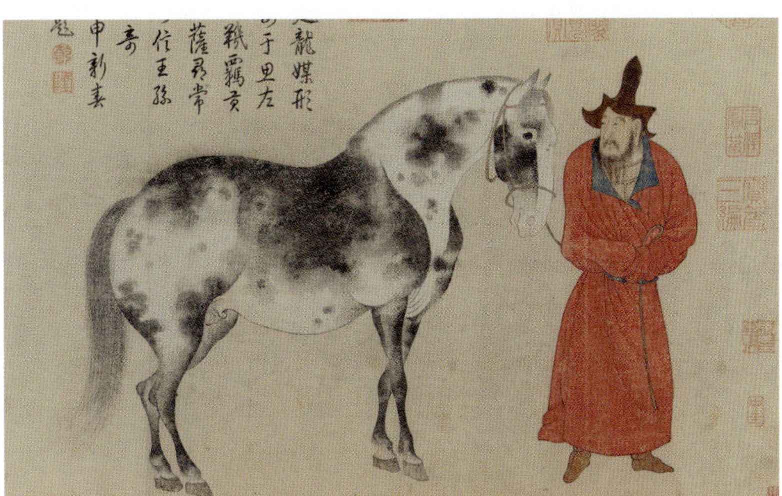

Figure 2. *Detail of Zhao Yong,* Horse and Groom, after Li Gonglin, *dated 1347, handscroll on paper. Freer Gallery of Art, F1945.32*

3. S1991.65. Unknown artist, *Portrait of Lady Hejia*, Qing dynasty, nineteenth century, hanging scroll on paper. (Stuart and Rawski no. 9, p. 194)
4. S1991.117. Unknown artist, *Portrait of Woman* (*Empress Dowager*), Qing dynasty, nineteenth century, hanging scroll on paper. (Stuart and Rawski no. 25, p. 196)
5. S1991.122. Unknown artist, *Portrait of Woman* (*Tao Kuang's fifth daughter*), Qing dynasty, late nineteenth century. (Stuart and Rawski no. 39, p. 197)
6. S1991.133. Unknown artist, *Man with Two Women and Child* (*Relative of Yung Cheng under House Arrest*), Qing dynasty, eighteenth century. (fig. 3)
7. S1991.135. Unknown artist, *Portrait of Man* (*Ching Shou*), Qing dynasty, late nineteenth century, hanging scroll on silk. (Stuart and Rawski no. 44, p. 198)

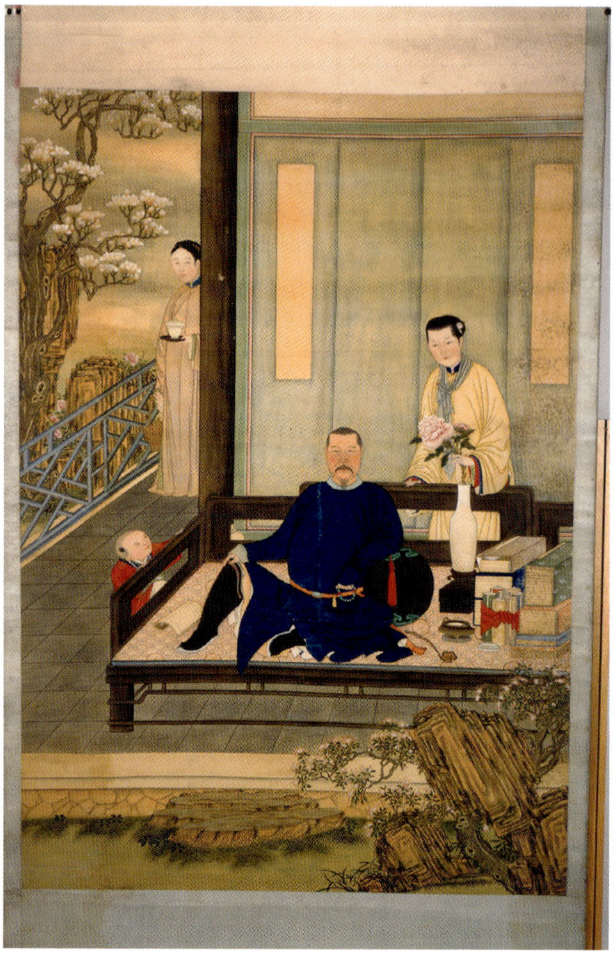

Figure 3. Man with Two Women and Child (Relative of Yung Cheng under House Arrest), *eighteenth century, hanging scroll on silk. Arthur M. Sackler Gallery, S1991.133*

Methods

The examination protocol was based on a survey of the colored areas using stereomicroscopic examination followed by the first two (both noninvasive) methods. Sampling for specific identifications was then done as necessary.

X-Ray Fluorescence (XRF)

This technique identified all elements heavier than argon and supplied evidence for most inorganic pigments. The instrument is an Omega 5 Museum and Industrial Object Analyzer (a modified Spectrace 6000 spectrometer) by Data Acquisition and Control.

Fiber Optic Reflectance Spectroscopy (FORS)

This technique was applied in the visible and near infrared region to distinguish indigo from other blue pigments and for the identification of organic red colorants. The instrument is a Cary 50 UV-visible spectrometer (Varian Analytical Inc). Measurements were taken with either the addition of a fiber optic reflectance probe (Leona and Winter 2001) or a newly designed fiber optic probe–remote detector attachment (fig. 4).

Fourier Transform Infrared Spectroscopy (FTIR)

The instrument was a Nicolet Nexus 670 Fourier transform infrared spectrometer with the addition of a SpectraTech IR-Plan Advantage infrared microscope. The technique

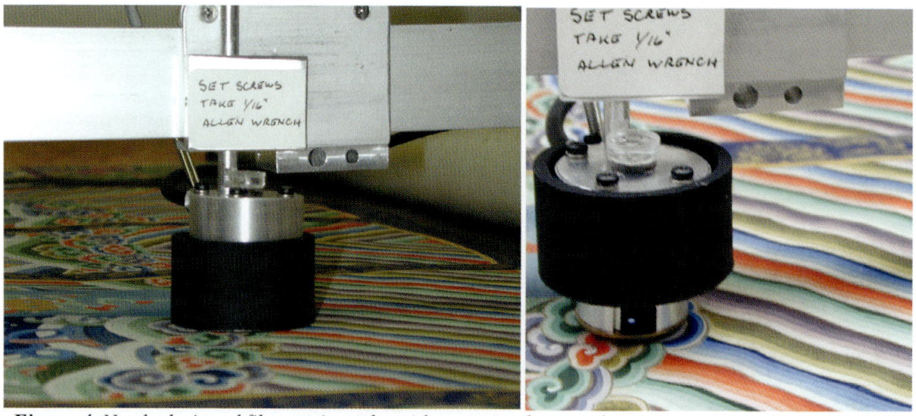

Figure 4. *Newly designed fiber optic probe with a remote detector for noninvasive vis/NIR reflectance spectroscopy: (left) examining S1991.65; (right) dot of white light showing area to be examined*

required the removal of a microscopic sample, which was compressed between diamond surfaces and analyzed in the microscope; it was applied specifically to the identification of gamboge and Prussian blue.

Infrared Fiber Optic Reflectance Spectroscopy (mid-IR FORS)

A chalcogenide optical fiber reflectance probe (Remspec Corporation) was coupled to the Nicolet instrument described above (Fabbri et al. 2001). This technique was used on one painting and is useful in providing midinfrared data noninvasively.

X-Ray Diffraction (XRD)

This technique used Gandolfi-pattern cameras (a modified Debye-Scherrer technique) on a standard Philips x-ray generator operated at 40 kV and producing copper K± radiation. A microscopic sample was mounted on the tip of a glass fiber. This technique identified crystalline inorganic pigments.

Polarized Light Microscopy (PLM)

A Leitz Orthoplan rotating stage microscope was used to study removed samples by standard methods. This technique was especially useful in the case of mixtures.

Scanning Electron Microscopy (SEM)

A Philips Electron Optics XL-30 microscope with EDAX x-ray detector was used for element identification.

Results

Identifications followed standard methods for the most part. The use of FORS in distinguishing indigo from other blue pigments has been covered (Leona and Winter 2001). The same technique was used on several paintings to identify organic red colorants as "insect dyes" (see below) by the presence of a characteristic pair of peaks at 530 and 575 nm. Gamboge was identified using the FTIR microscope on a removed sample. Inorganic pigments were initially identified by XRF, then in most cases confirmed by XRD on a removed sample. For convenience, results are presented for each individual painting listed in order of accession number, and with identification tabulated approximately in order of the color in the spectrum. Mixtures of pigments are shown with a +; colors that are formed by painting one color over another or back painting are noted in the tables. Lead white, when clearly used only as an extender or as a ground layer, is not mentioned in the pigment identifications for the sake of brevity.

F1911.222, Garden Scene, *Ming Dynasty*

Results are summarized in table 1. This rather brightly colored painting owes much of its color to organic, rather than inorganic, pigments. While the brown and white colors are iron earths and lead white, respectively, all the red shades make use of insect dye, and, in addition, insect dye appears in a number of darker shades in the painting. Violet stripes on the flowers are a mixture of insect dye and indigo. The double peak of insect dye is clearly observed, together with a prominent peak at 665 nm with a triangular shape, characteristic of indigo (fig. 5). The dark color of the eggplants, however, shows the insect dye peaks but has nothing to suggest indigo. It does have marked absorption in the near infrared region, suggesting the admixture of Chinese ink.

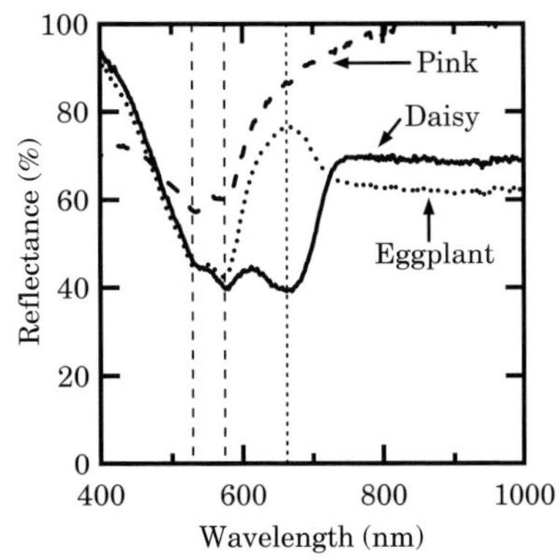

Figure 5. *Vis/NIR reflectance spectra from violet stripe on daisy, pink flower, and dark violet of eggplant on F1911.222 (fig. 1)*

Table 1. *Pigments in F1911.222,* Garden Scene, *Ming dynasty*

Color (motif)	Identification	Methods
White	Basic lead carbonate	XRD
Pink	Insect dye	FORS
Yellow	Gamboge	FTIR
Green (leaves)	Unknown copper compound (back painted) + indigo	XRF, FORS
Green (dark spots on ground)	Gamboge + indigo	FTIR, FORS
Opaque blue	Azurite	XRF
Blue wash	Indigo	FORS
Violet	Insect dye + indigo	FORS
Violet, dark (eggplant)	Insect dye + (probably) Chinese ink	FORS
Brown	Iron oxide earth	XRF

Gamboge was found alone and as a mixture with indigo to give a green (dark spots on ground). Another green color used for plant leaves showed the presence of copper as a major element by XRF, but the surface appearance of the material was not what one would expect for an inorganic pigment. The painting was also x-radiographed: the green areas were moderately radio-opaque. Our interpretation is that a copper-based green pigment was applied to the back of the silk support, a known technique for paintings on silk. Indigo, identified in the same areas by FORS, appears to have been applied to the front.

F1945.32, Zhao Yong, Horse and Groom, *Dated 1347*

Results are summarized in table 2. This painting had been studied in 1970 by R. J. Gettens, who identified vermilion by PLM, and by John Winter in 1980, who identified vermilion and azurite by XRD (object records, Freer Gallery of Art).

Many details on this painting reveal, on careful examination, the use of thin washes of color. The thin wash of pink defining the nose and lips of the horse, and possibly the groom's lip, was identified as insect dye from the characteristic pair of peaks in FORS (fig. 6). An orange wash near the horse's hoof appears to be an iron oxide–based earth pigment with the typical broad absorption peak and sharp absorption edge of an inorganic pigment spectrum; a red wash on the groom's robe was also inorganic.

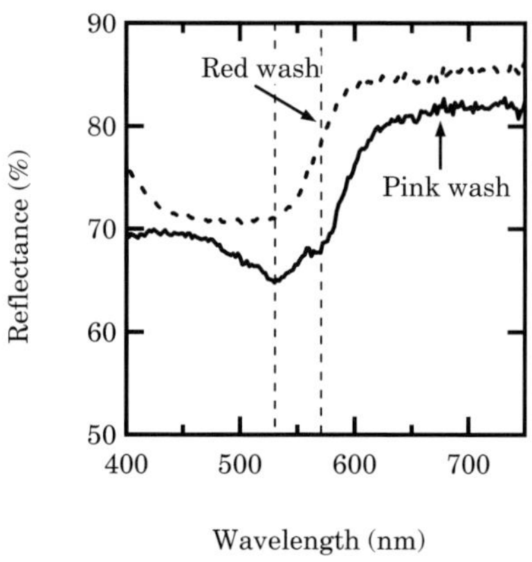

Figure 6. *Vis/NIR reflectance spectra from organic pink wash and inorganic red wash on F1945.32 (fig. 2)*

Table 2. *Pigments in F1945.32, Zhao Yong,* Horse and Groom, *dated 1347*

Color (motif)	Identification	Methods
White	Basic lead carbonate	XRD
Red	Vermilion	PLM, XRD
Pink wash (horse lips, nose)	Insect dye	FORS
Orange wash (horse hoof)	Iron oxide earth	XRF
Blue	Azurite	XRD
Brown	Iron oxide earth	XRF

Table 3. *Pigments in S1991.65, unknown artist,* Portrait of Lady Heija, *Qing dynasty, nineteenth century*

Color (motif)	Identification	Methods
White	Basic lead carbonate	XRD
Red	Vermilion	XRF
Pink	Insect dye	FORS
Orange-yellow	Arsenic sulfide, pararealgar	XRD
Yellow	Gamboge	FTIR
Yellow-green (flowers on rug)	Emerald green + gamboge + Prussian blue	XRD, FTIR
Green, medium	Emerald green	XRD
Green, dull (stripes on robe)	Indigo + gamboge	vis FORS, mid-IR FORS
Green, dark (stripes on robe)	Copper based green (probably emerald green) + ultramarine	PLM
Green, dark (carpet)	Prussian blue + gamboge	FTIR
Blue, medium	Azurite	XRF
Blue, bright (darkest stripe on robe)	Artificial ultramarine	FTIR, PLM
Blue, dark (robe)	Prussian blue + vermilion	XRD, FTIR, PLM
Violet, medium	Insect dye + indigo	FORS
Violet, dark	Insect dye + Prussian blue	FORS, FTIR
Brown	Iron oxide earth	PLM

S1991.65, Unknown Artist, Portrait of Lady Heija, *Qing Dynasty, Nineteenth Century*

Results are summarized in table 3. The orange-yellow pigment (pararealgar) showed marked x-ray fluorescence in XRD pattern acquisition and had to be run with an aluminum foil shield inside the film to render the pattern readable. Infrared spectroscopy on the organic yellow (gamboge) was done in this case by extracting the yellow compound from accompanying fibrous material with acetonitrile and evaporating the solution onto a barium fluoride disc. Emerald green was identified by XRD in a number of different shades of green. The proportion of vermilion in the dark blue (Prussian blue + vermilion mixture) appeared to be very small from an optical microscope slide, but was enough to give a faint mercury sulfide (HgS) pattern in x-ray diffraction. The same area contained a small proportion of a blue insoluble in aqueous alkali, which may have been ultramarine.

S1991.117, Unknown Artist, Portrait of Woman, *Qing Dynasty, Nineteenth Century*

Results are given in table 4. The arsenic sulfide in the mixture with vermilion (region of throne) was not identified more specifically. In the case of the orange-yellow of the

Table 4. *Pigments in S1991.117, unknown artist,* Portrait of Woman, *Qing dynasty, nineteenth century*

Color (motif)	Identification	Methods
White	Basic lead carbonate	XRD
Red	Vermilion	XRF
Pink	Insect dye	FORS
Orange (throne)	Arsenic sulfide + vermilion	XRF
Orange-yellow (belt)	Arsenic sulfide, alacranite	XRF, XRD
Orange-yellow (dots on carpet design)	Arsenic sulfide	PLM
Yellow	Gamboge	FTIR
Green, medium	Emerald green	XRD
Green, dark	Prussian blue + gamboge	FTIR
Blue, dull	Prussian blue + unidentified blue	FTIR, FORS
Blue, bright (bead)	Artificial ultramarine	XRD, PLM
Blue, medium (robe)	Prussian blue over organic red over azurite–emerald green mixture	FTIR, XRF
Blue, dark (stripe of rug, shading on robe)	Prussian blue	XRF, FTIR
Violet, all shades	Insect dye + Prussian blue	FORS, FTIR
Brown (decoration on robe, all shades)	Gamboge + finely divided dark brown, probably an iron oxide earth	FTIR, PLM
Brown (fur)	Iron oxide earth	XRF

subject's belt, the XRD pattern corresponded well to the international powder diffraction file d-spacing listing no. 25-57; this has been described both as "unnamed mineral" and as "alacranite, synthetic." The colorants in the dull blue color found on beads and thread of the subject's necklace defied identification.

S1991.122, Unknown Artist, Portrait of Woman, *Qing Dynasty, Late Nineteenth Century*

Results are given in table 5 and, though complex, most identifications were straightforward.

Table 5. *Pigments in S1991.122, unknown artist,* Portrait of Woman, *Qing dynasty, late nineteenth century*

Color (motif)	Identification	Methods
White	Basic lead carbonate	XRD
Red	Vermilion	XRF
Red, dark	Insect dye (over light violet-blue)	FORS
Pink	Insect dye	FORS
Orange (rug)	Iron oxide earth	XRF, PLM
Orange-yellow (belt)	Arsenic sulfide, pararealgar	XRD
Yellow (rug)	Iron oxide earth	XRF, PLM
Green (headdress)	Basic copper chloride, atacamite	XRD
Green (robe design)	Basic copper chloride, atacamite	XRD
Green, dark	Prussian blue + gamboge	FTIR
Blue, dark	Prussian blue	FTIR
Blue, dark (robe)	Azurite + organic red	XRF, PLM
Blue, medium	Azurite	XRF
Blue, light	Prussian blue	FORS
Violet, all shades	Insect dye + Prussian blue	FORS, FTIR
Olive brown (carpet)	Prussian blue + lead white + iron oxide earth	XRD, FTIR, PLM
Red-brown (edge of throne)	Vermilion + iron oxide earth	XRF, PLM
Brown (fur)	Iron oxide earth	XRF, PLM

S1991.133, Unknown Artist, Man with Two Women and Child, *Qing Dynasty, Eighteenth Century*

Results are given in table 6. The yellow component of the dark green color remains unidentified. Indigo dominated the infrared spectrum obtained from IR microscopy, and while iron oxide earths could not be identified in the small sample taken for polarized light microscopy, they would be a likely source of yellow.

Table 6. *Pigments in S1991.133, unknown artist,* Man with Two Women and Child, *Qing dynasty, eighteenth century*

Color (motif)	Identification	Methods
White	Basic lead carbonate	XRD
Red	Vermilion	XRF
Pink	Insect dye	FORS
Orange-yellow	Arsenic sulfide, realgar	XRD
Pale orange (carpet)	Iron oxide earth	XRF
Yellow	Gamboge	FTIR
Dark green	Indigo + unidentified yellow	FORS, FTIR
Light green	Malachite	XRD
Medium blue (fence)	Azurite	XRF
Dark blue (robe)	Azurite	XRF
Light blue (scarf)	Indigo	FORS
Violet	Insect dye + azurite	FORS, XRF
Brown (vase stand)	Iron oxide earth	XRF
Brown (cushion)	Iron oxide earth + vermilion	XRF

S1991.135, Unknown Artist, Portrait of Man, *Qing Dynasty, Late Nineteenth Century*

Results are in table 7. In the case of the orange-yellow arsenic sulfides, lines for pararealgar were identified in the XRD pattern, along with other lines. Since the SEM sample showed only arsenic and sulfur as major elements, this identification is returned as pararealgar plus an unidentified arsenic sulfide species. In the olive brown from the carpet,

Table 7. *Pigments in S1991.135, unknown artist,* Portrait of Man, *Qing dynasty, late nineteenth century*

Color (motif)	Identification	Methods
White	Basic lead carbonate	XRD
Red	Vermilion	XRF
Pink	Insect dye	FORS
Orange, dark (dots on carpet)	Gamboge + insect dyes	FTIR, FORS
Orange-yellow	Arsenic sulfides, pararealgar, + unidentified	XRD, SEM
Yellow	Gamboge	FTIR
Yellow-green (stripe on underskirt)	Gamboge + Prussian blue	FTIR
Green, dark	Gamboge + Prussian blue	FTIR
Green	Malachite	XRD
Green-blue	Malachite + azurite	XRD
Gray-blue, light	Azurite (coarsely ground) and lead white	FTIR
Gray-blue, medium	Finely ground azurite	FTIR, PLM, SEM
Blue, medium	Azurite + Prussian blue	XRF
Blue, dark (rug details)	Prussian blue	FTIR
Blue, dark (robe)	Azurite + organic red	XRF, PLM
Violet, light	Lead white + Prussian blue + azurite	XRD, PLM, SEM
Violet, medium	Insect dye + indigo	FORS
Violet, dark	Insect dye + indigo	FORS, FTIR
Olive brown (carpet)	Malachite + arsenic sulfide	PLM, SEM
Brown (purse tie)	Arsenic sulfide + copper pigment	XRF
Brown (fur)	Iron oxide earth + arsenic sulfide	XRF, PLM

the XRD pattern was complex, but malachite lines were identified among one or more unknown species. SEM on the same mixture showed copper, arsenic, and sulfur at all locations; we conclude that malachite is mixed with an unidentified arsenic sulfide. The medium gray-blue areas contained, by SEM, localized iron-rich areas, which may have been either Prussian blue or an iron-based impurity. No Prussian blue peak was found by examining a separate sample of this pigment by FTIR.

Discussion

All of the paintings had black design elements in what was plausibly assumed to be conventional Chinese ink; this substance was not investigated experimentally. Several more well-known pigments occurred on all or nearly all of these paintings and are unremarkable: mercuric sulfide (vermilion or cinnabar), iron oxides, azurite, and lead white. The consistent use of the last of these may be worth noting; in Japan during the period covered by these works, lead-based whites became replaced largely by oyster shell white (Yamasaki and Emoto 1979; FitzHugh 1979).

The use of "insect dyes" on all paintings calls for comment. These products originate from a number of species of scale insect and are chemically rather similar, being based on heavily substituted anthraquinones. Although the specific constituents do differ chemically between different insect species, reflectance spectra by the FORS technique allow us only to assign a red pigment to this class without distinguishing between different members. The latter can be done using high performance liquid chromatography (HPLC), but this method currently requires, in our hands, unacceptably large samples. For this reason, we identify these pigments here as insect dyes without further distinctions being drawn. The two most likely possibilities, on historical grounds, are cochineal and lac dye. Cochineal is a product of the insect *Dactylopius coccus*, native to Central and South America, and is expected to be found only in more recent centuries. Lac dye, from *Kerria lacca*, is well known in southern Asia and has evidently been imported into China since at least the Tang dynasty (Schafer 1957); it may well be the product corresponding to the "ant-ore of Nan-hai" described by Zhang Yanyuan in the ninth century. Clearly it would be of historical interest to be able to distinguish reliably these different natural products on paintings, but currently the very low surface density of these highly colored compounds frustrates such an endeavor; it remains a hope for the future.

All five of the Qing dynasty portraits used an arsenic sulfide for yellow to orange shades. Realgar, pararealgar, and alacranite (all AsS) were identified in addition to at least one for which the crystallographic species could not be discovered. Orpiment (As_2S_3) was not found, even though it (though not other arsenic sulfides) has occasionally been identified on Japanese paintings (FitzHugh 1997). We have no evidence concerning whether these compounds are minerals or synthetic, but realgar occurs as a mineral in parts of China (Golas 1999) and would have been available. Yu Feian, in *Chung-kuo hua yen-setiyen-chiu* (Chinese Painting Colors) lists pigments that "folk painters" or portrait painters used and includes realgar, with the note that the realgar was made by the painters themselves (Silbergeld and McNair 1988). The crystal chemistry of arsenic sulfides is complex, and the occurrence of more than one version, whether the pigment is natural or artificial, is not inherently surprising.

The yellow organic resin gamboge was identified on all but one of the paintings, being found both as a yellow colorant and in green areas with a finely divided blue (see below). Its practicality for the latter is undoubtedly helped by its transparency (Winter 1997). It should be noted that the one painting on which gamboge was not found (F1945.32) contains no yellow or green design elements. Gamboge is well known as a pigment in East Asia, though its known areas of origin are outside the region and presumably it has always been imported.

Three copper-based greens were found on the portraits: malachite, atacamite (basic copper chloride), and emerald green (copper acetoarsenite). This last dates from the early nineteenth century in Europe, and the two paintings on which it occurs are evidently quite late. On these paintings emerald green appears to have displaced the other copper greens entirely.

Several blue pigments were found in addition to the azurite already noted. Artificial ultramarine (invented in 1826) was found in addition to azurite and Prussian blue on two ancestor portraits (S1991.65 and S1991.117). All but one of the paintings had indigo or Prussian blue (two contained both), the latter appearing on four of the five portraits. While it is unclear when Prussian blue was first available to the Chinese, it is known to have been imported into Japan from China from as early as 1782 (Smith 1998). These four paintings are all conventional "ancestor portraits," unlike the formal portrait of a man with his family (S1991.133), which used indigo and which may have been painted earlier. The two paintings containing both Prussian blue and indigo (S1991.65 and S1991.135) used Prussian blue as a blue pigment and in a violet paint combined with insect dye. A second violet shade (a lighter violet in the case of S1991.65, medium and dark violets in the case of S1991.135) was made with indigo and insect dye (figs. 7 and 8).

The five Qing dynasty portraits, and especially the four ancestor portraits, afforded many examples of mixed color effects, either in the sense of pigments being premixed or by painting one color over another. As mentioned above, violet

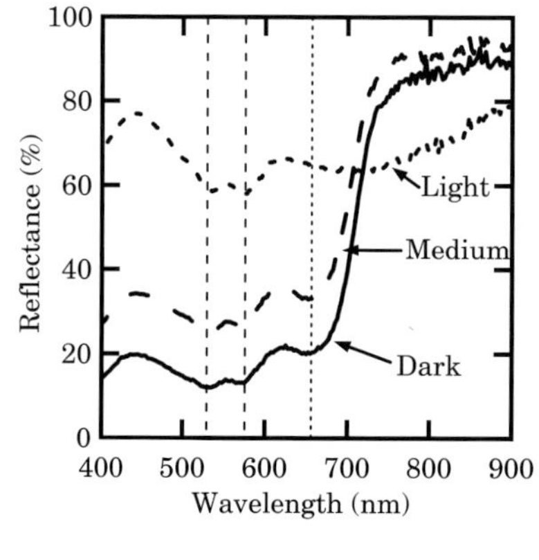

Figure 7. Vis/NIR reflectance spectra from light, medium, and dark violet shades of S1991.135 (fig. 8)

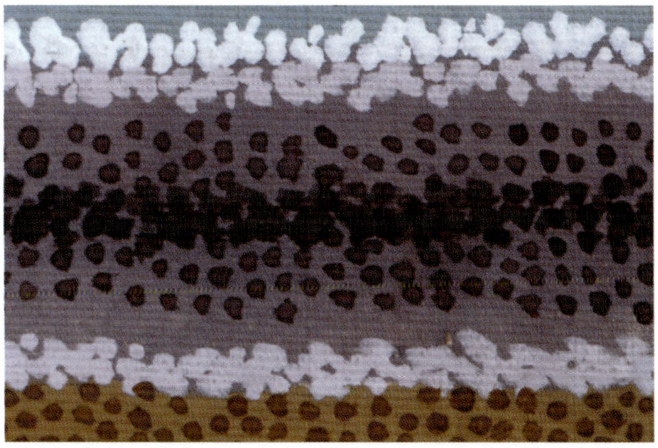

Figure 8. Detail showing light, medium, and dark violet shades in S1991.135

and purple colors were generally a combination of insect dye (a transparent red) with one or other of the blue pigments. There were repeated examples of a green achieved by a mixture of gamboge, a transparent yellow, with a finely divided blue such as indigo or (in the later works) Prussian blue. This mixture was indeed found on the Ming dynasty handscroll (F1911.222) as well as on the portraiture. There was considerable variation in the recipes for browns, not only between paintings but often on the same painting. A conventional iron oxide brown is found, sometimes mixed with vermilion or arsenic sulfide to adjust the color. In other cases, arsenic sulfide mixed with malachite and/or azurite was encountered.

Many light to dark shades of green and blue were noted. Apart from variations between the pigments themselves, this shade was achieved by mixing azurite or a copper-based green with lead white as well as by the well-known method of grinding malachite or azurite more or less finely. These two approaches were apparently used for different design elements. In one painting (S1991.133) a rather bright shade of light blue proved to be coarse azurite mixed with lead white, with a duller shade being based on more finely ground azurite.

The portraiture described here is undoubtedly the product of professionally trained studio painters, with, probably, more than one individual contributing to a given painting. A detailed discussion of them is provided by Stuart and Rawski (2001). The contrast between these and the literati or amateur painters has been described earlier. Whatever aesthetic judgments may be made, or whatever views the traditional connoisseur may have had, it is clear that they had a sophisticated expertise in the choice and application of color.

References

Acker, W. R. B. 1954 [1979]. *Some T'ang and Pre-T'ang texts on Chinese painting.* Vol. 1. Leiden: E. J. Brill; reprint, Westport, Conn.: Hyperion Press, 1979.

Acker, W. R. B. 1974. *Some T'ang and Pre-T'ang texts on Chinese painting.* Vol. 2, *Chang Yen-Yüan, Li Tai Ming Hua Chi chapters IV-X.* Leiden: E. J. Brill. In two parts; part 2 (unpaginated) is Chinese texts.

Benedetti-Pichler, A. A. 1937. Microchemical analysis of pigments used in the fossae of the incisions of Chinese oracle bones. *Industrial and Engineering Chemistry, Analytical Edition* 9:149–52.

Bush, S., and H.-Y. Shih. 1985. *Early Chinese texts on painting.* Cambridge, Mass.: Harvard University Press.

Fabbri, M., M. Picollo, S. Porcinai, and M. Bacci. 2001. Mid-infrared fiber-optics reflectance spectroscopy: A non-invasive technique for remote analysis of painted layers. Pt 1, Technical setup. *Applied Spectroscopy* 55:420–27.

FitzHugh, E. W. 1979. A pigment census of ukiyo-e paintings in the Freer Gallery of Art. *Ars Orientalis* 11:27–38.

FitzHugh, E. W. 1997. Orpiment and realgar. In *Artists' pigments: A handbook of their history and characteristics*, vol. 3, ed. E. W. FitzHugh, 47–79. Washington, D.C.: National Gallery of Art.

FitzHugh, E. W., and L. A. Zycherman. 1983. An early man-made blue pigment from China: Barium copper silicate. *Studies in Conservation* 28:15–23.

FitzHugh, E. W., and L. A. Zycherman. 1992. A purple barium copper silicate pigment from early China. *Studies in Conservation* 37:145–54.

Franke, H. 1950. Two Yüan treatises on the technique of portrait painting. *Oriental Art* 3:27–32.

Golas, P. J. 1999. *Chemistry and chemical technology.* Pt. 13, *Mining.* Science and civilisation in China 5. Cambridge: Cambridge University Press.

Hsu, W.-Y., K.-H. Chou, and Y.-H. Li. 1983. X-ray analysis of the inorganic Dunhuang pigments. *Dunhuang Yanjiu* 1:187–97. (also identified as issue no. 3)

Kuchitsu, N., ed. 1997. *The conservation of Dunhuang Mogao grottoes and the related studies.* International Symposium on the Conservation and Restoration of Cultural Property, 1–3 February 1996. Tokyo: Tokyo National Research Institute of Cultural Properties.

Laufer, B. 1919. *Sino-Iranica. Chinese contributions to the history of civilization in ancient Iran.* Anthropological Series 15: 3, publication 201. Chicago: Field Museum of Natural History.

Leona, M., and J. Winter. 2001. Fiber optics reflectance spectroscopy: A unique tool for the investigation of Japanese paintings. *Studies in Conservation* 46:153–62.

Leona, M., and J. Winter. 2003. The identification of indigo and Prussian blue on Japanese Edo period paintings. In *Studies using scientific methods: Pigments in later Japanese paintings.* Freer Gallery of Art Occasional Papers, n.s., vol. 1. Washington, D.C.: Freer Gallery of Art, Smithsonian Institution.

Maeda, R. J., trans. 1970. *Two twelfth century texts on Chinese painting.* Translations of the *Shan-shui ch'un-ch'üan chi* by Han Cho, and chapters 9 and 10 of *Hua-chi* by Teng Ch'un. Michigan Papers in Chinese Studies 8. Ann Arbor: University of Michigan.

Schafer, E. H. 1957. Rosewood, Dragon's blood and lac. *Journal of the American Oriental Society* 77:129–36.

Schafer, E. H. 1963. *The golden peaches of Samarkand: A Study of T'ang exotics.* Berkeley and Los Angeles: University of California Press. 2nd printing 1981.

Shen, Z. 1959. *Jie Zhou xue hua bian* [Jie Zhou's painting studies]. Beijing: Ren ming mei sha chu ban she.

Silbergeld, J., and A. McNair, trans. 1988. *Chinese painting colors: Studies of their preparation and application in traditional and modern times.* Translation of Yu Feian, *Chung-kuo hua yen-se ti yen-chiu.* Hong Kong, Seattle, and London: Hong Kong University Press and University of Washington Press.

Smith, H. 1998. Ukiyoe ni okeru "burû kakumei" [The "Blue Revolution" in ukiyoe]. *Ukiyoe Geijutsu* 128:3–26.

Soper, A. C. 1958. T'ang Ch'ao Ming Hua Lu: Celebrated painters of the T'ang dynasty by Chu Ching-hsüan of T'ang. *Artibus Asiae* 21:204–30.

Stuart, J., and E. S. Rawski. 2001. *Worshiping the ancestors: Chinese commemorative portraits.* Washington D.C.: Freer Gallery of Art and Arthur M. Sackler Gallery.

Sung, Y.-H. 1966. *T'ien-kung K'ai-wu: Chinese technology in the seventeenth century*, trans. E-tu Zen Sun and Shiou-chuan Sun. University Park and London: Pennsylvania State University Press.

Sze, M.-M. 1977. *The Mustard Seed Garden Manual of Painting.* Translation of *Chieh Tzu Yüan Hua Chuan*, Nanking, 1679–1701, Princeton, N.J.: Princeton University Press. Paperback reprint of publication by Bollingen Foundation, New York, 1956 and 1963.

Wang, S. 1975. X-ray analysis of the pigments of a printed and colored gauze (N-5) from Han tomb no. 1 at Ma wang dui (Chang sha). *Huaxue Tongbao* 4:246–49.

Winter, J. 1997. Gamboge. In *Artists' pigments: A handbook of their history and characteristics*, vol. 3, ed. E. W. FitzHugh, 143–55. Washington, D.C.: National Gallery of Art.

Yamasaki, K., and Y. Emoto. 1979. Pigments used on Japanese paintings from the protohistoric period through the 17th century. *Ars Orientalis* 11:1–14.

Painting Materials and Deterioration Phenomena in a Yuan Dynasty Wall Painting

John Twilley and Kathleen M. Garland

ABSTRACT As part of a comprehensive effort to document the present condition and past restorations of an important Yuan wall painting in the collection of the Nelson-Atkins Museum of Art, Kansas City, Missouri, laboratory analyses of the paints have been undertaken. The sectional removal of the painting on its mud plaster support in the early twentieth century necessitated extensive reassembly and reintegration of losses upon its re-installation in the museum. Analysis is complicated by at least two major campaigns of work that occurred during the relocation. Prior historical refurbishments in China were also suspected at the outset. Analysis has shown that the most widely used original green pigment was one of the copper trihydroxychlorides that has undergone significant alteration. The micromorphology and trace components of the green suggest that it was originally produced from corroded bronze. Deterioration of this pigment has resulted in significant chromatic shifts in the colors of the mural. Gypsum, a minor component of the white paint layers whose principal pigments are clays, shows widespread evidence of dissolution and recrystallization, including possible adverse interaction with the copper green. While evidence of dissolution and dispersal of the green is readily found, no products of the process that would explain the fate of the copper have been identified. The kaolin clay white is of considerable morphological complexity by itself. However, other white pigments, including gypsum and lead white of a very fine particle size, seem to have been sparingly used to augment the properties of this paint. The paint compositions, and their conditions, have important implications for conservation treatment planning.

History of the Mural

The mural, *Tejaprabha Buddha and Attendants* (acc. no. 32–91, figs. 1 and 2), in the Nelson-Atkins Museum of Art, Kansas City, Missouri, is one of several wall paintings in North American collections that share similarities of construction and followed similar paths from China through the art market in the 1920s and 1930s. As a result of the time that has passed since their initial treatment and installations, some of these have required conservation attention in the last two decades. This attention has included varying types of technical investigation that have sought to achieve a better understanding of their materials, structures, and problems. The most closely related work, that of the Metropolitan Museum of Art, New York (*Paradise of Bhaisajyaguru*, acc. no. 54–88), originally faced the Nelson-Atkins mural

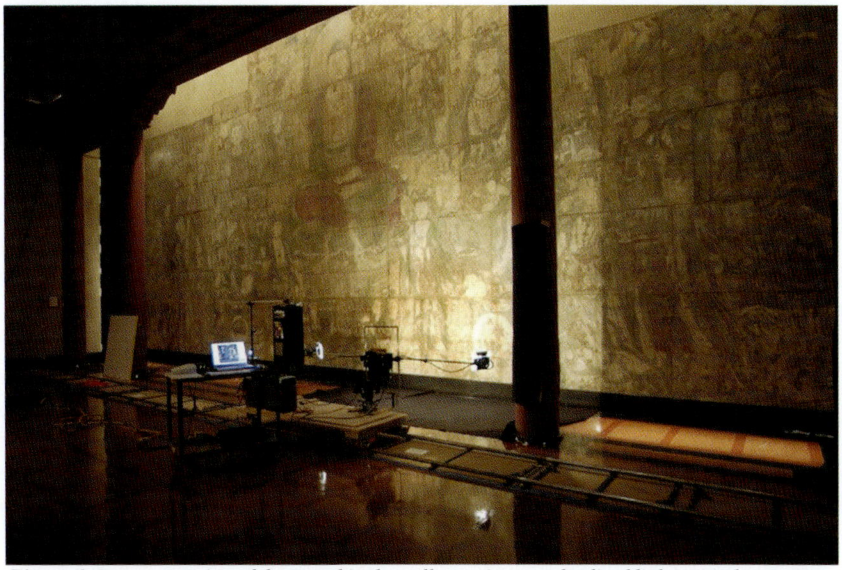

Figure 1. *Documentation of the mural in the gallery using standardized lighting and a carriage-on-track system for dimensional control*

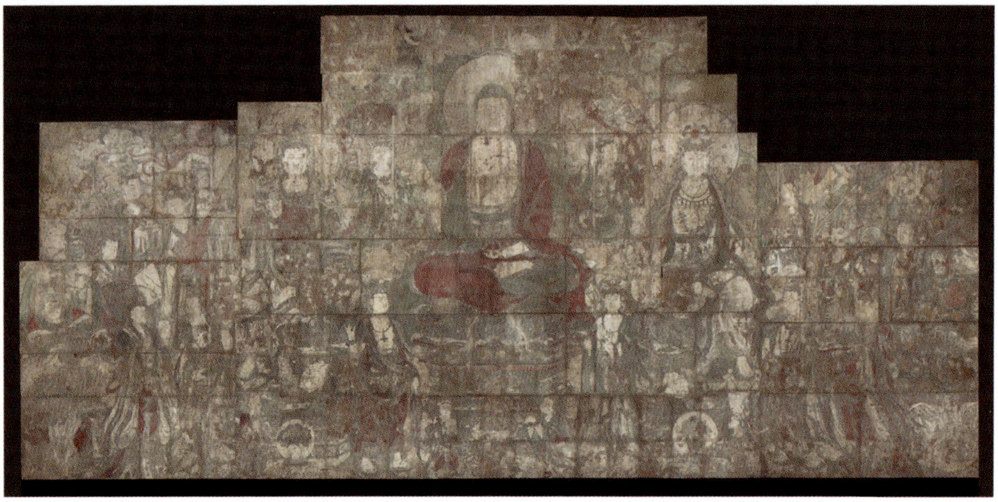

Figure 2. Full composite of the Tejaprabha Buddha and Attendants *assembled from 88 digital images comprising 1 gigabyte in total size. The subdivisions seen here are the individual panels of the mural as de-installed in China, not the boundaries of the documentation images that are invisible in the montage.*

across a large hall in the Guangsheng Lower Monastery, Shanxi, that was constructed in the years immediately following the destruction wrought by an earthquake in 1303 (Steinhardt 1984). The two paintings are generally thought to date to 1309–19 (Jing 1991). A later painting from another building of the same monastery complex is in the collection of the University Museum, University of Pennsylvania, Philadelphia. Other Yuan dynasty (1279–1368) works that have been studied in detail, less closely related in place but sharing technical attributes with these, include one from Henan province residing in the Philadelphia Museum of Art (Malenka and Price 1997) and three from southern Shanxi province in the Royal Ontario Museum, Toronto (Moffat et al. 1985; Gordon and Phillimore 1984). The paintings are all executed in tempera paints on mud plaster and share many of the materials and technical problems of mud plaster paintings from throughout China, the Himalayas, and Central Asia.

The Nelson-Atkins mural was acquired through Paris art dealer C. T. Loo, who arranged its removal and shipment to Paris in small sections. Loo is known to have maintained a staff of restorers, and it appears that an initial assembly of the small sections into frame-mounted groups of five to eight units was carried out in his shop so that the subject matter of the huge painting (over 15 meters in length) could be made comprehensible to potential buyers. This procedure seems to have entailed the filling and retouching of the intervening sectional divisions. These retouchings extend well beyond the losses of original color and make use of brush spatters to camouflage transitions between the original and filled areas. After the acquisition by the Nelson-Atkins, Rutherford J. Gettens of Harvard's Fogg Museum, who was enlisted as an adviser, examined the grouped sections there in 1932 and recommended various details of treatment and mounting procedures, including the use of a dilute solution of Vinylite A resin (believed to be polyvinyl acetate) as a consolidant. Indeed, by this time Gettens and his colleague George Stout had employed the material to treat similar works and are believed to have treated the unmounted units comprising the Metropolitan Museum of Art mural, which remained stored in Boston until the 1950s, when it was acquired by the Metropolitan. However, the installation and additional restorations at the Nelson-Atkins were carried out by local personnel and seem not to have employed some of these suggestions.

The objectives of the recent project were to perform the first detailed condition assessment of the wall painting using digital imaging and annotation techniques to serve as a baseline for monitoring and future treatment planning. Analyses of the painting materials were undertaken to assist in determining the extent of original and restoration paints, to identify past and current deterioration mechanisms, including those influencing the color interpretation of the mural, and to provide guidance for future treatment. While the differing emphases placed on the various analytical techniques employed by the authors of publications on the related works cited above lead to an uneven body of data on the murals, it is nonetheless clear that both differences and similarities exist in the aging phenomena reported for this group of paintings.

Pigment Alteration Processes

An accurate characterization of the original pigments is fundamental to the identification of their deterioration processes. The most difficult pigment in this regard was the green. The components of the original palette of the Kansas City mural include white kaolin clay, lead white, azurite, minium, cinnabar, hematite, hydrous iron oxide yellow, lampblack, charcoal, and green based on one of the copper trihydroxychlorides. Where it is sufficiently well preserved, the green has been identified as atacamite. The kaolin clay is of considerable morphological complexity, with large variations in both particle size and shape, and may have been used in differing grades. However, other white pigments, including gypsum and lead white of a very fine particle size, seem to have been used to augment the properties of this paint. Most of the kaolin was in the common flat morphology, but some areas included a significant portion of the tubular, halloysite morphology.

The alteration phenomena observed may be ranked in terms of their increasing complexity as follows.

Gypsum

Gypsum occurs widely as a component of whites and is sometimes found to have undergone recrystallization recognizable by differences in particle morphology. This process is an example of simple dissolution and redeposition, perhaps accompanied by some short-distance transport. Gypsum crust development, in terms of a continuous layer of interlocking crystals, was not found. The only gypsum source identified was the paint layer itself; the mineral was not found to play any role in the mud plaster beneath.

Lead White

Lead white, in the form of hydrocerussite, was found to have undergone an advanced degree of recrystallization in some locations. Examples were found of the complete conversion of the pigment particles into a continuous matrix of this mineral in which individual particles could no longer be recognized and particles of other pigments are engulfed. This process is also an example of simple dissolution and redeposition, unaccompanied by chemical change or any appreciable transport over distance.

Minium

Red lead oxide, minium, was also found to exhibit signs of recrystallization. Examples were found of extensive bridging by secondary lead deposits between particles, leading to aggregates that far exceed the dimensions of the original particles. These, too, appear to be examples of simple dissolution and redeposition without appreciable transport. It is notable that, while other investigators have often reported the oxidative formation of black plattnerite as a consequence of the alteration of red lead, no such transformation was found here. Whether morphological changes typically accompany such transformation of minium to plattnerite has not been reported.

Atacamite Green

Microstructural changes to the atacamite green take the form of its loss, apparently by incongruent dissolution, and simultaneous deposition of gypsum within the confines of the original particle. The destruction of the green can be observed at different stages to span a range of compositions from that of the copper chloride alone to that of calcium sulfate alone with examples of all four ions commonly found. The transformation takes place on a very fine scale resolvable only with scanning electron microscopy (SEM) at magnifications over 4,000×. The ultimate fate of the copper has not been found in the form of any secondary copper compounds and so may involve its dilution into the surrounding materials via adsorption of copper ions on particle surfaces. Why gypsum precipitation would be nucleated on the damaged atacamite in the absence of more abundant gypsum deposition is not clear.

Moolooite

Copper oxalate, in the form of moolooite, has been found to result from alteration of the atacamite green in areas of high oxalate content. The source of the oxalate is unclear, as it is found not only in immediate association with fungal activity. Although clearly identifiable from Fourier transform infrared spectroscopy (FTIR) spectra, particles of moolooite were not recognizable via either polarized light microscopy (PLM) or SEM.

Calcium Oxalate

Calcium oxalates were widespread and frequently identified in FTIR spectra. Many examples contain contributions from both copper and calcium oxalates, complicating their identification. However, examples were found where the two compounds occur alone, and their identifications could be more securely made. The specific calcium oxalate that was present could not be discerned from the FTIR spectra due to overlap of the distinguishing hydroxyl absorptions by those of gypsum and other hydrous minerals. As in the case of the copper oxalate, calcium oxalate could not be identified morphologically either as particulate matter or as a component of any fungal mycelium.

Fungal Growth

While the association between fungal growth and oxalate formation was weak in this mural, fungal growth itself was frequently encountered in the form of hyphae invading the paint layer. These were not confined to specific paints and seem to survive in immediate contact with the potentially toxic compounds of copper and lead. They were most numerous, however, in kaolin whites.

Lead Chloride

Lead chloride formation and the deposition of lead chloride on particles of atacamite green represent a chemical transformation that may occur along several routes. This formation of lead chloride, probably in the form of cotunnite, may come about through the interaction of lead white and chlorides from seepage containing salt extracted from the wall (Ordonez and Twilley 1997). This mechanism probably accounts for the lead chloride occurrences in areas far removed from the green pigment. As noted below, the atacamite appears to have contained some intrinsic lead chloride. Additional lead chloride may have formed in green mixtures from chloride ions liberated during deterioration of atacamite since there are clear examples of its formation on unrelated particles in the layer (such as quartz). Lead, in addition to that which accompanied the atacamite, was also available in soluble form, as evidenced by the above cases of lead carbonate recrystallization.

Lead Phosphate

Authigenic formation of the lead analogues of apatite has occurred through the interaction of lead pigment with phosphate from some extraneous source. None of the pigments of the original palette serves as a source of phosphate, and yet lead phosphates could be found in some areas of hydrocerussite transformation.

Tin Compounds

Throughout the examination of examples of the atacamite green, morphologically distinctive tin compounds were found. These most often took the form of square prisms in which the only detectable elements were tin and oxygen. The tin compounds, also visible as opaque particles using PLM, were never found in pigments without the copper green. This finding is very strong evidence in support of the origin of the atacamite as a man-made product produced through the intentional corrosion of bronze. Tin is not a natural contaminant of copper ores where natural atacamite might be found. Furthermore, the growth habit of the atacamite is complex and chaotic, as if frequently disturbed. Cleavage fragments, as would be expected of the pulverized mineral, were not nearly so prevalent as concentric rind formations with very fine dimensions suggestive of a developing corrosion layer. The tin compounds were euhedral particles, seldom affected by fractures as would be expected to occur had they been initially present in mineral matter undergoing pulverization.

Analytical Evidence

Virtually all of the original paints are affected by deterioration and diagenetic changes that undoubtedly occurred while the mural was *in situ* and were fostered by moisture, some of it from roof leaks (fig. 3). Simple recrystallizaton of gypsum has already been described. In other cases, the dissolution has been accompanied by chemical reactions between nearby pigments. These are the primary causes of the single most influential change affecting the chromatic balance of the surviving original passages of the mural: the loss of atacamite green and consequent desaturation of the greens (fig. 4). Microstructural evidence suggests that irreversible dissolution of the atacamite and its replacement by gypsum occurs when these two materials are in intimate contact. The complex morphology evident in figure 5, consisting of irregular rinds made up of thousands of acicular crystals that are barely resolved at this magnification, gives

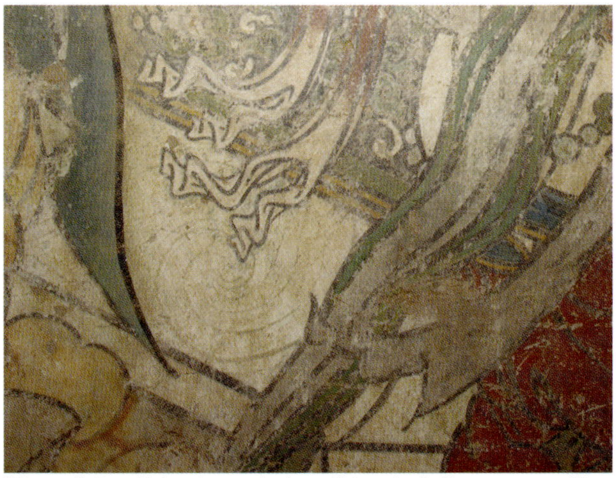

Figure 3. *Detail showing typical conditions including unrestored loss of red paint in the sash, graying and diminished saturation in many blues and greens, and extremely low contrast between a pale blue spiral and the underlying white background of an architectural border. The spiral line contains recrystallized lead white with a trace of azurite, while the background beneath contains kaolin.*

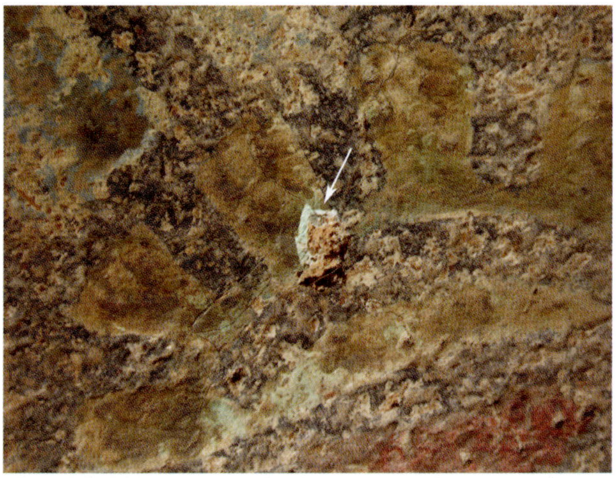

Figure 4. *Photomacrograph of a garment decoration showing darkening and lack of saturation in the green surface relative to better-preserved interior exposed by sampling (arrow). FTIR demonstrates the presence of both calcium and copper oxalates in this green.*

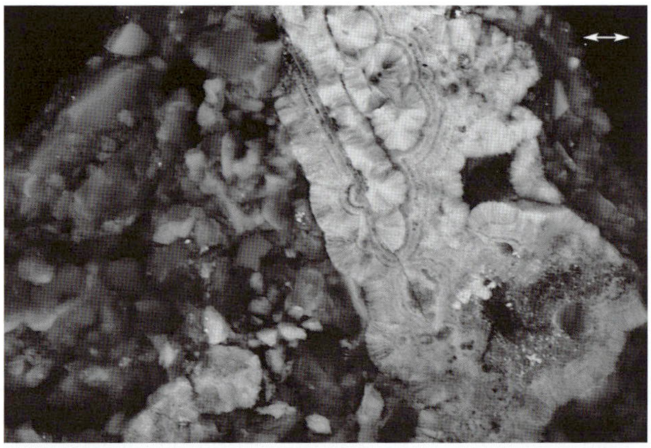

Figure 5. *Typical complex grain of atacamite exhibiting deterioration gradient from well-preserved upper end to eroded, low-density lower end. Backscattered electron image, scale bar = 2 microns*

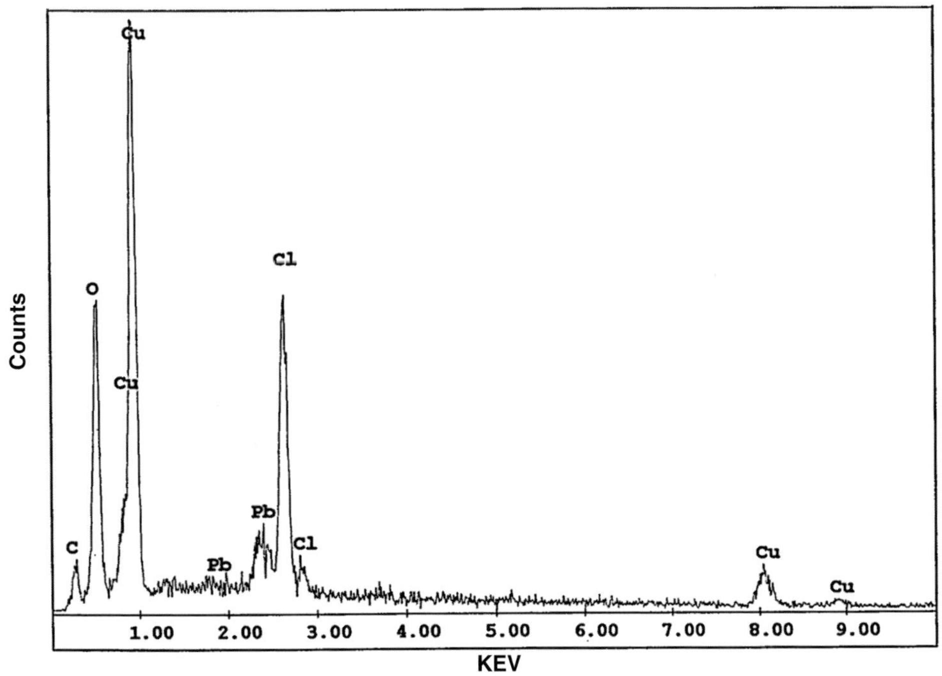

Figure 6. *Energy-dispersive x-ray spectrum from well-preserved upper end of the atacamite grain in figure 5*

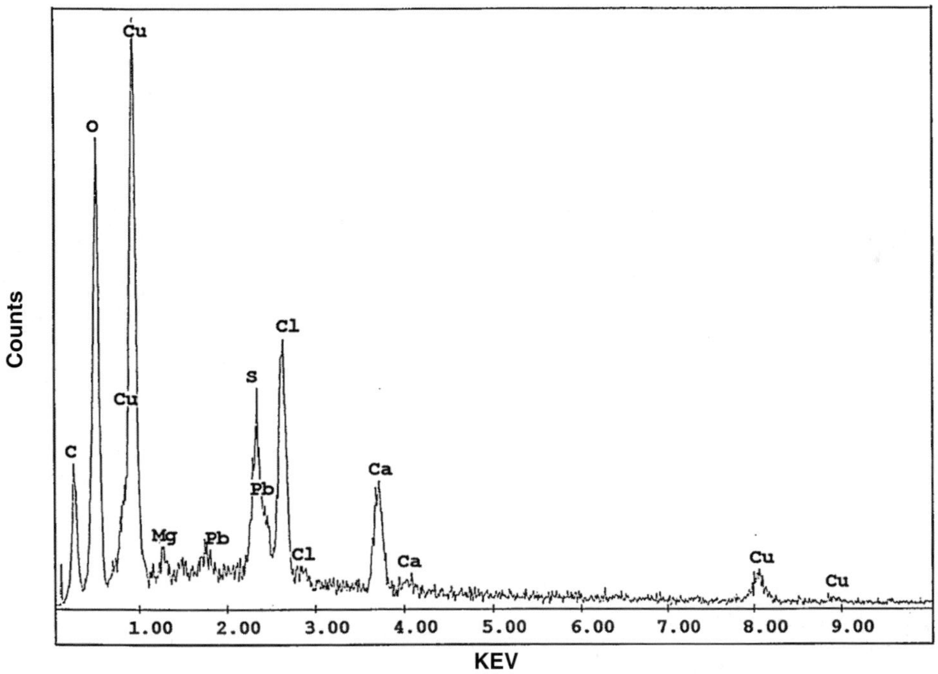

Figure 7. *Energy-dispersive x-ray spectrum for the deteriorated lower end of the atacamite grain in figure 5*

rise to varying optical properties. Well-preserved examples are birefringent and nearly colorless, with no consistency to their particle morphology. Highly disordered, deteriorated examples are almost invisible in PLM. Elemental analysis (fig. 6) of the well-preserved upper end of the atacamite grain shown in figure 5 yields only copper, chlorine, and oxygen as significant elements, consistent with the results expected for atacamite. (Traces of lead are ubiquitous in the atacamite, probably due to its origin in corroded bronze.) However, the deteriorated lower end of the grain yields a diminished intensity for the copper peak relative to oxygen, the presence of calcium and sulfur associated with gypsum, and an increased lead level associated with the bright, lead-rich features in the eroded zone shown in figure 5 (fig. 7).

The ultimate fate of the copper released by this interaction is unknown since individual particles of authigenic copper products could never be located for elemental analysis. It seems likely that the copper disperses into the adjacent materials, perhaps remaining only as a stain in extreme cases. This process may account for some examples that

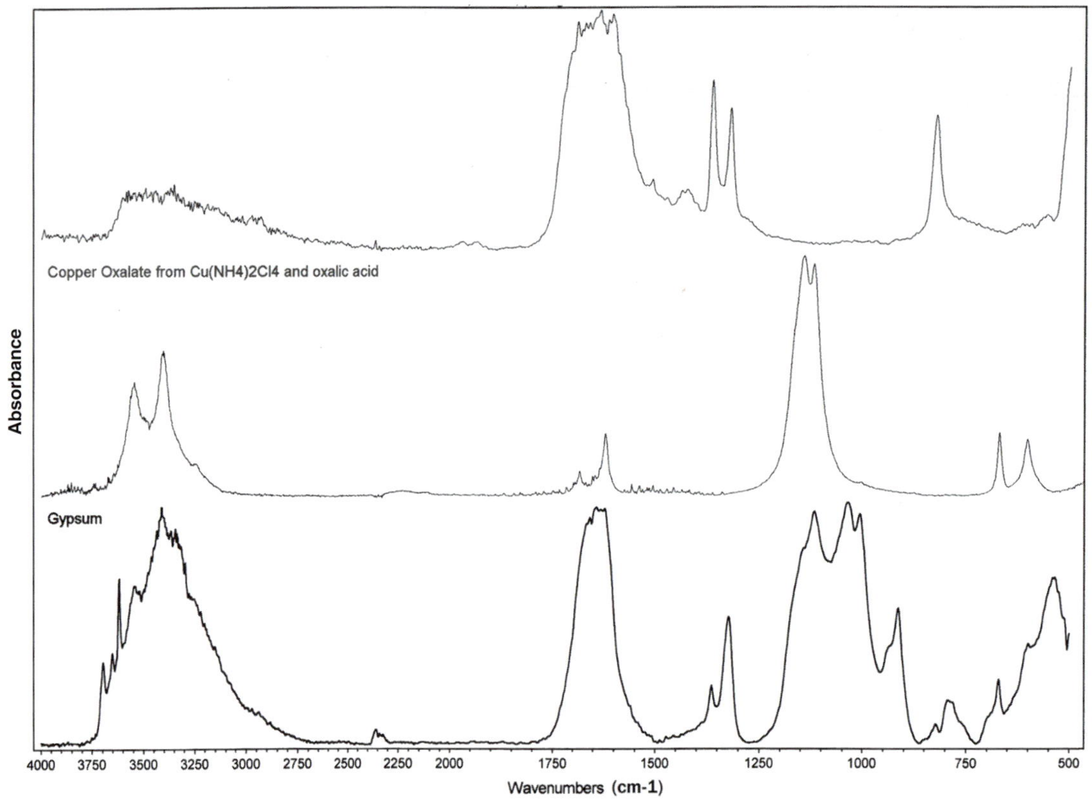

Figure 8. FTIR spectrum of degraded green decoration from an architectural member compared with gypsum and copper oxalate

exhibit pale green color on the wall but in which green pigments could seldom be recognized in the laboratory.

Figure 8 depicts an FTIR spectrum for the degraded green paint of an architectural decoration. A high content of gypsum is present. The relative intensities of the oxalate peaks at 1320 and 1360 cm^{-1} indicate that much of it occurs as calcium oxalate in addition to the copper compound. Peaks attributable to kaolin, atacamite, and lead carbonate are also present. The latter could not be assigned to the neutral or basic forms due to overlaps of the distinguishing features. SEM examination demonstrated that much redeposited lead chloride was present. Other evidence from infrared spectroscopy shows that in some cases the copper has reacted with oxalate to form moolooite in large amounts. However, there are also cases in which the green has degraded to products other than the oxalate even in the presence of that anion (as demonstrated by the presence of calcium oxalate but absence of moolooite in the damaged material).

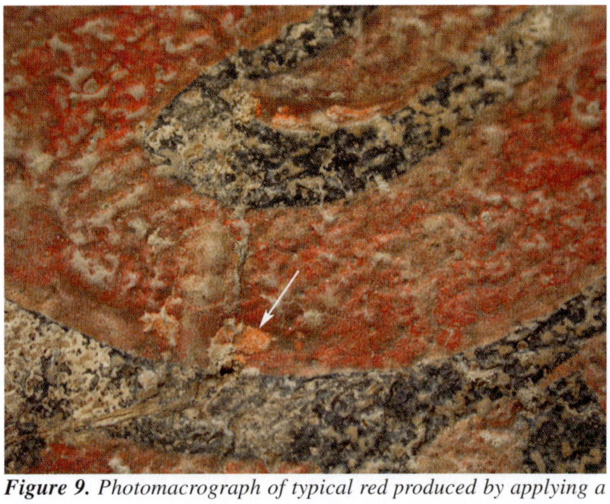

Figure 9. Photomacrograph of typical red produced by applying a thin coat of cinnabar over a base application of minium. Note orange interior at sample location, arrow.

The blackening of red lead by conversion to plattnerite, frequently reported by others in Chinese wall paintings (Kuchitsu et al. 1997; Pique 1997; Wainwright et al. 1997), was not encountered in this work. While it is true that much of the red lead serves as an undercoat for a minute layer of cinnabar and might be slightly protected (fig. 9), there are areas in which it is directly exposed and even these remain unaffected. SEM examination of the minium here disclosed widespread recrystallization and matrix formation involving this pigment, along with the development of secondary lead chlorides. Lead carbonates were not detected via FTIR or PLM, suggesting that not all of the recrystallization phenomena involve the carbonate.

While the alteration processes affecting the original paints have generally been detrimental to the mural, one aspect has been beneficial. There are numerous areas in which the dissolution and recrystallization of pigment, particularly of the lead white, has resulted in a hardening of the pigment layer in a sort of "self-consolidation." Dramatic evidence for the recrystallization of lead white as the cause of this phenomenon was found in many locations. Figure 10 shows an example in which the lead has not only recrystallized throughout the layer in which it was originally applied but also has begun to reinforce clays in the adjacent

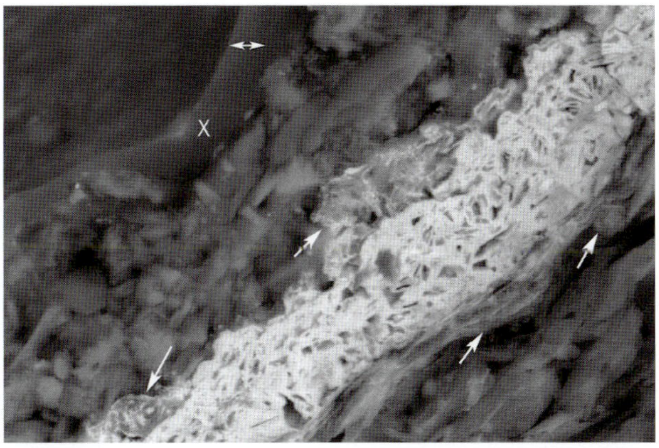

Figure 10. *Fracture section of white paint from the face of an attendant near floor level showing continuous matrix of hydrocerussite formed through recrystallization accompanied by the loss of discrete pigment particles. Note lead coatings in the interstices of clays above and below the pigment layer itself (arrows) and an organic coating thought to be polysaccharide on top surface (* X *). Backscattered electron image, scale bar = 1 micron*

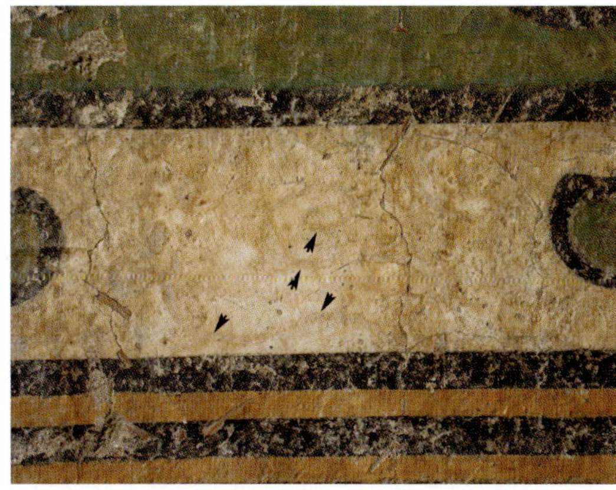

Figure 11. *White kaolin architectural border bearing spiral designs (arrows) in hydrocerussite. Field of view is approx. 25 cm. The lead white here exhibits the uptake of phosphorus, probably from water seepage.*

white ground and atop the paint surface by cementing those grains.

The widespread presence of phosphorus (not associated with any identifiable particle type) in areas where pigments have undergone alteration is suggestive of the introduction of phosphates in water seepage (fig. 11). Potential sources of the phosphate include bird droppings along the path of prior water seepage that reached the mural. Figure 12 shows a fracture section through one of the typical lead white architectural lines that has undergone recrystallization so nearly complete that the original particles of lead white are no longer recognizable.

This sample was selected for diffraction based on elemental analysis results suggesting that the content of secondary phosphate was sufficiently high that the end product of this transformation might be identifiable via x-ray diffraction (XRD). Phases that may be identified based on the

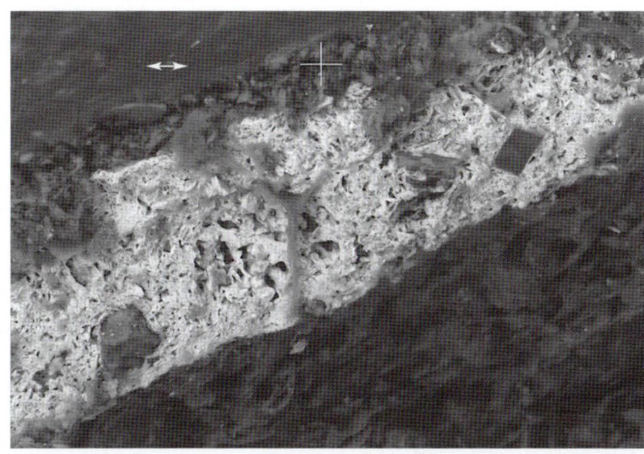

Figure 12. *Fracture section of cool white line from architectural decoration showing continuous matrix of lead compounds. Darker grains suspended in the lead layer are azurite. Note the organic film on top surface and fungal hyphae in pigment layer. BEI, scale bar = 3 microns*

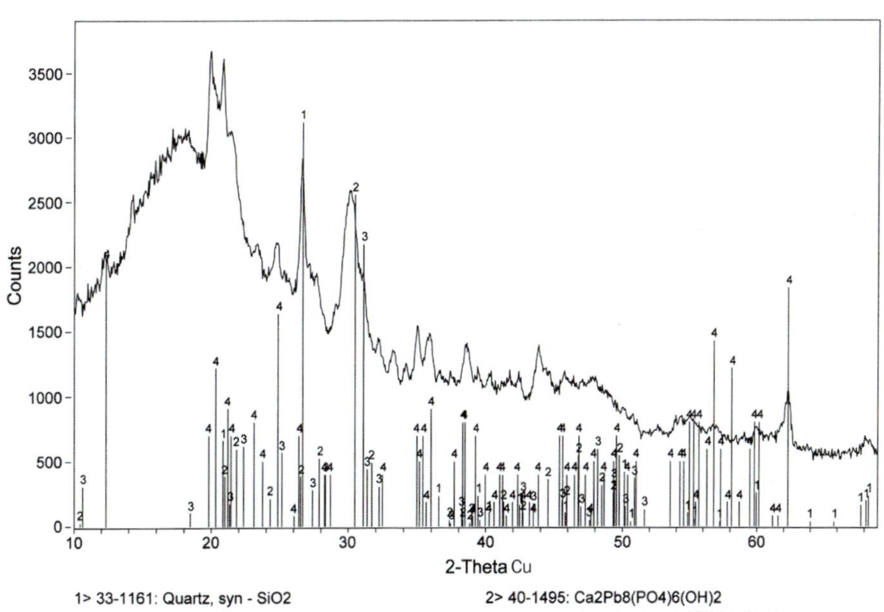

Figure 13. *X-ray diffraction pattern for a lead white decorative line atop a kaolin white architectural element obtained from a Brucker diffractometer equipped with an area detector (GADDS)*

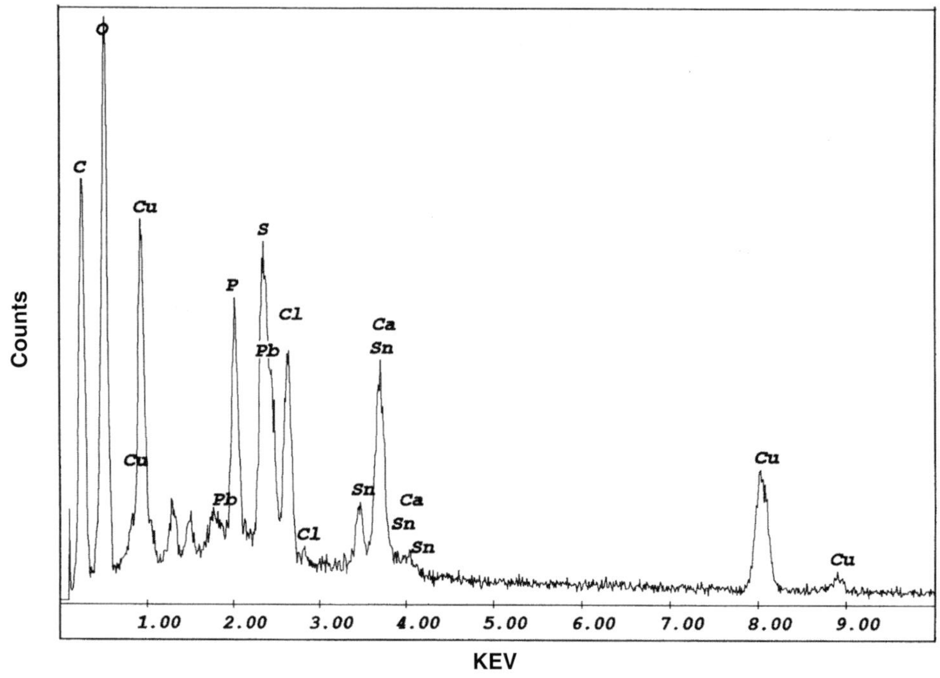

Figure 14. *Energy-dispersive x-ray spectrum for wisps of hair (painted in green) on the Buddha's face showing the complex assemblage of elements in this highly altered paint from an area subject to water seepage*

resulting pattern (fig. 13) and confirmatory PLM and FTIR data include quartz, kaolin, and lead-substituted hydroxylapatite. Two representative members of the latter class provide typical examples of the patterns with strong broad lines with a 2θ value near 30 degrees. The high background is due to scatter from the air path in the diffractometer and a high proportion of amorphous material—a commonly observed trait of authigenic hydroxylapatites.

The formation of hydroxylapatite and its highly substituted analogues is an underreported phenomenon in the conservation literature due to the experimental difficulties in its identification (Twilley 1996, 2002). The high degree and multiple forms of substitution allowed by the minerals of this class lead to a wide range of compositional variation and optical and crystallographic behaviors.

More complex examples exist in which several of the processes described above may be seen to have occurred, giving rise to a complex assemblage of elements. Figure 14 shows an x-ray spectrum obtained from wisps of hair (painted in green) on the Buddha's face in an altered paint that was subject to water seepage. Copper and chlorine are associated with atacamite pigment. Locally high concentrations of tin oxide are due to carryover from corroded bronze used as a source for the atacamite. Calcium and sulfur are primarily associated with gypsum, which has been seen to increase systematically in areas of atacamite decomposition. The high level of phosphorus, probably derived from water seepage contaminated by bird or bat droppings, is a constituent of secondary lead analogues of apatite, as are at least part of the lead and calcium. Carbon has multiple sources but is a common substituent in apatite in the form of carbonate anion.

The micromorphology and trace components of the green atacamite suggest that it was produced from corroded bronze. Tin oxide that occurs exclusively as a minor component of the green is the signature compound in this regard, but the conclusion is also supported by the micromorphology of the atacamite, which appears to have formed as a very thin reaction "rind" on a surface rather than being pulverized from a lump of the solid mineral or precipitated from solution (fig. 15).

In a majority of cases the infrared spectra were too complex for complete interpretation in the areas where critical distinctions occur between the various polymorphs of the copper trihydroxychlorides. Both original pigments and their alteration products often interfere. However, a sufficient number of clear cases were obtained where point-by-point comparison excludes the alternative species other than

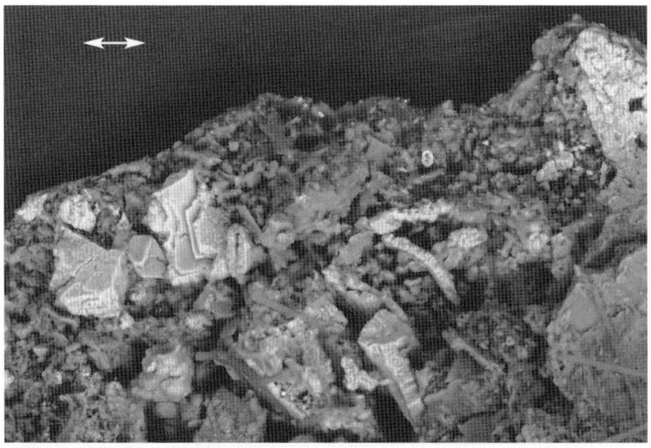

Figure 15. *The higher atomic number particles in this fracture section all consist of atacamite. Note the chaotic differences in morphology and size ranging from a small hollow "walnut," single and multiple acicular rinds, fragments, ovoid structures, and combinations of the above. Fungal hyphae are visible as well. BEI, scale bar = 10 microns*

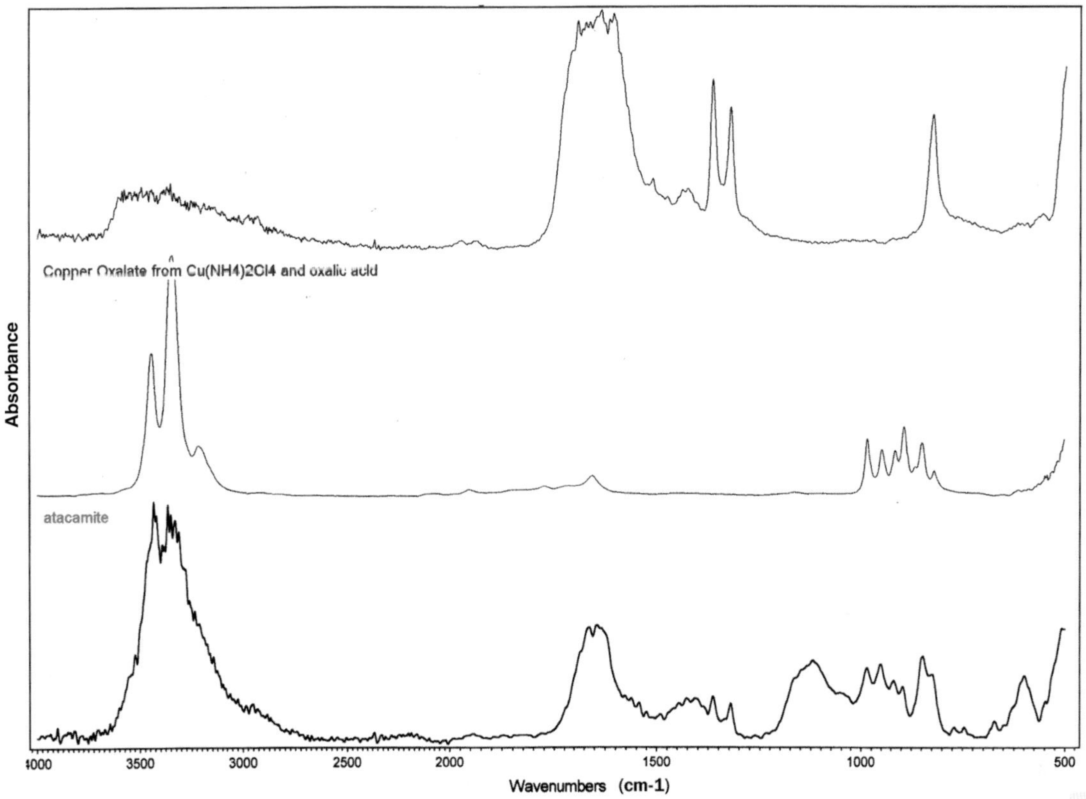

Figure 16. FTIR spectrum of partially degraded green from the hair of an attendant showing the correspondence between the distinctive peak positions in the region from 750 to 1,000 cm^{-1} with those of a superimposed spectrum for atacamite that serve to differentiate it from paratacamite. Peaks attributable to copper oxalate (top spectrum) are present as well, along with those of gypsum.

atacamite (fig. 16). No example, no matter how well preserved, conformed to malachite or other copper compounds from which atacamite could have been derived by alteration.

The signature species by which a bronze origin for the atacamite could be identified was tin oxide that occurs in distinctive square prisms only in the green paint. These were seldom recognizable in PLM but were frequently encountered in the SEM (fig. 17). Optical microscopy suggests that more of the tin oxide, in the form of opaque particles, may reside in the center of atacamite agglomerates, but this could not be confirmed by any instrumental means.

Ramifications of the Work

The discovery that the green originated as a copper alloy demands that the interpretation of the original pigment compositions be broadened to consider that some of the lead may have been introduced not in the form of lead white pigment but in the form of lead compounds resulting from corrosion of lead that, along with the tin, may have been a constituent of the bronze. For this reason we must consider that some of the lead compounds in green layers may not be due to the intentional addition of lead white. Furthermore, since the production of atacamite from leaded bronze would unavoidably lead to the formation of lead chlorides in the product, some of the lead chloride encountered may be original and not due to the introduction of chloride salts into the mural or to the interaction of lead white pigment with chloride released from the deterioration of the atacamite.

The very fine particle sizes, deterioration of the atacamite green, and diagenetic transformations involving changes in pigment morphology and composition greatly complicated the pigment analyses. However, their study has led to information about the mural that has important implications not only for the history, present appearance, and future care of this mural, but for similarly constituted murals in Asia and elsewhere. Electron microscopy of fracture sections was important for nearly every sample, while conventional cross-sections were often of little value due to the optical translucency of the clay-based layers and the extremely small dimensions of the features that are diagnostic of pigment alteration.

The instability of the principal green pigment and the numerous instances of pigment alteration—both microstructural and compositional—are indicative that tight constraints on the methods of future treatment must be observed if further damage is to be avoided.

The lead chloride, which, like the tin oxide derived from tin in the bronze, would be derived from the lead of a leaded bronze, accounts for at least a portion of the lead chloride as a contaminant in the atacamite. The discovery of a bronze source for the green also points to the role of chlorides as stimulants for the bronze corrosion in a manner analogous to many of the documented European recipes for the production of copper pigments (Scott 2002). If this is the case, it complicates the interpretation of some of the alteration

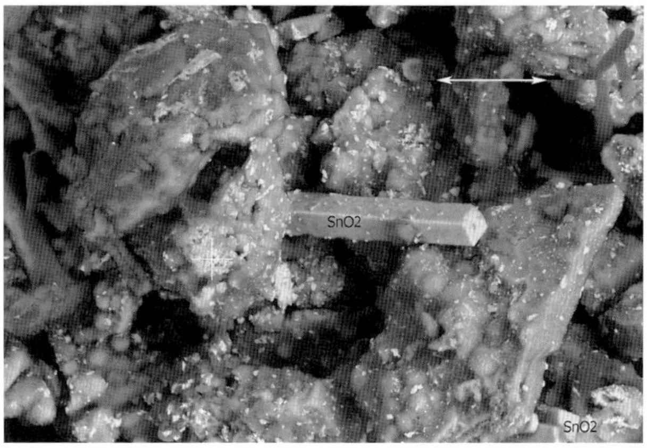

Figure 17. *Square prisms of tin oxide among particles of atacamite flecked by particles of lead chloride. The wide occurrence of tin oxide associated only with the copper green is strong circumstantial evidence for the production of atacamite by the intentional corrosion of bronze. BEI, scale bar = 10 microns*

phenomena cited above. For example, it suggests that at least part of the chlorides widely found to decorate the surface of other pigments originated with the atacamite and need not have an extraneous source in the wall or temple.

The origin of the mixed layer of clay and gypsum that often lies atop the pigments is an unresolved issue at the conclusion of this work. The consistency of the kaolin here is remarkable for airborne dust, and it is compositionally indistinguishable from the clay used in the execution of the mural. The gypsum that occurs along with it also serves as a widely used white ground layer for the mural. While it would seem on first consideration that the surface residue was the result of gypsum recrystallization bonding airborne clay to the surface, the microstructure contradicts this interpretation. The material on the surface of the mural does not appear to share the structural traits of redeposited gypsum or gypsum resulting from sulfation. Perhaps it played a role in the facing and removal of the mural that is obscure to us from our present vantage point.

Obscuring coatings on the colors were consistently found to contain both kaolin and gypsum. The gypsum did not occur in crust formations such as those typical for both gaseous sulfation phenomena and gypsum efflorescence. It seems, therefore, that at least part of the gypsum and the clay with it are the result of soiling. A history of microbiological colonization of the mural is apparent from fungal hyphae that are found sporadically inside the pigment samples. The fungi were found not to discriminate in terms of the presence of potentially toxic metals, occurring in areas of copper and lead compounds as well as those including only gypsum and clay. It was noted, however, that the example with the most prolific growth included only clay. Microbial overgrowths that might play a role in obscuring the colors were not found. Oxalates without associated fungal hyphae were more prevalent at the surface of the pigment than in its interior when the layer could be subdivided for analysis and often occur together with gypsum in dull coatings atop the pigments.

It has been suggested that peach gum (polysaccharide exudate of *Prunus persica*) would have been a likely choice of facing adhesive in China,[1] and it seems reasonable to suppose that locally available materials would have been used to facilitate the removal of the wall painting in China. This suggestion is consistent with our frequent observation of an organic layer uppermost atop the pigment and overlying clays (see, e.g., figs. 10 and 12). Infrared spectroscopy failed to disclose absorptions unique to this layer that were not hidden by those of the mineral matter, a result that is consistent with its identification as a plant polysaccharide gum. The spectra of such gums are dominated by broad hydroxyl peaks in areas where clays and hydrous silicates also absorb strongly (Twilley 1984). However, the results of separate gas chromatography–mass spectroscopy (GC-MS) analyses for free sugars and for carbohydrates in one location were inconclusive, yielding low overall levels of free mono- and disaccharides in the first case, and a low yield of hydrolysate in the second, for which the monosaccharide profile did not correlate well with that of any specific polysaccharide.

Although the paints are clearly tempera and unrelated to Western techniques of fresco involving lime, the organic medium was never identifiable in infrared spectra. Analysis by GC-MS at one location has demonstrated that the paint layer contains a protein whose amino acid profile is consistent with that of an animal glue binder. The history of water damage and microbiological growth has probably reduced the media content substantially over the centuries. This condition contrasts with what has been found in the analysis of structurally similar murals from temples in western Tibet (Twilley 1998) where proteinaceous medium was still abundant and organics of other classes could be readily found in specialized applications. The difference between these cases is probably partly a function of the differing climates—the generally dry conditions of Tibet tending to favor the survival of all classes of organics by inhibiting biological deterioration. It should also be noted that some ambiguity surrounds positive media analysis results obtained on the related works in Toronto, since the treatment of these works on the art market differed. The application of a proteinaceous facing material is certain in one case (Moffat et al. 1985).

Only one sample contained a small thallus- or mycelium-like structure that was recognizable using PLM. It consisted of a clear layer that conformed closely to the surface of some pigment agglomerates but that was lifted slightly from the agglomerate surface by highly birefringent crystals in between. This arrangement is characteristic of the structures observed on a coarser scale in the colonization of stone surfaces by calcium oxalate-producing microorganisms (Twilley 1993) and similar to examples illustrated for copper oxalate by Moffat et al. (1985). However, none of the samples specifically identified as copper oxalates through instrumental analyses in the Nelson-Atkins examples were found in this form. The finding that oxalates occur in the twentieth-century restorations to the Toronto example suggests that at least some of that oxalate is not ancient and may not have a biological origin.

Acknowledgments

Funding for the documentation project, of which this analytical study forms a part, was provided by the Getty Grant Program. Conservator Jerry Podany participated in initial examinations and helped to frame the scope of activities for this project prior to funding. Director Marc Wilson and curators Dr. Xiaoneng Yang and Ling-en Lu allocated staff time and assisted in gathering information on the prior history of the painting. Joe Rogers and Steve Bonham undertook much of the detailed documentation work using digital photographs prepared by Louis Meluso and Jamison Miller. Conservator Eric Gordon provided comparative information, on site, based on his participation in the prior treatment of murals in the Royal Ontario Museum, Toronto. Access to the Brucker GADDS x-ray diffractometer was generously provided by Dr. John Parise of the Earth and Space Sciences Department, SUNY Stony Brook. GC-MS analyses were conducted by Richard Newman of the Scientific Research Laboratory, Museum of Fine Arts, Boston. The FTIR reference spectrum for atacamite was contributed to the Infrared Users Group (IRUG) database by Beth Price, Philadelphia Museum of Art.

Note

1. We thank Lucienne van Valen for this suggestion derived from her experience in China.

References

Gordon, E., and E. Phillimore. 1984. The treatment of two thirteenth-century Chinese wall paintings in the collection of the Royal Ontario Museum. *ICOM preprints*, 8415.12–14. Seventh Triennial Meeting, 10–14 September 1984, Copenhagen. Paris: International Council of Museums in association with the J. Paul Getty Trust.

Jing, A. 1991. The Yuan Buddhist mural of the *Paradise of Bhaisajyaguru. Metropolitan Museum Journal* 26:147.

Kuchitsu, N., D. Xiuye, C. Sano, G. Hong, and L. Jun. 1997. Color change of pigments in the Mogao grottoes of Dunhuang. In *Conservation of ancient sites on the Silk Road*, ed. N. Agnew, 329–33. Los Angeles: Getty Conservation Institute.

Malenka, S., and B. Price. 1997. A Chinese wall painting and palace hall ceiling: Materials, technique and conservation. In *Conservation of ancient sites on the Silk Road*, ed. N. Agnew, 127–38. Los Angeles: Getty Conservation Institute.

Moffat, E. A., N. T. Adair, and G. S. Young. 1985. The occurrence of oxalates on three Chinese wall paintings. In *Application of science in examination of works of art*, ed. P. A. England and L. van Zelst, 234–38. Seminar, Boston, 7–9 September 1983. Boston: Research Laboratory, Museum of Fine Arts.

Ordonez, E., and J. Twilley. 1997. Clarifying the haze: Efflorescence on works of art. *Analytical Chemistry: News and Features* 69:416A–22A.

Pique, F. 1997. Scientific examination of the sculptural polychromy of cave 6 at Yungang. In *Conservation of ancient sites on the Silk Road*, ed. N. Agnew, 348–61. Los Angeles: Getty Conservation Institute.

Scott, D. 2002. A review of copper chlorides and related salts in bronze corrosion and as painting pigments. *Studies in Conservation* 45:39–53.

Steinhardt, N. 1984. *Chinese traditional architecture*. New York: China Institute in America, China House Gallery.

Twilley, J. 1984. The analysis of exudate plant gums in their artistic applications: An interim report. In *Archaeological Chemistry III*, ed. J. B. Lambert, 357–94. American Chemical Society Symposium Series 205. Washington, D.C.: American Chemical Society.

Twilley, J. 1993. The lichen/rock interface: Structural and compositional studies. In *Biodeterioration of cultural property 2*, ed. K. Toishi, H. Arai, T. Kenjo, and K. Yamano, 80–97. Proceedings of the 2nd International Conference on Biodeterioration of Cultural Property, Yokohama, 5–8 October 1992. Tokyo: Organizing Committee and International Communications Specialists, Inc.

Twilley, J. 1996. Three panels with relief carving. In *Ancient art from the Shumei family collection*, ed. E. Shultz, 142–43. New York: Metropolitan Museum of Art.

Twilley, J. 1998. Materials of the monastic wall paintings of western Tibet. Paper presented at Art et Chimie la Colour, Ecole du Louvre, Paris, 16–18 September 1998.

Twilley, J. 2002. Pigment analyses for the grave stelae and architectural fragments from Chersonesos. In *Color in ancient Greece: The role of color in ancient Greek art and architecture (700–31 B.C.)*, ed. M. A. Tiverios and D. S. Tsiafakis, 171–78, 302–3. Proceedings of the Conference, Thessalonica, 12–14 April 2000. Thessalonica: Aristotle University.

Wainwright, I., E. Moffat, P. Sirois, and G. Young. 1997. Analysis of wall painting fragments from the Mogao and the Bingling Temple grottoes. In *Conservation of ancient sites on the Silk Road*, ed. N. Agnew, 334–40. Los Angeles: Getty Conservation Institute.

Research Plan for the Restoration of the Multicolored Paintings in Lungshan Temple at Lu-Gang

Tung Kuo-king and Tsai Yu-lin

ABSTRACT As a result of recent developments in both materials and techniques, as well as cultural changes, the number of craftsmen capable of producing traditional colored paintings has been declining. Since new materials and techniques are popular in modern paintings, professional craftsmen using traditional methods are now rare in Taiwan. Thus it is difficult to find knowledgeable people for restoration work. During a building restoration project at Lungshan Temple, Lu-Gang, Taiwan, the traditional architectural paintings by Kuo Shin-lin required restoration, but few architects were willing to investigate the colored paintings in detail. Scientific methods were used to investigate Kuo's work, and the results of scientific analysis were combined with traditional techniques to develop an appropriate restoration process. Ultimately, we hope our work will help preserve not only the original colored paintings created by Kuo Shin-lin but also the treasured traditional art of painting.

Introduction

This project was organized by the National Center for Research and Preservation of Cultural Properties, Taiwan, which assigned the authors to study the paintings by Kuo Shin-lin at Lungshan Temple at Lu-Gang, Taiwan. The project also was assisted by the Japan Conservation Project Specified Nonprofit Corporation (JCPNOP) and, in particular, Kaloku Miwa, Tsutomu Watanabe, and restorer Kentaro Obayashi. The center was in charge of researching previous historical references, performing an investigation, recording the colored paintings with slides, mapping the damage, and searching for the causes of damage. The JCPNOP and the center worked together to take pigment samples, examine the restoration methods, and discuss restoration plans. A symposium was held to present our research and the progress of the investigation in Lu-Gang on 11 November 2002. Kentaro Obayashi and Tsutomu Watanabe were invited speakers.

The aim of this project is to help the caretakers of Lungshan Temple at Lu-Gang use scientific methods to investigate its paintings and to provide results that can aid the restoration work. The range of this project is limited to the multicolored paintings on the trigram roof of the Stage of Lungshan Temple.

History of Lungshan Temple at Lu-Gang

Lungshan Temple was constructed on its present site in Lu-Gang village, Chang-Hua County, Taiwan, in 1786 (Wang and Chen 2002). It was a branch of the Lungshan Temple in Chuen-Jou, Fu-Chien province. Reconstructed in 1829, it remains intact today. Lungshan Temple is a building occupying approximately 5,000 square meters and constructed with joined wooden beams and brackets. It consists of three courtyards and a temple gateway, a Five-Gate Hall, a Stage, a Great Hall, a Worship Hall, and a Posterior Hall. In 1983, Lungshan Temple was listed as a Treasure of Taiwan's Cultural Heritage, and for the last three years it has been nominated as one of Taiwan's most popular cultural tourist attractions.

The Stage in Lungshan Temple is a place for giving thanks to God. The six columns (originally four) occupy 5.73 × 5.73 square meters. The roof of the stage as shown in figure 1 consists of twelve groups of *tou-kung* (trapezoidal brackets). In the middle these form a trigram shape, and from this trigram are formed another eight groups of *tou-kung*, which are concentrated in the center of the roof. The surfaces of the *tou-kung* and the beams are covered with

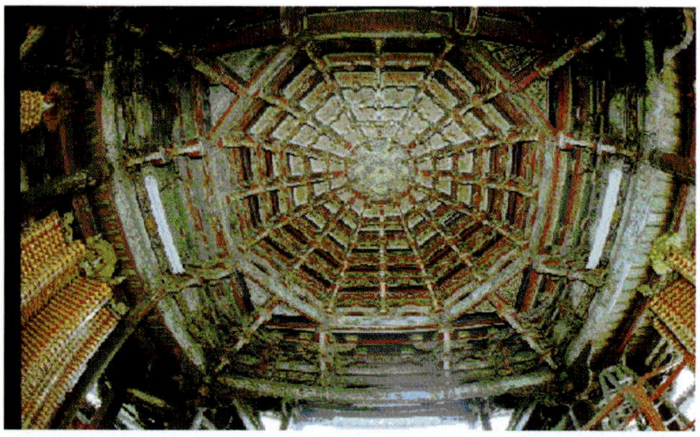

Figure 1. *Trigram roof of the stage at Lu-Gang Lungshan Temple*

many beautiful and elegantly colored designs, Chinese paintings, and calligraphies.

The upper and lower parts of the trigram roof consist of 2,650 pieces. Each piece has been numbered and photographed, focusing particularly on the beams with Chinese paintings and calligraphies. The total area with color painting is about 1,116,465 square centimeters. The Chinese painting works occupy about 171,544 square centimeters. According to the computer aided drawing (CAD) damage maps, the damaged area of the Chinese painting is about 88,074 square centimeters.

Colored Painting

The multicolored painting in the temple is of a type called *tsai-hui*. *Tsai* indicates the use of a ground layer on the wood or wall to prevent the wood oil from coming to the surface and also to prepare a smooth surface to paint. *Hui* refers to the artist showing artistic talent through the painting of beautiful works. All of the traditional ancient–styled paintings in Lungshan Temple were painted by Kuo Shin-lin, a famous Taiwanese folk artist (Tsai 1980). Kuo's ancestor came from Mainland China, and for three generations his family has worked as *tsai-hui* craftsmen. Kuo Shin-lin is the most talented and most famous *tsai-hui* craftsman of his generation in the Kuo family.

History of the Restoration of the Painting of the Stage's Trigram Roof

The following information was obtained on the history and prior restorations of the paintings (Hai 1980; Hai-Kuang 1985):

1. The columns used with the Five-Gate Hall were replaced with cement columns.
2. In 1964 the temple invited Kuo Shin-lin to paint the colored paintings.
3. The addition of two extra columns to the original four columns was considered necessary for additional support due to problems with the weight of the roof.
4. In the structural restoration work done in 1986, the colored paintings were washed with water and sprayed with WP-20 (see the FTIR section for the analytical results for this material).

Except for the application of WP-20, the paintings have not had any formal restoration or preservation work performed on them. The paintings were flaking very severely. The temple was damaged seriously by the earthquake that occurred on 21 September 1999. Presently, structural conservation work is being done on the Great Hall, the Worship Hall, and the Posterior Hall. The Worship Hall was moved temporarily beside the Stage. The second phase of conservation work will include conservation of the paintings on the wooden structure.

Environmental Conditions

Lu-Gang is located at 24°–24°10″ N and 120°22″–120°30″ E. Summers are hot, and it is warm with little rain in winter. According to the Central Weather Bureau of Taiwan at Wu-Chi station, the nearest station to Lu-Gang, in 2002 the average temperature was 23.7°C, and the average humidity was 75.8% RH. Temperature and relative humidity values at the temple were recorded every ten minutes from the Rotronicag Hygrolog during a two-week period in late summer for comparison with the Weather Bureau's measurements. The highest temperature was 32°C in the afternoon and the lowest temperature was 28°C in the early morning, whereas the highest humidity was about 78% RH in the early morning and the lowest humidity was about 62% RH in the evening. Although the temperature and humidity are relatively high, they are regular. The change in temperature is less than ±4°C, and the relative humidity has about a 15% fluctuation. The moisture content of wood of the temple was also measured at this time and it was found to be 10.5–14.5% RH.

Although the trigram roof is exposed to outside air, the painting on the interior does not receive direct sunlight. However, in the temple three fluorescent lamps hang above the third floor. In the daytime they block the tourists' view of the beautiful colored paintings. At night, the light can illuminate only small areas. Moreover, the light shines in the eyes of the tourists. The wavelength emitted by the fluorescence lamps is about 217 nm.

Degradation of the Painting

Investigation revealed various forms of damage.

1. The fading of the color on the cement column is more serious than that on the wooden column.
2. The beams of the third floor ground *tou* are thicker than other parts, and the loss of pigment and flaking is more serious than in other areas.
3. Some of the paintings have unacceptable retouching.
4. In the south part of the roof, the paint losses are significant. This area seems to have been destroyed by outside factors.
5. The paint layers flaked together with the ground layer, indicating that the adhesive force between the ground layer and paint layer is stronger than that between the ground layer and the wooden surface.
6. The smoke of the incense sticks from a temporary Worship Hall built beside the Stage will make the cleaning of the paintings more difficult.

The condition of each painting layer is listed in table 1.

Table 1. *Damage to the painting layer*

Layer	Damage
Wooden base layer	Warping, displacement, bare wood surfaces, insect damage
Ground layer	Flaking, losses, cracking
Paint layer	Flaking, losses, cracking, fading
Paint surface	Dust, spotting

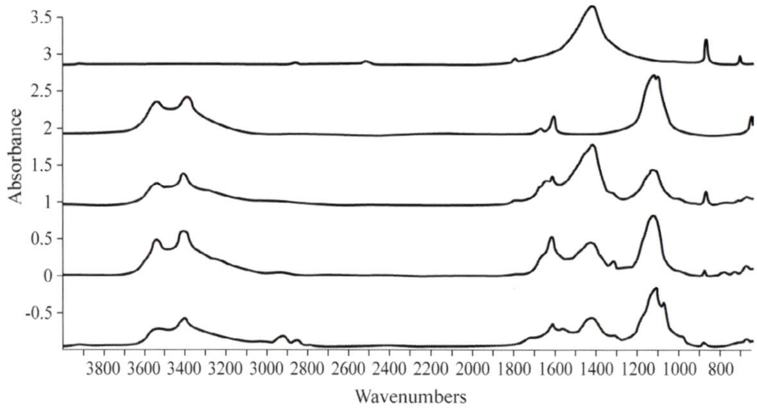

Figure 2. *FTIR spectra of calcium carbonate, calcium sulfate, and the ground layer of the samples*

Scientific Examination

In order to investigate the composite materials of the colored layer, Fourier transform infrared spectroscopy (FTIR) and microscopic examination of cross-sections were performed.

FTIR

The restoration record of Lungshan Temple states that "the surface of the colored paintings was sprayed with WP-20" (Hai-Kuang 1985, 69), but we could not find any detailed information about this treatment. Therefore, a cotton stick with ethyl acetate was used to collect a sample from the painting surface. After evaporating the ethyl acetate, the residue was redissolved with hexane, then deposited on the detection disc and analyzed using FTIR. Absorption at 1750 cm^{-1} in the spectrum reveals that the material on the painting surface contains a carbonyl functional group, indicating that it is coated with a material such as an acrylic resin or some other species. (The binder of WP-20 is a modified acrylic resin, and the chemical structure of acrylic resin contains a carbonyl functional group.)

The FTIR spectra of conventional fillers in paint manufacture such as calcium carbonate and calcium sulfate were collected and compared with that of the white color on the reverse of the paint sample. Figure 2 shows the FTIR spectra. The characteristic absorptions of calcium carbonate and calcium sulfate are found in the white layer on the reverse, indicating that the ground layer contains both inorganic fillers. Figure 3 shows the FTIR spectra of some colored samples. Five color paint samples were found to contain carbonyl functional group species, and most of the color samples contain calcium carbonate as filler pigment in varying quantities.

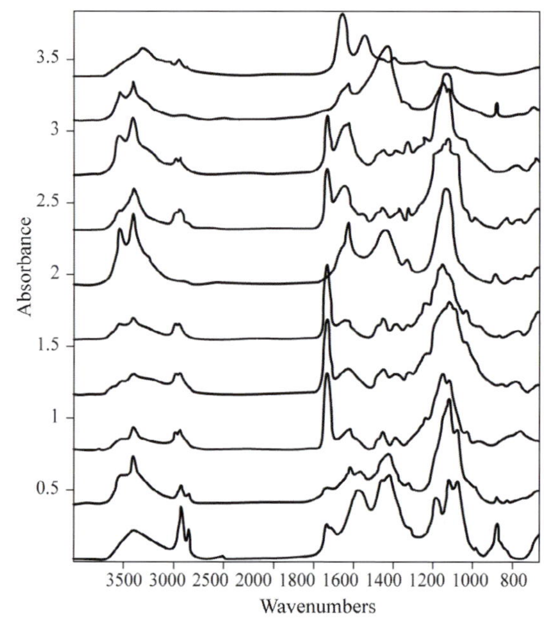

Figure 3. *FTIR spectra of color samples. From top to bottom: "pig blood", white background, China ink, black, white, gold, green, red, blue, and "thick" green*

Cross-Section Investigation

In order to understand the application methods of the paints and their condition, cross-sections were mounted in

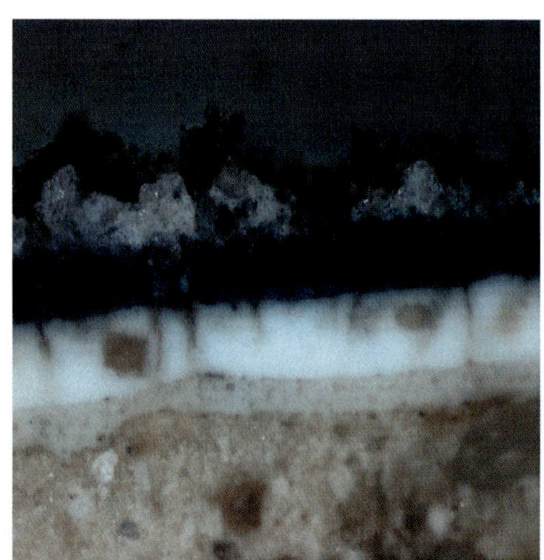

Figure 4. *Cross-section of blue paint*

polyester resin and analyzed by microscope. Cross-sections show that the painting is composed of a ground layer with one base paint and one surface paint. Only one colored layer was seen in each cross-section (see, e.g., fig. 4). Dust covered the surface of the paint in a thick layer, and the ground layer contained occasional deep cracks.

Test of Cleaning for Restoration

Following information from the scientific analysis, ethyl acetate ($CH_3COOC_2H_5$) was used to clean a small square of each color in one area. After being cleaned, the color of the squares was measured by using a spectrophotometer (Minolta cm-2600d D8) and with values calculated using CIE 94. Figures 5 and 6 demonstrate the condition of the colored area before and after cleaning. The color difference between clean and unclean pieces is listed in table 2. Obviously, the brightness (ΔL*) and chromaticity (ΔC*) of each color increased due to cleaning. The color change was Δb* = –6.41 for blue, Δa* = –7.87 for green, Δb* = 6.8 for yellow, Δa* = 5.68 for red, and Δb* = 3.05 for black, respectively (h and Δh refer to hue angle and the hue angle difference). Figure 7 shows the change in reflectance spectra for a green color area.

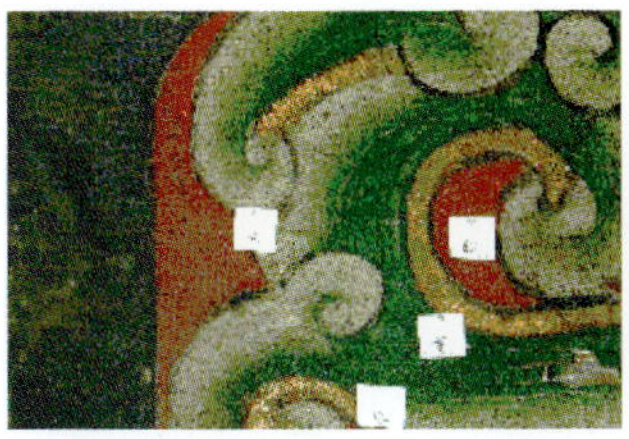

Figure 5. *Painting with areas to be cleaned marked*

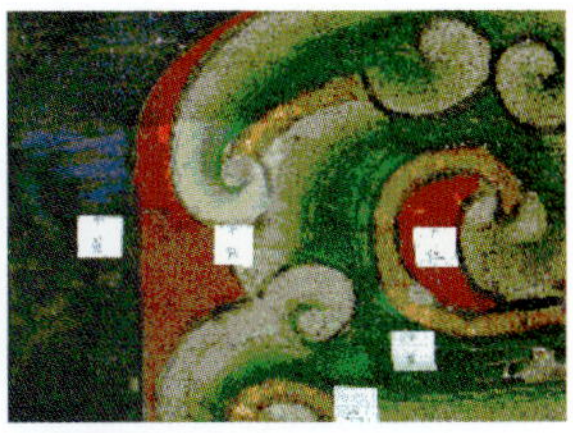

Figure 6. *Painting after cleaning marked areas*

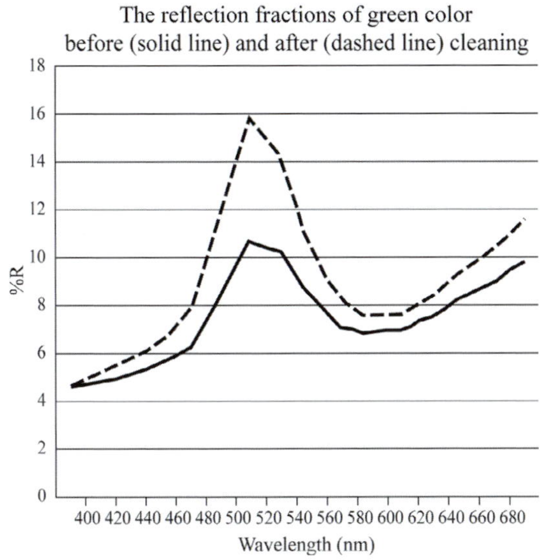

Figure 7. *The % reflectance of a green painted area before and after cleaning*

Conclusions

1. The fading of the colors is very serious. Almost half of the painting is damaged. Conservation is necessary.
2. The variation of humidity for one day is frequently about 15% and the moisture content of wood is between 10.5% and 14.5%. We expect the wood expansion and contraction to cause further separation between the paint layer and the support.
3. The artificial lighting should be redesigned, and until this is done, the fluorescent lamps should be covered with UV-absorbing plastic.
4. The temple should be cleaned regularly to reduce dust.
5. A coating on the painting has been identified as material having an infrared absorption of 1750 cm^{-1}, that of a carbonyl functional group, suggesting it is an acrylic resin or other polymer with plasticizer. Ethyl acetate is suggested for use as a solvent to remove the coating and the dust on it. The cracks in the colored layer are very serious. A nonaqueous material is suggested to stabilize the cracks.
6. The painting research should be extended to the Five-Gate Hall, the Great Hall, the Worship Hall, and the Posterior Hall.
7. Continual environmental monitoring should be used in the temple.

Acknowledgments

The authors would first like to thank our institution, the National Center for Research and Preservation of Cultural Properties, for giving us permission to use the material from this project. The authors appreciate the help of Japan Conservation Project Specified Nonprofit Corporation (JCPNOP), Mr. Kaloku Miwa, Restorer Mr. Kentaro Obayashi, Mr. Tsutomu Watanabe, National Chung-Hsing University chemistry professor Jen Jen-Fon, Tainan National College of the Arts wood conservation professor

Table 2. *Difference in color due to cleaning*

	Before cleaning					After cleaning					Change in L*a*b*C* h				
	L*	a*	b*	C*	h	L*	a*	b*	C*	h	ΔL*	Δa*	Δb*	ΔC*	Δh*
Blue	33.69	–0.46	–0.69	0.83	236.55	35.58	–0.56	–7.10	7.12	265.53	1.89	–0.10	–6.41	6.29	28.98
Green	34.81	–10.19	10.96	14.97	132.92	39.44	–18.06	14.92	23.43	140.44	4.63	–7.87	3.96	8.46	7.52
Yellow	47.44	12.00	25.47	28.16	64.76	56.92	10.89	32.27	34.05	71.35	9.48	–1.11	6.80	5.89	6.59
Red	34.41	19.30	10.53	21.99	28.61	34.78	24.98	13.83	28.55	28.97	0.37	5.68	3.30	6.56	0.36
Black	34.96	2.42	8.14	8.49	73.42	45.76	2.14	11.19	11.40	79.18	10.8	–0.28	3.05	2.91	5.76

Antoni Wimmerring, Feng-Chia University Architecture Department assistant professor Lin Wen-Chen, Chi Mei Culture Foundation, and many people associated with Lungshan Temple at Lu-Gang.

References

Hai, Pao-Te. 1980. *Study of Lungshan Temple at Lu-Gang*. Taipei.

Hai-Kuang Architecture Office. 1985. *The restoration record and report of Lungshan Temple at Lu-Gang*. Taipei: Hai-Kuang Architecture Office.

Tsai, Ya-Hui. 1980. Study of color painting craftsman Chin-lin Kuo at Lu-Gang. Master's thesis, Department of Architecture, Taipei Art College.

Wang, Kang-Shou, and Chen Shih-Hsien. 2002. *The Lungshan Temple at Lu-Gang, Lu-Gang*.

Crossing the Line: The Interplay between Scientific Examination and Conservation Approaches in the Treatment of a Fifteenth-Century Nepali *Thangka*

Marco Leona and Sandhya S. Jain

ABSTRACT *Chakrasamvara and Vajravarahi*, a fifteenth-century *thangka* painted by a Newar artist in Tibet, was examined in detail prior to its inclusion in a large exhibition of Buddhist meditational art. The visual examination highlighted the need to consolidate flaking paint, in-fill several losses, and remount the *thangka*. During the examination and treatment sequence, conservators and scientists consulted regularly in a process that led to an improved understanding of this *thangka* and of *thangka* painting technique in general. Techniques such as polarized light microscopy, scanning electron microscopy, and Raman microscopy were used on pigment samples and paint cross-sections. Additionally, Fourier transform infrared microspectroscopy was used to identify binding media. Lastly, a deterioration phenomenon affecting the orpiment pigment present in certain areas of the painting was investigated. The results of the scientific study of the *thangka* influenced the formulation of the treatment proposal. The discovery of selective applications of a glue varnish over parts of the *thangka* to modulate the gloss of the paint constrained the conservators' choices in terms of consolidation approaches. In order to preserve the optical texture of the painting, broadcast glue applications were ruled out in favor of localized microapplication of isinglas.

Introduction

Chakrasamvara and Vajravarahi (fig. 1), a large painting on cloth representing the embrace of Chakrasamvara (the Buddha's mind of compassion) and his consort Vajravarahi (the nature of omniscient wisdom of a Buddha) is considered one of the finest examples of the Newar *bal bris* style (Huntington and Bangdel 2003). Dated to the middle of the fifteenth century and painted by a Newar artist in Tibet, or a Tibetan artist working in the Newar style for a Tibetan patron, *Chakrasamvara and Vajravarahi* is characteristic both for its stylistic quality and iconographic complexity, and for its exceptionally large size. The overall dimensions of the painting are 137.2 × 114.3 cm, making it two-to-three times larger than contemporaneous *thangka*s. The *thangka* is one of the most comprehensive visual representations of the generation and completion stage practices of the *Chakrasamvara Tantra*. The center of the *thangka* depicts the male god Chakrasamvara in embrace with his consort, the orange-colored female Vajravarahi. They stand within a red *prabha mandala* that emits multicolored flames, which rests on a lotus-petal throne. Surrounding the *prabha mandala* are the holy cremation grounds containing various symbolic figures. Also depicted in the upper and lower sections of the composition are specific holy figures symbolizing particular aspects of the Tantric practice.

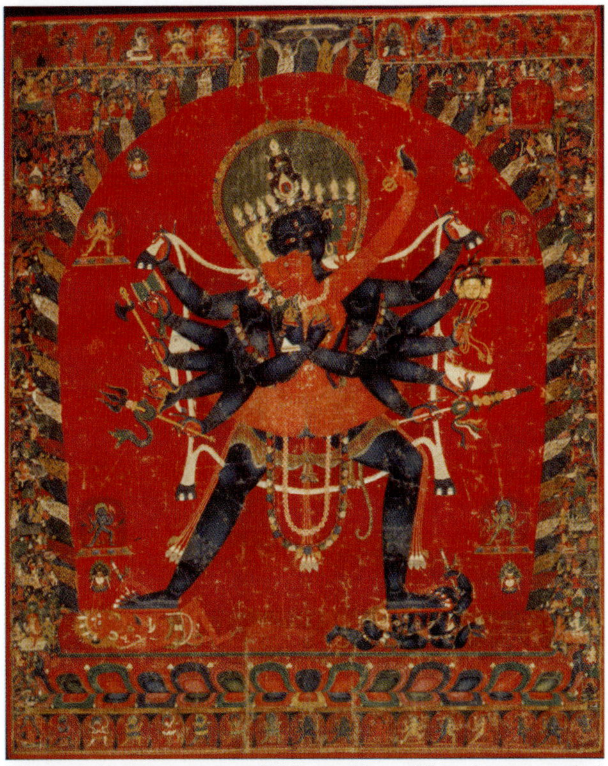

Figure 1. Chakrasamvara and Vajravarahi, *ca. fifteenth century, cloth with natural pigments, 137.2×114.3 cm. Los Angeles County Museum of Art, Nasli and Alice Heeramaneck Collection, M.70.1.3*

*Thangka*s were used by Buddhists practitioners to gain enlightenment through intense meditation and, as such, were often commissioned by monasteries to guide their *yoginis*, or tantric practitioners, through the four classes of Tantric Buddhism. During meditation the *yoginis* used this *thangka* in the last step toward realizing their own identity as a fully enlightened Buddha. Every detail of its complex iconography suggests several nuances of meaning, all of which must be completely internalized by the practitioner until they can be effortlessly invoked during meditation. Since these *thangkas* were used for secret rituals, only artists who had received formal initiation could paint them (Jackson and Jackson 1984). The artists might work with specific instructions or even under the direct guidance of the monastery. Prior to use, *thangkas* were consecrated with an inscription, usually after the painting was completed.

In preparation of its inclusion in the exhibition *The Circle of Bliss: Buddhist Meditational Art* (5 October 2003 – 4 January 2004, Los Angeles County Museum of Art; 6 February 2004 – 9 May 2004, Columbus Museum of Art, Columbus, Ohio) *Chakrasamvara and Vajravarahi* was examined at the Paper Conservation Laboratory of the Los Angeles County Museum of Art (LACMA). The painting was still in its 1970s mount, a honeycomb panel sandwiched within wooden strainers, with acid-free rag boards on either side, the whole wrapped with bast fiber paper. The *thangka* was fitted with silk brocade borders, in turn adhered to the backing board (Blyth-Hill 1989). As a result of shrinkage of the silk borders, tears had appeared in the painting support. Additionally, considerable loss to the paint and ground layers, with fibers and weave bare in places, was noticed. To avoid further damage and to stabilize the painting, the *thangka* needed to be remounted and appropriate treatment carried out to consolidate flaking paint and in-fill losses.

After removal of the painting from its mount, the primary support was found to consist of two pieces of plain weave fabric joined vertically near the proper left side, the larger part of the support measuring approximately 94 cm in width.

The primary support was still flexible, indicating that no overall consolidation step had been performed in the past, and the weave was relatively intact. Most of the losses of paint and ground were found to correspond to areas of physical manipulation, such as rolling or, as the lines of loss indicate, folding. Some of the patterns suggested by the losses, however, may be related to structural details of the *thangka*, such as flaps or coverings once in place. A detailed discussion of these features is outside the scope of this article.

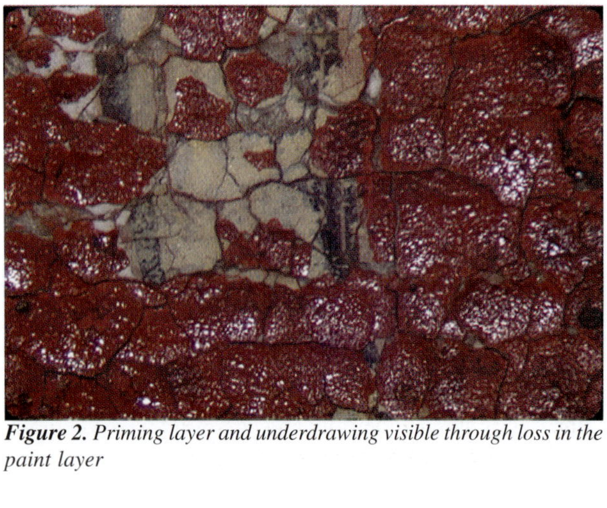

Figure 2. *Priming layer and underdrawing visible through loss in the paint layer*

The conservation treatment proceeded in parallel with a scientific investigation, initially aimed at surveying the range of pigments used in the paintings. As the conservation work continued, and following the indications of the conservators, the aim of the scientific analysis grew to include a study of the stratigraphy of the painting, the identification of translucent coatings selectively applied to areas of the *thangka*, and the identification of glittering particles noticed in some of the yellow painted areas.

Microscopic Examination

Examination of the *thangka* using stereomicroscopy showed that the losses were small and scattered fairly regularly all over the image and that they varied in severity. Some areas of partial paint loss revealed an ocher-colored preparatory layer and faint lines of what seemed to be underdrawing (fig. 2). In many areas, the losses exposed the ground layer and the primary support, with flakes often literally hanging on by a thread. These losses provided excellent sites for sampling without creating further damage to the paint surface.

Microscopic examination with raking light also revealed unexpected details of the painted surface. The first was the presence of a series of coatings in certain specific areas of the image. These coatings appeared to have been preferentially applied to saturate and/or highlight different colors. The most prominent example is in the flaming *prabha mandala* that frames Chakrasamvara and Vajravarahi. The difference in the light and dark areas of the dancing flames is created by the application of a darker, matte coating (compare figs. 3a and 3b). In another example, the skulls in the headdress of Chakrasamvara have a coating applied only to their red eyes and mouths (fig. 4). The black hair on the three-headed Brahma, held in Chakrasamvara's top left hand, also has a coating.

The second surface detail highlighted by the stereomicroscopic examination was the presence of microscopic crystals over part of the painting. At a close visual inspection, in raking light, areas of yellow pigment seemed to have a glittering quality. Subsequent examination at the stereomicroscope revealed the presence of crystals in the faces of the three-headed Brahma as well as in other areas of yellow pigment. These are the yellow face of Chakrasamvara; the body of the water urn in the *khatvanga*; the two leaf-shaped ornaments in the head necklace worn by Chakrasamvara; the hair of Kalaratri; one of the small figures in the lower part of the painting, in the charnel grounds next to Indra; Indra's green-colored attendant; the yellow *Rupini*; three of the flames encircling Chakrasamvara and Vajravarahi; above the upper proper right arm of Chakrasamvara; and three of the *ishtadevata* on the upper register of the painting (for information on the Tibetan terms used, a more detailed list, and references to the deities represented in the painting, see the Glossary, Catalogue entry no. 71, and Appendix 3B in Huntington and Bangdel 2003).

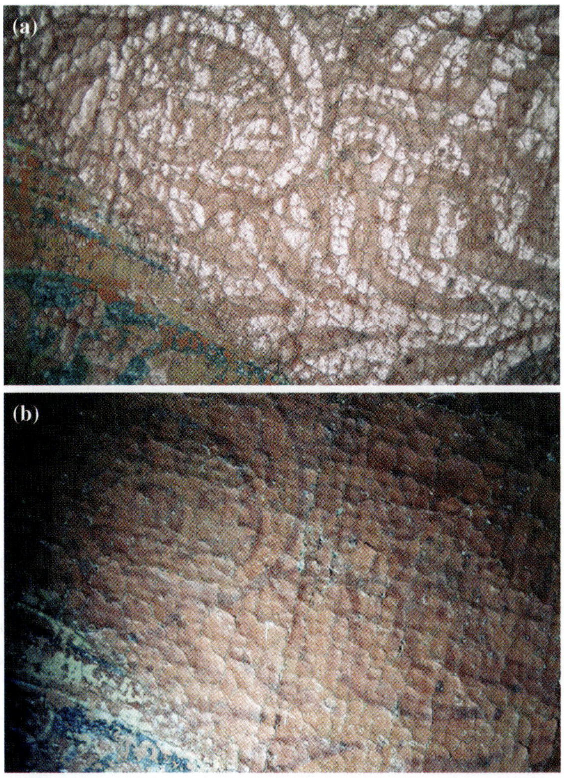

Figure 3 a. *The background of the central part of the painting, in direct light.* ***b.*** *The background of the central part of the painting, in raking light. Comparison with figure 3a highlights the effect obtained by the artists using transparent coatings.*

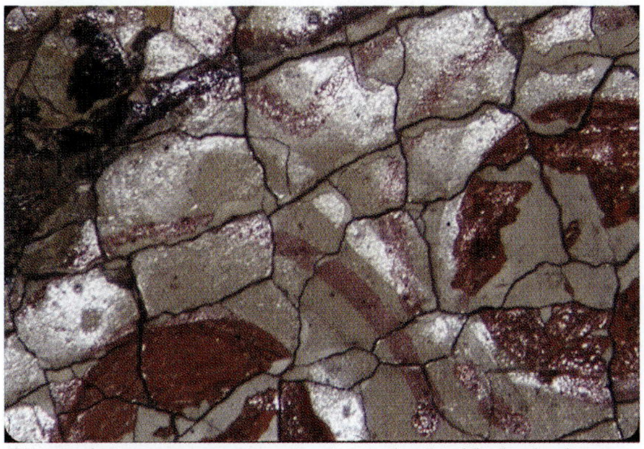

Figure 4. *A transparent coating is used to highlight the features in the garland of skulls.*

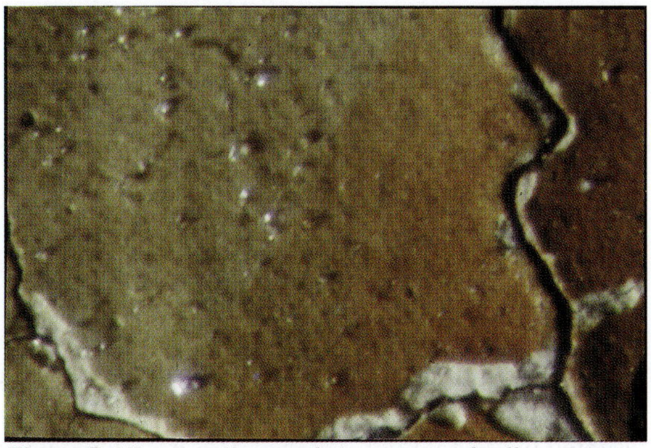

Figure 5. *Microscopic crystals in an area of yellow paint, as seen through the stereomicroscope*

With the exception of the one green area, all of the crystals were in yellow areas, but not all of the yellow areas of the painting showed their presence.

While a precise observation of the morphology of the crystals was not possible at the relatively low magnification of the stereomicroscope (40×), the presence of triangular faces was noted (fig. 5). Art historians, who have appreciated for years the glittering appearance of this and other *thangka*s, have suggested the use of mica (Pal 1984, 1990). Even at the magnification afforded by the stereomicroscope, it was apparent that the crystals were not mica, as their morphology differed conspicuously from that of mica platelets. It thus became extremely important to identify the crystals and to determine what function or origin they had. Additionally, subsequent examination of several other *thangka*s in the LACMA collection and in other public and private collections uncovered the presence of the same crystals, as determined from stereomicroscope examination or from careful visual examination. Six *thangka*s in the *Circle of Bliss* exhibition, spanning in period and origin from thirteenth-century Tibet to nineteenth-century Nepal, were found to present the glittering crystals.[1]

Infrared Reflectography Examination

An important component of the examination of any *thangka* is to determine the location of any consecration inscriptions. If inscriptions are present on the reverse, the object is considered a double-sided painting, a fact that can limit or alter the treatment and display options.[2] In this case, the examination of *Chakrasamvara and Vajravarahi* with the unaided eye revealed no marks or inscriptions on the bottom or reverse of the *thangka*. In many *thangka*s the inscription is beneath the painted surface; in such cases infrared reflectography (IRR) has been used to visualize it. In addition to consecration inscriptions, color notations intended as instructions for the master painter's assistants can often be found on *thangka*s when conducting an IRR examination (Duffy and Elgar 1995).

No writing was found on *Chakrasamvara and Vajravarahi* using IRR. The only marks found were beneath the paint layer, to the right and to the left of the *chattri*, near the blue deities in the red circles. The marks are three "squiggles"—two on the proper right (fig. 6) and one on the proper left. These squiggles may have something to do with a consecration inscription, color notations, or something entirely different. Their purpose remains open for investigation.

In addition to detecting hidden inscriptions, IRR can reveal changes in the underdrawing, reflecting composition changes or adjustments. In the case of *Chakrasamvara and Vajravarahi*, the examination did reveal changes made to the composition during painting. These include changes to the placement of Chakrasamvara's left arms (fig. 7), right arms (fig. 8), and left feet, and Vajravarahi's right hip, as well as Chakrasamvara's and Vajravarahi's faces (fig. 9), and the width of the *prabha mandala*. IRR confirmed the initial impressions about underdrawings that were formed during the visual examination of areas of paint loss.

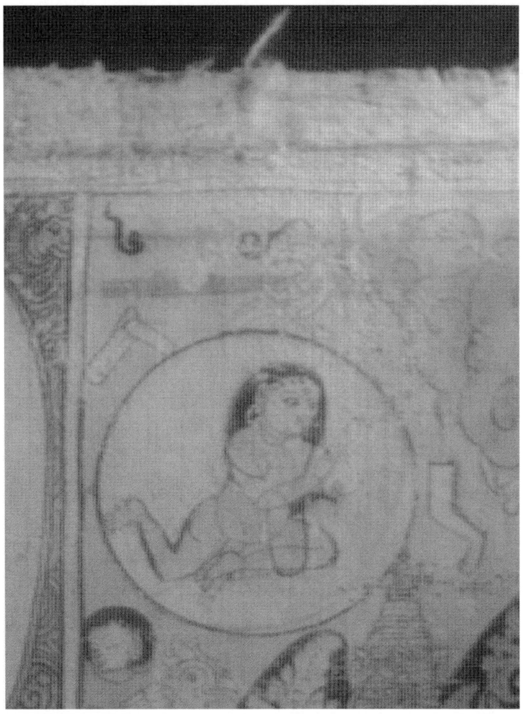

Figure 6. *Marks revealed by IRR in correspondence of the* chattri

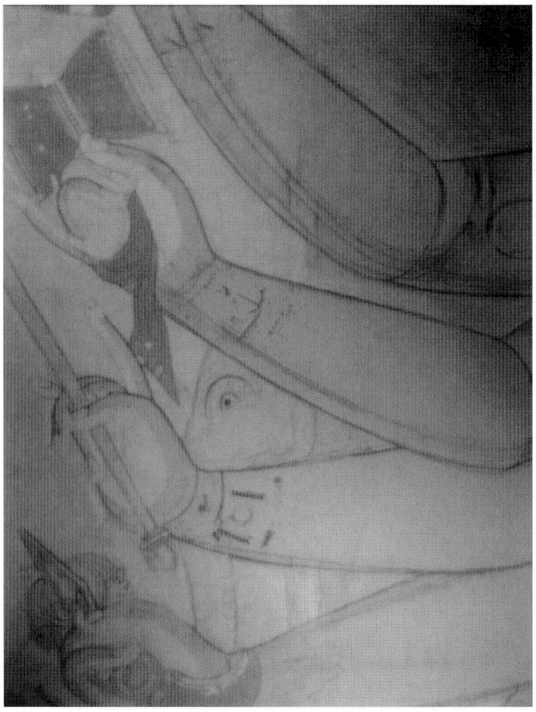

Figure 7. *Changes in the position of the proper left arms of Chakrasamvara*

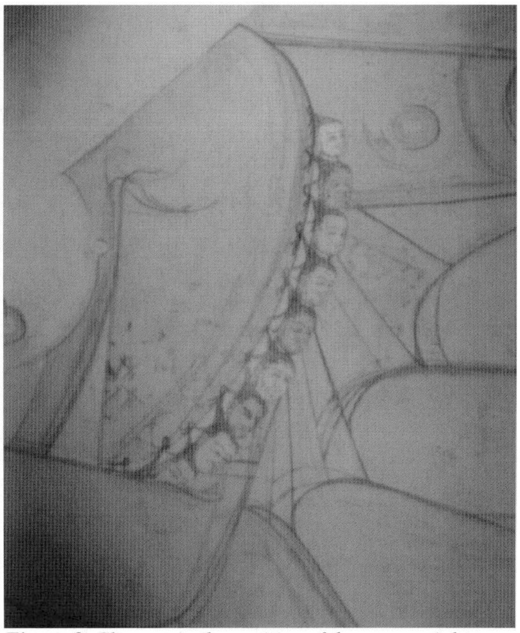

Figure 8. *Changes in the position of the proper right arms of Chakrasamvara*

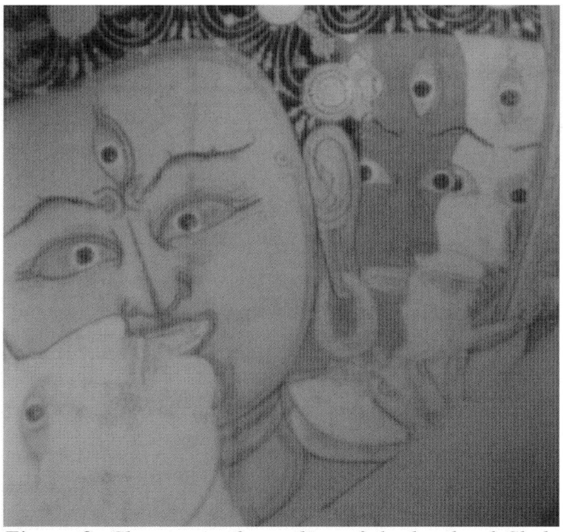

Figure 9. *Changes in the outline of the heads of Chakrasamvara and Vajravarahi*

The observations made by IRR also suggest that the large size of this *thangka* proved to be a challenge for the workshop. Traditional methods include planning and sketching the composition according to iconometric theory. This theory uses strict canons of proportions for specific subjects, which were available to the workshop from examples drawn by the master. The master's drawings were made on a grid background that the workshop scaled up or down to fit the new *thangka*. After the main figures and iconographic elements were sketched and the proportions were checked, the painting began. Since *Chakrasamvara and Vajravarahi* is nearly three times are large as most contemporaneous *thangka*s, it is possible the workshop had difficulty finalizing the design before painting. In particular, the artists had trouble predicting how much room would be required to fully realize the implements Chakrasamvara holds in his hands. Many of the changes create space for these attributes, which are essential to the practice of the Chakrasamvara tantra.

The subject, size, and attention to detail in this *thangka* make it clear that it was very expensive to commission. It would be very strange not to have consecrated this *thangka* for use in meditative practices. It is very possible that an inscription does exist, but we were unable to detect it with IRR.

From the Visual Examination to the Scientific Analysis

To fill the losses in the paint layer, more information was needed about the materials and techniques of the *thangka*'s construction. A precise identification of the grounds, pigments, and binders can assist in determining the most appropriate materials for the fills. The fills must be similar to the original paint, both in terms of color matching and also in terms of mechanical properties. Thus, knowing what the original materials are allows the conservator to formulate a fill material having similar color, granulometry, and adhesion properties to the original, with a microscopic marker added to differentiate it from the original paint under appropriate conditions (for instance, a fluorescent marker, invisible under normal illumination, but clearly evident under UV illumination).

The visual and stereomicroscopic examination revealed a sophisticated treatment of the surface. The artist appeared to have preferentially applied coatings to highlight and saturate different colors. This situation affects conservation choices, as the application of consolidants could potentially unsettle the balance between glossy and matte areas. Further research into the painting's stratigraphy was deemed necessary before decisions could be made regarding the consolidation of the friable paint. Finally, more research was needed on the crystalline material observed on the yellow areas: the nature of the crystals, as well as their origin, needed to be investigated more thoroughly.

Scientific Examination

Methods

Microscopic samples for pigment and media analysis and for cross-section examination were taken with a tungsten needle. Cross-sections were embedded in Bioplast polyester resin, polished on Micromesh cloth, and finally cut with a diamond knife on Reichert Autocut microtome to obtain both thin sections and "microplaned" cross-sections. This treatment proved to be very effective for highlighting clear coatings.

Pigment identification by polarized light microscopy (PLM) and cross-section examination were conducted on a Nikon Microphot microscope equipped with a Nikon DMX 1200 digital camera. Fourier transform infrared microspectroscopy (FTIR) was conducted on a SensIR IlluminatIR system mounted on an Olympus BX51 microscope.

Loose samples, mounted pigments, and cross-sections were analyzed by Raman microspectroscopy (RM) using a Chromex Senturion Raman microscope equipped with a 785 nm laser diode. Spectra were acquired at 2 cm^{-1} resolution in the 150–1650 cm^{-1} spectral range, using 20× and 100× Olympus long working distance objectives.

Scanning electron microscopy with energy-dispersive spectrometry (SEM-EDS) was conducted on a Cambridge Stereoscan 200 system.

Pigment Identification

Pigment samples, representative of the chromatic range of the painting, were taken from various areas, according primarily to ease of sampling. Losses and abrasions were generally chosen as natural areas for sampling, with the exception of some samples for cross-sections, which were taken from areas containing the features required for the investigation. The samples obtained include the following colors: white (matte), white (glossy), blue (light, medium, and dark), green (rich green and dark green), yellow, orange, pink, red, deep red, and black. The ground was sampled in various locations and was also part of some of the cross-sections. Additionally, samples of the clear crystals were taken from the head of Brahma and from the hair of Kalaratri.

Cross-sections were prepared from samples of the red background with floral scroll motif behind Chakrasamvara and Vajravarahi: one section from the outline of the floral scroll (deep red color) and one from the transition from the glossy red background to the matte deep red outline. An additional cross-section was also prepared from a fragment from the yellow and orange Yamaduti, to investigate the use of the deep red glaze over pigments other than red. Sample locations, colors, and pigments identified are listed in table 1.

Overall, with the exception of two pigments never before documented on a *thangka*, the range of pigments found did not differ substantially from what was expected. Interesting variations from what was found by other investigators (Colinart 1995; Duffy and Elgar 1995) were discovered in the way pigments were mixed or applied for particular effects.

Whites

The examination of the ground and of the two whites showed that three different compositions were used. The ground was found by PLM to contain mostly sub-micron-size platelike particles, a morphology typical of clay minerals. Analysis by SEM-EDS showed only sodium, aluminum, magnesium, silicon, and potassium—elements compatible with a white clay. The identification was confirmed by FTIR spectroscopy (fig. 10). The glossy white had an appearance similar to the ground in polarized light, but the FTIR analysis revealed the typical spectrum of kaolinite. Finally, the matte white was shown to contain claylike particles and a fibrous material with appearance and optical properties similar to that of *gofun* or shell white. FTIR analysis confirmed the presence of kaolin and calcium carbonate, while RM analysis, conducted on the fibrous particles within the slide prepared for PLM, revealed the presence of aragonite (as detected by the presence of an intense Raman peak at 1085 cm^{-1} compared to 1088 cm^{-1} for calcite), an indication of the origin of the calcium carbonate from ground mollusk shells. The use of shell white rather than calcite has not previously been documented for *thangka* paintings.

Table 1. *List of samples and pigments identified*

Color	Location	Description	Identification	Methods[a]
Ground	Various areas	Off-white layer	White clay	PLM, FTIR, SEM-EDS
White (matte)	Decoration over throne of Mahasiddha Luipa (second throne-seated figure from top, proper right charnel ground)	Matte white paint	Calcium carbonate (*gofun*) and kaolin	PLM, RM, FTIR, SEM-EDS
White (glossy)	Toenail of elephant skin, proper right upper edge of the flayed skin	Smooth and glossy white	Kaolin	PLM, FTIR, SEM-EDS
Light blue	Acanthus leaf from mandala	Light blue paint with smooth surface	Indigo	PLM, RM
Medium blue	Throne petal	Grainy surface	Azurite	PLM, FTIR, RM
Dark blue	Shaded portion of throne petal	Grainy surface	Indigo over azurite	PLM, FTIR, RM
Rich green	Tree in middle of proper right charnel field	Rich green color	Indigo and orpiment	PLM, RM
Dark green	Throne petal	Heavily bound paint	Indigo, orpiment, and copper chlorides	PLM, RM, FTIR, SEM-EDS
Yellow	Yellow and red Yamaduti	Rich yellow	Orpiment	PLM, RM
Orange	Identified in cross-sections as priming layer	Priming layer	Iron oxide red and vermilion	PLM, RM
Pink	Near sampling area for matte white	Mauve stain over white paint	Organic red	PLM
Red	Background near Chakrasamvara's proper right foot	Flat red (with glossy coating)	Vermilion	PLM, RM
Deep red	Background near Chakrasamvara's proper right foot	Glazelike red	Organic glaze over vermilion layer	PLM (cross-section)
Black	Head of Brahma	Glossy black	Lampblack	RM
Crystals	Various locations (see text)	Highly light scattering	Arsenolite and sulfur	RM, SEM-EDS
Cross-section 1	Same as red sample	Deep red	Stratigraphy: ground, orange, red, glue varnish	
Cross-section 2	Same as deep red sample	Glossy red and matte deep red	Ground, orange, red, glue varnish, areas of red glaze	
Cross-section 3	Red and yellow Yamaduti	Red over yellow	Opaque red glaze over orpiment	

[a]PLM: polarized light microscopy; FTIR: Fourier transform infrared microspectroscopy; SEM-EDS: scanning electron microscopy–energy-dispersive spectrometry; RM: Raman microscopy

Blues

Examination of the light blue areas and of the medium and dark blue areas (the latter really being a dark blue shading of a medium blue layer) revealed the use of indigo for the light blue, of azurite for the medium blue, and of indigo over azurite for shading (the pigments were identified by PLM and RM, with the characteristic Raman spectra of indigo and azurite shown in fig. 11).

Greens

Two types of green paint were used. A rich green color, used in trees and leaves, was found to be a mixture of indigo and orpiment. A darker green, used to depict the throne petals, was found to contain, in addition to the mixture of indigo and orpiment, characteristic spherulitic particles. The presence of copper was determined by SEM-EDS and the FTIR spectrum identified the pigment as a mixture of the copper chlorides, atacamite, and paratacamite (fig. 12). The use of these chlorides as pigments in Japanese paintings has been previously described (FitzHugh 2003), and the appearance of the pigment on the *thangka* matches those descriptions. The use of copper chlorides in *thangka* painting has not been previously reported.

Yellow

The only yellow pigment present on the painting is orpiment, as identified by RM (characteristic peaks at 154, 202, 292, 309, 353, and 381 cm^{-1}).

Orange

An orange pigment is used both for details and as a priming layer (visible in all cross-sections). RM of the priming layer in the cross-section was hampered by fluorescence due to the clay ground. Using a new technique to suppress the interference from sample fluorescence—shifted excitation Raman difference spectroscopy (SERDS) (Zhao et al. 2002)—the components of the orange layer were identified as vermilion and iron oxide (fig. 13).

Reds and Pink

The main red pigment is vermilion, but an organic red (tentatively identified as such because of the crimson color and stainlike appearance in PLM) is used for pink-purple details (painted over white) and as an overglaze (painted over the mineral red), to obtain a deep red. This deep red is especially visible in the background. The deep red glaze will be discussed in more detail contextually with the discussion of the stratigraphy of the painting.

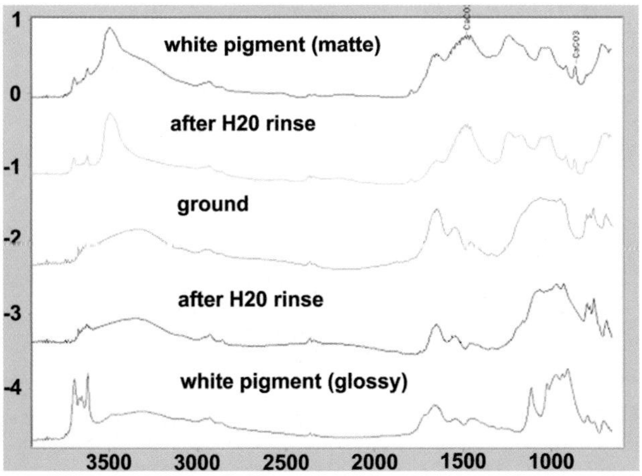

Figure 10. *FTIR spectra of whites: the spectra of the matte white and of the glossy white are shown before and after treatment of the samples with hot water to remove the binding medium, together with the spectrum of the ground. Kaolin is evident in the glossy white, while a generic clay and the same clay plus calcium carbonate were detected in the ground and in the matte white, respectively.*

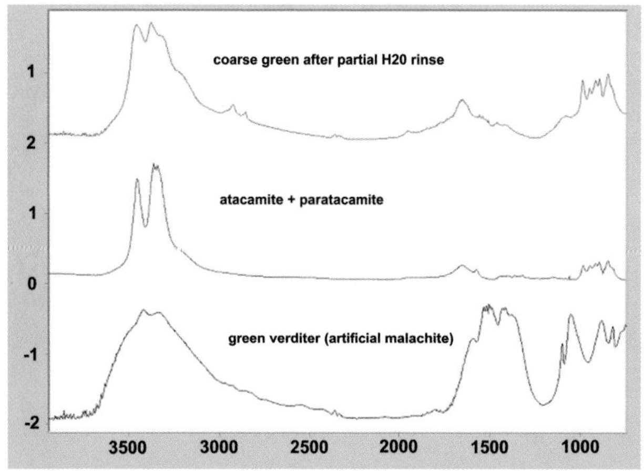

Figure 12. *FTIR spectrum of green pigments. Top: sample from* Chakrasamvara and Vajravarahi. *Middle: a reference spectrum of an atacamite plus paratacamite mixture. Bottom: green verditer (the pigment has a similar morphology to the chloride pigments observed in the sample, yet the IR spectrum allows clear differentiation)*

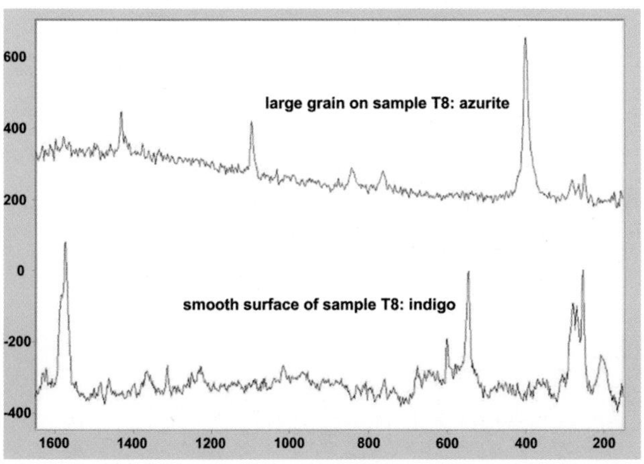

Figure 11. *Raman spectra of blues: the upper spectrum was obtained from a coarse grain, and the lower from the shaded part of the sample, using a 100× objective. Top: azurite; bottom: indigo*

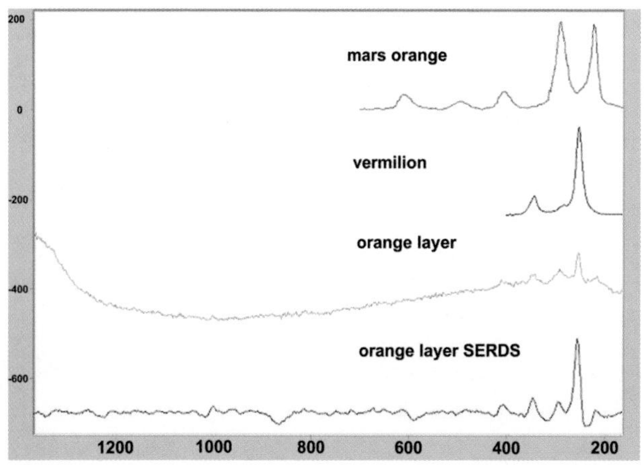

Figure 13. *The spectrum obtained from the priming layer on cross-section 1 suffered from interference due to the fluorescence of the clay in the ground. Use of SERDS allows identification of vermilion and an iron oxide red.*

Black

The pigment responsible for the glossy black paint is lamp-black, as identified by RM (broadband at 1580 cm^{-1}).

Binding Medium and Surface Treatment

The binding medium used throughout the painting, both for the paint layers and for the ground, is animal glue, as evidenced by FTIR. Besides its use in the formulation of the paint, glue was found to be present in the painting as a varnish and as a vehicle for an organic red used as a glaze. The use of a glue varnish and of the red glaze are the most salient characteristics of this *thangka*. Thin coatings of pure glue are responsible for the increased gloss of some details (see fig. 4), while a combination of glue varnishing and glazing is responsible for the matte-glossy contrast in the background. The sophisticated manipulation of the optical texture is clearly illustrated by the cross-sections in figures 14 and 15. The cross-section in figure 15 is taken from the background within the *prabha mandala*, in correspondence of the transition between red background and deep red outline (see figs. 3a and 3b). A clear coating over the mineral red makes the background layer glossy, while the addition of a crushed mineral (a mixture of silicates, as evidenced by FTIR attenuated total reflection [ATR] analysis on the deep red layer of a cross-section) gives body to the glaze, making it more opaque and deepening the red color (the figure shows both the brightfield reflected light image and the fluorescent light image, to assist the viewer in identifying the separation between the glaze and varnish layers). The use of an organic red layer over vermilion has been described by Duffy and Elgar (1995), who also identified the color notation used for it (*ma* for *dmar po* [dark red] as opposed to *ga* for *rgya mtshal* [bright red]). The layering sequence observed here, however, and the addition of ground minerals to the red glaze have not been reported previously.

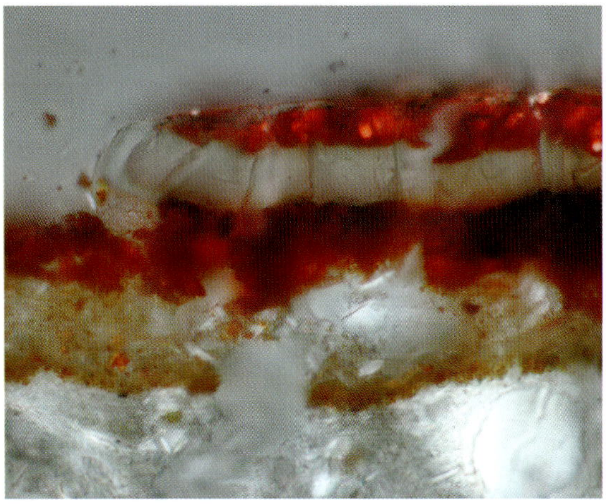

Figure 14. *The thin section (microplaned from a cross-section) shows the stratigraphy (ground, orange priming layer, mineral red, glue varnish, and organic glaze) and highlights the addition of ground-up clear mineral (highly scattering irregularly shaped fragment) to the glaze. Transmitted light, 1,000× original magnification (oil immersion objective)*

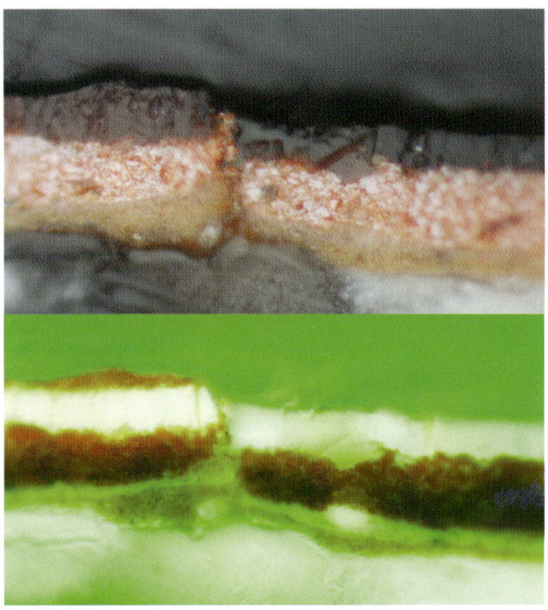

Figure 15. *The sample was taken in an area of transition between the red base and the dark red outline of the scroll decoration on the background of the two central figures. The use of glue varnish and of the deep red glaze to create a contrast between glossy and matte areas is evident when the brightfield image (top in the composite) and the fluorescent image (bottom) are compared. Reflected brightfield and fluorescent illumination, 40× original magnification*

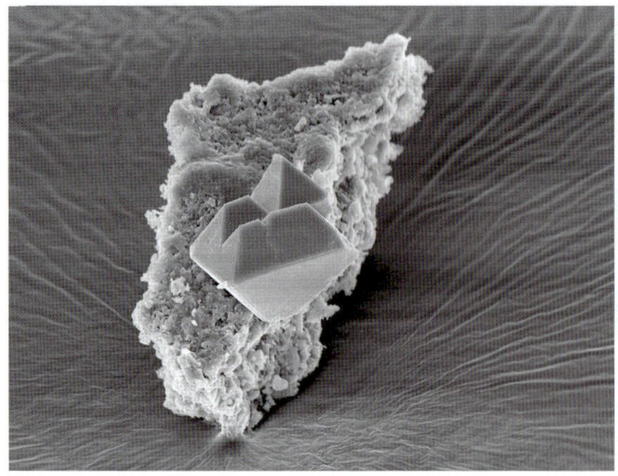

Figure 16. *A relatively large euhedral arsenolite crystal*

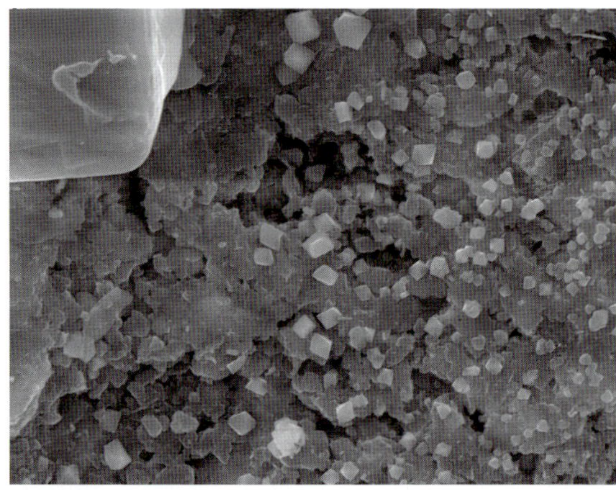

Figure 17. *A number of* in situ *grown arsenolite crystals*

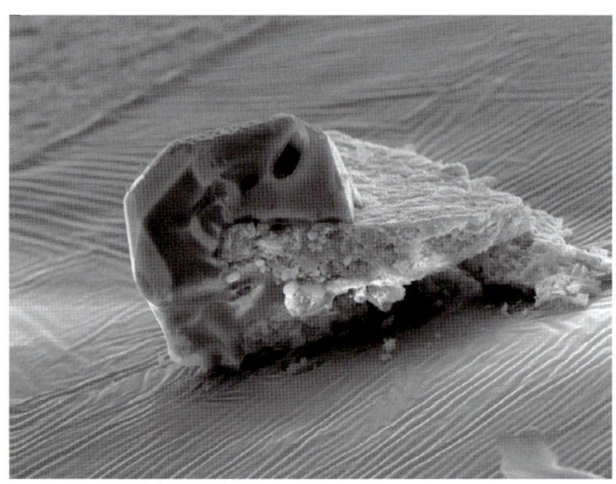

Figure 18. *Sulfur crystal*

The Glittering Crystals

Stereomicroscopic examination of the *thankga* revealed that crystals on the surface of some of the yellow painted areas (and one painted in green) were responsible for the glitter effect observed macroscopically. Their morphology, even as observed at the limited magnification afforded by the stereomicroscope, clearly hinted at something other than mica. Removal of some of the crystals and examination by SEM and RM led to their identification as arsenolite, the cubic form of arsenic trioxide. The Raman spectrum of arsenolite is characterized by peaks at 184, 268, 370, 417, 470, 560, and 780 cm^{-1}. The crystals are euhedral and display octahedral morphology; their triangular faces are responsible for the characteristic reflection, and the high refractive index of arsenolite adds to its effectiveness in generating the glitter effect. In addition to the octahedral crystals, occasional larger crystals displaying a more complex morphology were also found. They were identified as sulfur by SEM and RM (Raman peaks at 187, 216, 219, 247, 438, and 474 cm^{-1}).

Both types of crystals appear to have formed *in situ*, as evident from the electron micrographs (figs. 16–18). The absence of a meniscus around the edges of the crystals, the variety of sizes, and the relation of each crystal to the substrate exclude the possibility that the sulfur and arsenolite had been applied to the painted surface (their unbroken appearance is also in sharp contrast to the smooth surface of the burnished paint layer). The presence of sulfur and arsenolite is consistent with the documented photodegradation reaction affecting orpiment, with oxidation of the sulfide (generally to sulfate) and conversion to the oxide. More interesting, while this reaction is generally known to yield large white masses of microcrystalline arsenolite with disfiguring results for works of art (Daniels 1983; Leach and Green 1995; Green 1995; Trier 1972), in the case of this painting the reaction rate appears to be extremely slow, so that equilibrium growth is predominant. The presence of sulfur crystals would seem to hint at a relatively oxygen-poor environment for the orpiment crystals: it is possible that the glue used to bind the paint provided some protection from the atmospheric oxygen. Indeed Leach and Green (1995) describe experiments in which gum arabic–based orpiment paints subjected to accelerated aging did not degrade if coated with isinglas.

A likely explanation for the present appearance of the *thangka* would see the glue binder play two roles: that of a protective, relatively oxygen-impervious coating, and that of a gel (when swollen under high humidity condition) in which the slightly water-soluble arsenic trioxide could diffuse. Upon fluctuations in the relative humidity level and moisture content of the paint layer (witness the ubiquitous water damage signs evident in almost all *thangkas*), the arsenic trioxide would effloresce to the surface. The crystals thus formed would be relatively few and large, leading to the observed glitter phenomenon, rather than the white powdery deposits described by Trier (1972) or Green (1995). It is interesting that Green also noted a "sparkly" (1995, 89) appearance of some of the orpiment in one of the papyri she examined but attributed it to lathlike cleaved particles of orpiment.

It is important to note that freshly abraded areas of the yellow paint did not show a higher concentration of white crystals compared to adjacent glue-varnished areas, as would be expected if the reaction rate for the degradation process were so high as to pose a concern for the current well-being of the painting. Certainly though, light exposure for *thankga* paintings should be carefully monitored and limited to avoid additional damage.

Conclusions

What started as two separate projects—the conservation examination and treatment and a technical art-history study of materials—immediately became a highly integrated research project, with the direction of the scientific analysis strongly influenced by the conservators' needs and observations.

By applying noninvasive and microsampling techniques, it was possible to completely characterize the pigment range of the painting, including pigments such as *gofun* and copper chlorides that had never been previously documented on Himalayan paintings on cloth. Moreover, the application of a technique typical of the examination of easel paintings—that is, the study of cross-sections—allowed clarification of the stratigraphy of the painting and illustration of a very sophisticated manipulation of optical texture through the use of contrasting glossy and matte coatings. The analytical results directly influenced the art-historical perception of this and other *thangkas*, redressing long-standing and erroneous ideas concerning the use of mica to obtain a glitter effect: it is quite clear that a natural photodegradation of the orpiment pigment is responsible for the glitter occasionally observed in *thangka* paintings. Likewise, the documentation of the use of coatings to modulate the visual appearance of the painting was a determining factor in the conservators' choice of a consolidation treatment, dictating the use of localized microinjections instead of broadcast consolidant application.

Acknowledgments

We wish to thank Victoria Blyth-Hill, who originated this project, pointed out the need for a thorough scientific study of the *Chakrasamvara and Vajravarahi*, and lent her guidance and support throughout the examination and treatment of the *thangka*. Several other members of the Conservation Center at the Los Angeles County Museum of Art were involved in the treatment and contributed countless hours of work and precious observations: Soko Furuhata, Chail Norton, Liz Werden, Grace Jan, and Terry Schaeffer. Colleagues from other institutions were likewise generous with their time and commented on the orpiment deterioration phenomenon: these were John Winter, Elisabeth FitzHugh, Richard Newman, John Twilley, and Richard Ernst. To all of them, our sincerest gratitude. Lastly, we would like to thank the Andrew W. Mellon Foundation for its support of scientific research at the Conservation Center of LACMA and the Irvine Foundation for establishing the Camilla Chandler Frost Internship in Conservation, which supported Sandhya S. Jain's work at LACMA.

Notes

1. Richard Ernst (personal communication 2003) also noticed the crystals in two *thangkas* in a private collection; two additional *thangkas* in LACMA's collection were found by the authors to present the phenomenon, as well as one on display at the San Francisco Museum of Asian Art.
2. For a more complete discussion of ethical, theoretical, and technical considerations in the conservation of *thangkas*, the reader is referred to articles by Huntington (1970), Mehara (1970), Shaftel (1991), the book by Agrawal (1984), and a review on the history of the development of *thangka* conservation by Blyth-Hill (1993). An updated survey of *thangka* conservation principles and practices is offered by Blyth-Hill (1996).

References

Agrawal, O. P. 1984. *Conservation of manuscripts and paintings of South East Asia*. London: Butterworths.

Blyth-Hill, V. 1989. Conservation Center condition report, 7/89. Los Angeles County Museum of Art, Los Angeles, Calif.

Blyth-Hill, V. 1993. A history of thangka conservation in Western collections. *WAAC Newsletter* 15(2):17–21.

Blyth-Hill, V. 1996. The conservation of thangkas. In *On the path to void*, ed. P. Pal., 270–81. Mumbai: Marg Publications.

Colinart, S. 1995. Un pas vers le Tibet ou dix thang-ka peints du muse Guimet. *Techne* 2:93–104.

Daniels, V. 1983. Chinese realgar figurines: A study of their deterioration and method of manufacture. *Masca Journal* 2(6):170–73.

Duffy, K. I., and J. A. Elgar. 1995. An investigation of palette and color notations used to create a set of Tibetan thangkas. In *Preprints: Historical painting techniques, materials, and studio practice*, ed. A. Wallert, E. Hermens, and M. Peek, 78–84. Symposium, University of Leiden, The Netherlands, 26–29 June 1995. Los Angeles: Getty Conservation Institute.

FitzHugh, E. W. 2003. Pigments in Japanese ukiyo-e paintings in the Freer Gallery of Art. In *Scientific research in the field of Asian art: Proceedings of the First Forbes Symposium at the Freer Gallery of Art*, ed. P. Jett et al., 150–55. London: Archetype Publications.

Green, L. R. 1995. Recent analysis of pigments from ancient Egyptian artefacts. In *Conservation in ancient Egyptian collections*, ed. C. E. Brown, F. Macalister, and M. M. Wright, 85–91. London: Archetype Publications.

Huntington, J. 1970. The iconography and structure of the mountings of Tibetan paintings. *Studies in Conservation* 15(3):190–205.

Huntington, J., and D. Bangdel 2003. In *The circle of bliss: Buddhist meditational art*. Columbus, Ohio: Columbus Museum of Art and Serindia.

Jackson, D. P., and J. A. Jackson. 1984. *Tibetan thangka painting: Methods and materials*. Boulder: Shambhala.

Leach, B., and L. Green. 1995. Removal of unsuitable linings from illustrated papyri: An investigation into suitable consolidants and facings. In *Conservation in ancient Egyptian collections*, ed. C. E. Brown, F. Macalister, and M. M. Wright, 29–35. London: Archetype Publications.

Leona, M. 2003. Appendix 3B: Conservation report on the chemistry of the sparkles in the surface of the Ratnasambhava painting (cat. 14). In *The circle of bliss: Buddhist meditational art*, J. Huntington and D. Bangdel, 512–13. Columbus, Ohio: Columbus Museum of Art and Serindia.

Mehara, V. R. 1970. Note on the technique and conservation of some thang-ka paintings. *Studies in Conservation* 15(3):206–14.

Pal, P. 1984. *Tibetan paintings: A study of Tibetan thangkas, 11th to 19th centuries*. London: Sotheby Publications.

Pal, P. 1990. *Art of Tibet: A catalogue of the Los Angeles County Museum of Art Collection*, expanded ed. Los Angeles: Los Angeles County Museum of Art, in association with Mapin Publishing.

Shaftel, A. 1991. Conservation treatment of Tibetan thangkas. *Journal of the American Institute for Conservation* 30(1):3–11.

Trier, J., 1972. *Ancient paper of Nepal*. Vol. 10. Copenhagen: Jutland Archaeological Society Publications.

Zhao, J., M. M. Carrabba, and F. S. Allen. 2002. Automated fluorescence rejection using shifted excitation Raman difference spectroscopy. *Applied Spectroscopy* 56(7):834–45.

Kalighat *Pats*: An Examination of Techniques and Materials

Christine Mackay and Aditi Nath Sarkar

ABSTRACT Kalighat *pats* were produced in India in the thousands through most of the nineteenth century and were the most popular art form of their time. Viewed now with the perspective of time, they appear to have a greater affinity with the art movements of the future than those of the past, yet they were the work of migrants from an impoverished caste of picture showmen. The few people to have written about them in depth have considered two prime factors essential in their genesis and development: the advent of machine-made paper and the availability of prepared watercolor pigments. However, the investigation of the history of the early papermaking industry in colonial Bengal and the analysis of some of the pigments suggest the painters' selection of materials was more complex than previously assumed.

Introduction

During the last decade of the eighteenth century, the Belgian artist F. B. Solvyns (1760–1824) embarked on an ambitious project to record life in the native quarter of Calcutta, preparing *A Collection of Two Hundred and Fifty Coloured Etchings: Descriptive of the Manners, Customs and Dresses of the Hindoos*. He published the first plates in 1796; included is a "Puttooa" (a village artisan) (fig. 1). Leaving Calcutta in 1803, Solvyns added text to the volumes printed in Paris from 1808 to 1812, and his entry for the *patua* reads:

> The Hindoo painters are also sculptors; they carve[1] and colour the images of their Gods, and, as there is a great consumption of them, both because they are everywhere exposed to public view, and because some of their feasts terminate by throwing them into the water, the *Puttooas* [*sic*] never are in want of employment. They also make toys for children [fig. 2], of earth which they dry in the sun, and sell in the bazaars, particularly on their feast days. Their varnishes are much superior to ours and are very well suited to their climate. . . .
>
> There are also *Puttooas* who make only pictures and drawings, but always upon the same subjects; M Solvyns having never seen any other thing represented than their deities. They have some good copiers, but that is the utmost extent of their art; if some original painters are met in Hindoostan, they are Mahometans, Persian, Greeks, and other foreigners.
>
> The *Puttooas* are degraded to the class of *Parriahs*, because they handle grease and all sorts of colours. It might be added too that they seem to wish to merit their appellation by their dissolute conduct: consequently there are few true Hindoos among them. (Solvyns 1808–12, Vol. 1, No. 11, Pl. [5])

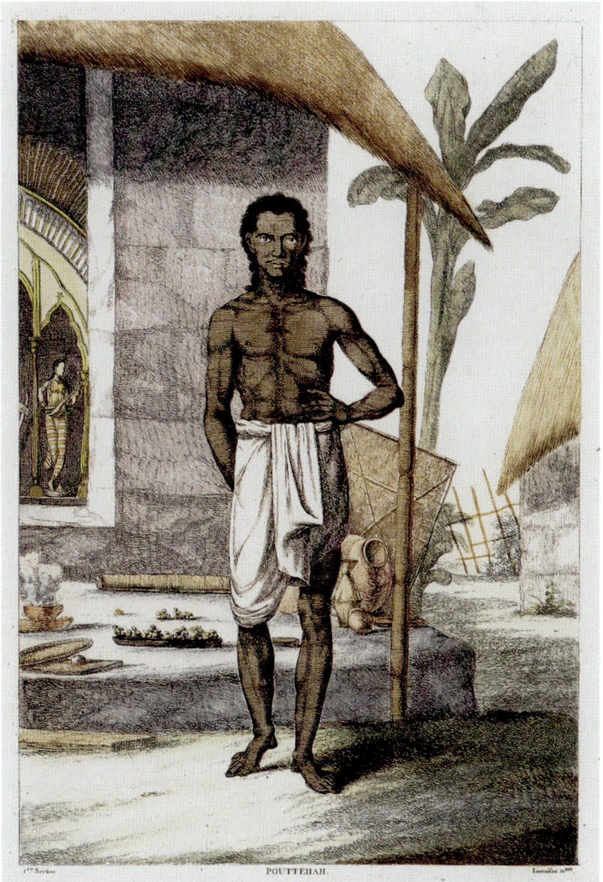

Figure 1. *F. B. Solvyns,* Pouttetah. A Collection of Two Hundred and Fifty Coloured Etchings: Descriptive of the Manners, Customs and Dresses of the Hindoos *(Calcutta: Mirror Press, 1796–99), 159. Reproduced by permission of the British Library, Oriental Collections, X 803.*

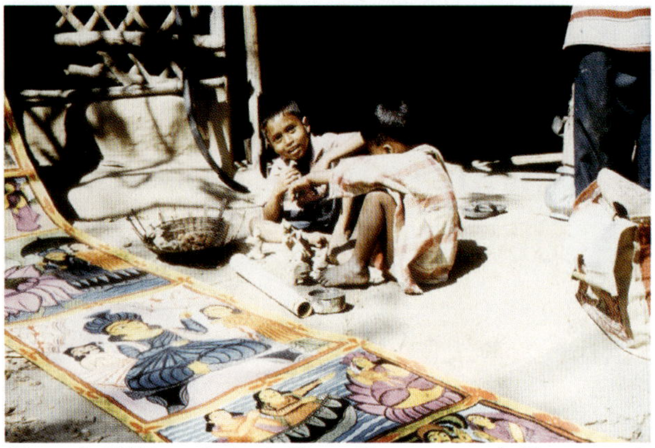

Figure 2. *Children playing in Naya village, surrounded by toys made from air-dried clay and a recently completed scroll*

About eighty years later, T. N. Mukharji, an official of the India Museum, Calcutta, wrote:

> Until recently a superior kind of water-colour paintings were executed in Bengal by a class of people called the *patua*s, whose trade also was to paint idols for worship. These paintings were done with minute care and considerable taste was evinced in the combination and arrangements of colours. The industry is on the decline owing to cheaper coloured lithograph representations of Gods and Goddesses turned out by the ex-students of the Calcutta School of Art having appeared in the market. A painting in the old style can still be had, by order, at a price of Rs 10 and upwards. The *patua*s now paint rude "daubs" which are sold by thousands in stalls near the shrine of Kalighat in the neighbourhood of Calcutta as also in other places of pilgrimage and public fairs. The subjects are usually mythological, but of late they have taken to making pictures representing a few comical features of Indian life. Such pictures are generally sold at a price ranging from a pice to an anna. (Mukharji 1888, 20)

Solvyns witnessed the arrival of the *patua*s to the city in order to gain a living, and Mukharji describes the reason for their decline, but during the intervening period, when an extraordinary innovative art—paintings called *pat*s—erupted in the city with tens of thousands of pictures being sold, it was all but ignored by the social and artistic commentators of the time.[2] Not until the second half of the twentieth century was the contribution of the Kalighat *patua*s assessed and their legacy to Bengali art evaluated. Their pictures were bought by rich and poor alike. The most avid purchasers were the pilgrims to the shrines of Calcutta who could for the first time afford an image of their chosen deity to worship at home. Few examples survive in India today, as the tropical climate destroyed the inherently weak support on which they were painted. Most of the sizable collections are now in Western museums, brought back by Europeans, the first datable examples being the Underhill Collection in the British Library, purchased before 1859. Unfortunately, none of these nineteenth-century collectors left a record of where or why they made their purchases.[3]

The Scroll Painters

According to Coomaraswamy (1929), picture showmen such as the *patua*s in Bengal have an ancient ancestry in the history of Indian arts. The earliest mention of picture showmen in Indian texts is in the *Bhagavati Sutra* of the Jaina sect in which the rival religious leader Gosala is called *mankhaliputta*, or "the son of one whose hand is kept busy by the picture board," which Basham (1951) interprets as an exhibitor of religious pictures and a singer of religious songs. The Jaina sect predates the foundation of Buddhism in India, suggesting picture showmen were part of the culture by 600 B.C.E.

The *patua*s of West Bengal are professional painters or *chitrakar*s (picture makers) whose occupation is to paint stories on rectangular sheets of paper (*pata*s) stitched together to form a vertical scroll, which they take from village to village, and especially to the markets and fairs, where, squatting on the earth, they unroll the sequential frames while reciting the narrative of the story in a high-pitched chant. Their designs and songs are commonly owned; once a song is memorized, it may be sung by anyone. While modern themes are now included in the repertoire, the great Hindu epics like the *Ramayana* and the *Mahabharata* have never lost their popularity. Their art is complete only when it is performed, and, never having been a landowning caste, they entertain in exchange for uncooked rice, lentils, or vegetables and sometimes for money.

The *patua*s survive to this day centered in rural areas like the Medinipur district, especially around Tamluk, as well as in Birbhum, Bankhura, Haora, and the Twenty-Four Parganas (fig. 3). As Solvyns observed, they were on the fringes of society; the *Brahmavaivartta Purana*, a text of the thirteenth century, indicates that the *chitrakar* caste had been one of the nine "pure" craftsmen castes but had been expelled for deviation from canons in their religious imagery. The *patua*s have interesting religious affiliations, almost all having two names, one Muslim and one Hindu. They perform Muslim birth and death rites but also worship the local Hindu village deities and make images of Hindu gods and goddesses. It is likely that the *patua*s originally became Muslims, like many other lower castes, to avoid paying taxes levied on non-Muslims by the Islamic rulers. But whereas other converted groups have become wholly absorbed into Islam, the *patua*s remain betwixt and between as they continue to ply their "idolatrous" craft.

Originally, the *pata*s were painted onto fabric (*patta*); today the *patua*s use machine-made paper, which they often back with worn-out clothing like saris to withstand the thousands of times the scrolls are rolled and unrolled. The painting of the scroll is a family affair: the male *patua* designs the main part, while women and children fill in the colors and borders, leaving no part of the paper unpainted (fig. 4). Although admitting to the occasional use of manufactured pigments like vermilion or zinc white, mostly they continue to use traditional pigments: powdered burnt rice or the burnt ground roots of the *Gaab* tree, *Diopyros peregrina*, for black; scrapings from the interior of terracotta ovens for browns; powdered turmeric for yellow; the extract of bean leaf (either *Dolichos lablab* or *hinche*) for green

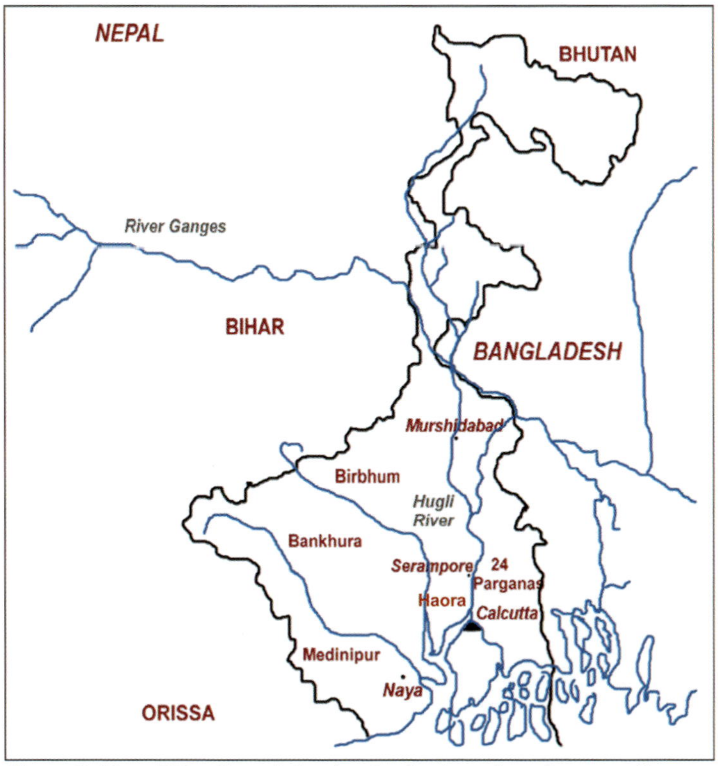

Figure 3. *Map of West Bengal showing the districts where the* patua*s live*

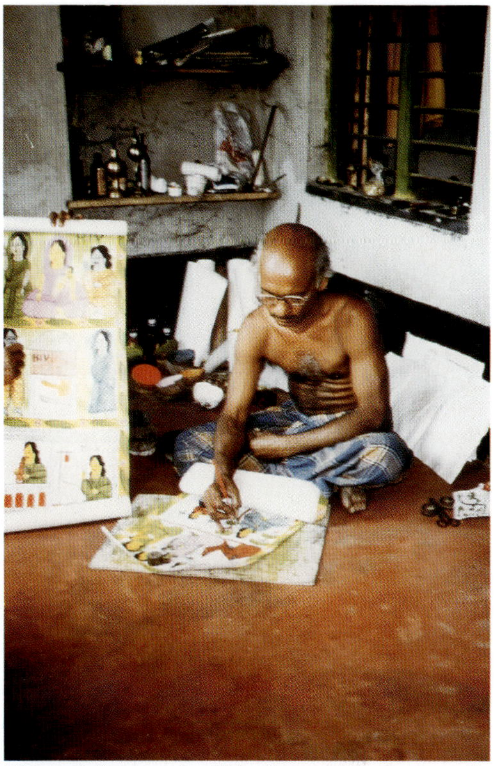

Figure 4. *Syamsundar Chitrakar working on a scroll*

(Palchoudhury 1988). The blues come from the *nilmoni* flower or *aparajita* berries and occasionally indigo. Reds are prepared from the hibiscus or the *palash* flowers, while purple is the result of mixing mud from the Ganges with the ground seeds of *pui saag*, a member of the spinach family, the flowers of this plant providing a very delicate pink.[4] The *patua*s make the colors opaque with the addition of ground conch shell, and for the binder they usually use gum from the *Acacia catechu*, which belongs to the same taxonomic class as *Acacia senegal*. Another binder is made from the strained, pressed fruits of the *Bael* tree, *Aegle marmelos*, referred to by the British as the Bengal quince (Usher 1974). Of interest is the instruction of English painter, Thomas Gainsborough (1727–1788) to his fellow artist Ozias Humphry (1742–1810), in a letter dated 24 March 1771 on the preparation of paper for varnished watercolors, "If you want to paint on Paper you must prepare yr paper first; either with thin Gum Arabic Water; or [which] is better the Mucilage of Quince Seed" (quoted in Sloman 2002, 111).

The Urban Artist

It was almost certainly the widespread famines of the second half of the eighteenth century as well as the political upheavals after the Battle of Plassey (1757) that led some *patua*s to attempt to take advantage of the rapidly growing economic opportunities in the city of Calcutta. It is not known when they discovered that "performance art" was not viable in the urban hustle and that to survive they needed to produce easily portable pictures for instant sale (fig. 5). However, they would have seen around them that "the commonest bazaar is full of prints—and Hodges' *Indian Views* are selling off by cart loads," as the artist Thomas Daniell (1749–1840) wrote in 1788, being another correspondent of Humphry (quoted in Hardie and Clayton 1932, 26).

Archer (1971) suggests while the prime purpose of the East India Company in developing Calcutta was for trade. Industrial enterprises were also started, and among the commodities that came to be manufactured was an article of special importance to these artists—paper that was both cheaper and thinner than the indigenous handmade paper. Without this steady supply, Kalighat paintings could hardly have come into existence, claims Archer. Yet recent investigation by the authors into the history of early paper mills appears to contradict this statement.

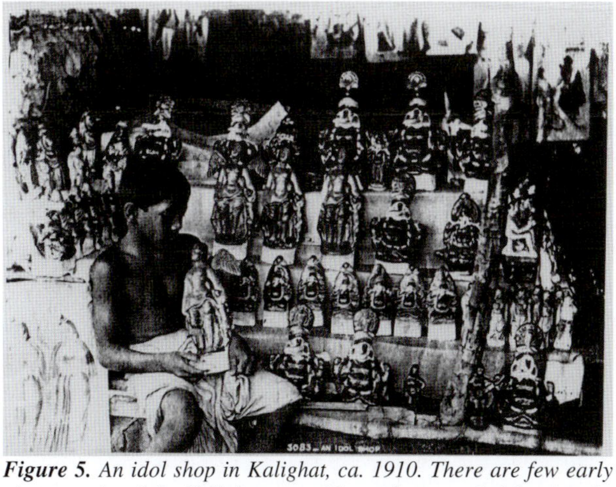

Figure 5. *An idol shop in Kalighat, ca. 1910. There are few early photographs of the Kalighat area. The authors are indebted to the late R. P. Gupta, a great authority on Kalighat* pats*, for permission to copy this postcard.*

For a city that took to the printing press with such enthusiasm after its tardy introduction in 1777, it is extremely surprising to read that more than a hundred years later, on Tuesday, 11 March 1884, the principal newspaper, the *Statesman* "had been printed for some weeks past on paper made at the Bally Mills and we congratulate Messrs Henderson and Co and the country on their success. The time cannot be far distant when India will produce for herself a large portion of the 'goods' which she now imports from Europe" (quoted in Ray Choudhury 1988, 85). According to Shaw (1981), probably no other problem of materials so vexed these early printers as much as that of paper. Two kinds were commonly in use: Indian handmade paper usually referred to as Patna paper after one of the chief centers of manufacture, characterized as "a dingy, porous, rough substance" or the imported "Europe" paper, higher in quality, but also higher in price (Shaw 1981, 35).

By 1790 there was a mill in Calcutta advertising for sale "paper, far superior to that made at Patna and of various sizes" (Shaw 1981, 36) but the business does not seem to have prospered. A more enduring mill was set up at Serampore in 1809, 48 kilometers north of the city, with the specific purpose of supplying the Serampore Baptist Missionary Press, founded by William Carey and his fellow missionaries in 1800. When working toward his plan to translate and print the Bible, Carey bemoaned the lack of any paper suitable for printing except for imported paper, which was 200% more expensive in Calcutta (Carey 1836). Production must have been slow, as he was still asking for shipments of paper in 1812. Shortly, however, an engineer, William Jones (known as "Meckanick" or "Guru" Jones to distinguish him from William "Oriental" Jones, the founder of the Asiatic Society), dramatically improved output at the mill by introducing steam power in 1820—the first use of steam in India for any kind of manufacture. But demand continued to outstrip supply (Ghosh 1990).

The East India Company controlled all the trade, and winning one of its contracts was highly lucrative. In 1782, Jacob Bosanquet became chairman of the company; one of his brothers-in-law happened to be James Whatman the younger. It became "one of the primary objects of Mr Whatman's early attention to supply such paper as would stand the Indian climate in exclusion of the Dutch papers which till then the Company had exclusively used" (Balston 1957, 111). He was successful, as during the first few years of the nineteenth century, supplying the company accounted for a large percentage of the mill's production; by 1814 his order was worth over £3,000. Other European suppliers vied with Whatman, and there are regular advertisements in the Calcutta press over the next decades about the arrival of all sorts of paper, cards, and boards at the docks.

Although none of the prolific diarists like William Hickey or the artist William Daniell (1796–1837), living and working in the city during this time, mention a shortage of paper, it is questionable as to whether supplies were as steady and easily affordable to the *patuas* as Archer (1971) suggested, until the introduction of wood pulp papers, which arrived at about the time that the art died out.

There are some paintings on extremely good quality imported paper. Others, particularly when the artists started to portray the social life about them rather than keeping to strictly religious subject matter, are on paper of lesser quality, as if the artists laid their hands on whatever they could to meet demand. Mostly, they seem to have obtained a thin, off-white smooth paper, and analysis reveals a high percentage of recycled European linen fibers mixed with a small proportion of what are believed to be jute fibers.[5]

Artistic Techniques

What the *patuas* painted on the paper was a complete divergence from their traditional *pats*. Focus was given to the main figures, while the background was left bare. Designs were kept simple, to be repeated as often as required according to the popularity of a picture. There was no longer the detailed draftsmanship of the scrolls; figures were sketchily outlined in pencil before the base color was swiftly applied in broad, wet strokes. Using vivid colors and translating the skills gained from making clay images, the artists added a darker hue to create a sculptural volume before the base coat was dry, to avoid tidemarks. Faces were mostly drawn in three-quarter profile. Finally, on the better examples, "silver" (in fact, tin ornamentation)[6] was added with such precise expertise that it and the paintings' detailed brushwork belie the notion that these paintings were always hurriedly produced.

The painters' greatest influence was probably what they saw around them in the bazaars. The British penchant for watercolor had also crossed the oceans, and the invention in 1780 by Reeves, the artists' colormen, of moist watercolor cakes transformed the life of the watercolorist outside the studio (Goodwin 1966).

Although the East India Company was most strict about who was allowed passage, several professional artists made the voyage and almost certainly brought their materials with them. Far more numerous were those going out as military officers in the company service who were given drawing lessons as part of their basic training, as it was essential for reconnaissance of this vast new territory. Reeves, displaying smart business acumen, supplied the drawing masters of these military academies, thus instigating a brand loyalty in the alumni wherever they might be posted. In the 1820s a large number of their color boxes, which included stationery and drawing instruments, were shipped to India labeled as military stores. Reeves supplied private trade as well; among its customers were the owners of the menageries of Calcutta, such as the Marquis Wellesley, the governor-general; Sir Elijah Impey, the chief justice of Bengal; and John Fleming, a surgeon. These people employed artists to record all types of flora and fauna of scientific interest as well as specimens in their own collections.[7] Some of the artists were British, but most were Indian, migrating from the fading centers of Mughal painting in Patna and Murshidabad. They had to be taught the different techniques required with transparent watercolor and the use of gum arabic to enhance darker hues, and the intensely colored results are far removed from the tinted landscape drawings of the English school seen in the bazaars. The way they placed the subject on the sheet of paper, often running over the edge, is very similar to a typical feature of the Kalighat *pats*.[8] How closely these two different groups of artists would have communicated is unknown, but it seems that the *patuas* might have found the work of these natural history painters a fertile ground for eye-catching ideas.[9]

While there was no doubt that manufactured, transparent watercolors were available in Calcutta in the first half of the nineteenth century, it may not have been so easy for the *patuas* to purchase them. However, in 1842 the situation changed when N. C. Dutt opened the first specialist paint shop in the city on Dharumtola Street, with his son-in-law, Aukhoy Coomar Laha, taking over in 1872. At first supplying the domestic and industrial trades, the third generation subsequently opened a shop next door selling artists' materials, becoming the first importers of Winsor & Newton brands. Besides selling all types of prepared paints, the shop also stocked dry pigments of all grades. Since that time, the Laha stores, trading as A. C. Laha and G. C. Laha, have been the place where Calcuttans needing pigment for any purpose have shopped (G. C. Laha ca. 1950, also personal communication 2001).

The *patuas* had earned the term Kalighat painters because they sold most of their pictures at the temple of Kali in Kalighat, a neighborhood in the south of the city; furthermore, most of them lived there. As Mukharji (1888) points out, they also traded at other places of pilgrimage like the Chitpore Temple in the north, and when the pilgrim trade was slack, they would return to their other profession of coloring images. Any journey away from home took them along the Pilgrim Road, now known as Chowringhee, and past the Laha stores, then as now conveniently placed close to the central crossroads of Calcutta.

Analysis of Pigments

To establish some facts about the paints used by the urban *patuas*, analysis was carried out on some of the inorganic colors on works in the National Museums and Galleries of Wales collection.

The *patuas* used two shades of blue. A sample from the dome of the fairly rare Muslim image of the *Tajia* (fig. 6) was analyzed using a scanning electron microscope (SEM) with an energy-dispersive x-ray spectrometer (EDS).[10] It was found to be ultramarine, characterized by the presence of sodium, aluminum, and silicon on the spectrum. The same pigment was also confirmed using polarized light microscopy (PLM) on the very popular, hence often repeated, image of *Hanuman Baring His Breast to Show Rama and Sita on His Heart* (fig. 7). In the latter example, the pigment was applied more densely, proving difficult to brush out evenly, as is often the case with this pigment in the watercolor medium.

A blue, identified by a small iron peak and a large organic (carbon) content as Prussian blue, was found on the figure of Krishna in *Krishna and Balaram Attack Kangsha* (fig. 8) using EDS. A large amount of calcium sulfate was found as well, possibly indicating the mixing in of some gypsum to gain intensity without losing transparency, although the result is a little dusty and granular. Gypsum, unlike the reactive calcium carbonate, would be more compatible with the alkali-sensitive blue. A similar Prussian blue has been found on other figures of Krishna using PLM.

The scarlet seen on Hanuman's heart and frequently used for clothing has been identified as red lead, using PLM.

Two yellows form part of the *patuas*' palette. One, of a golden hue, was widely used for skin tones, with a light tan

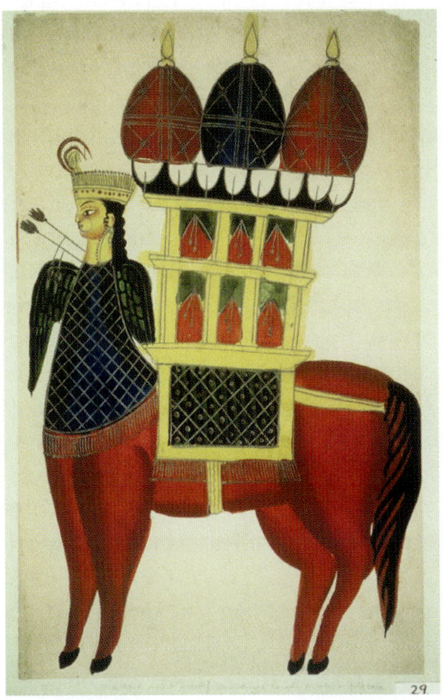

Figure 6. A Tajia of Burak. *National Museums and Galleries of Wales, 11646. Blue and yellow samples were taken from the domes where the pigment has overrun the outline.*

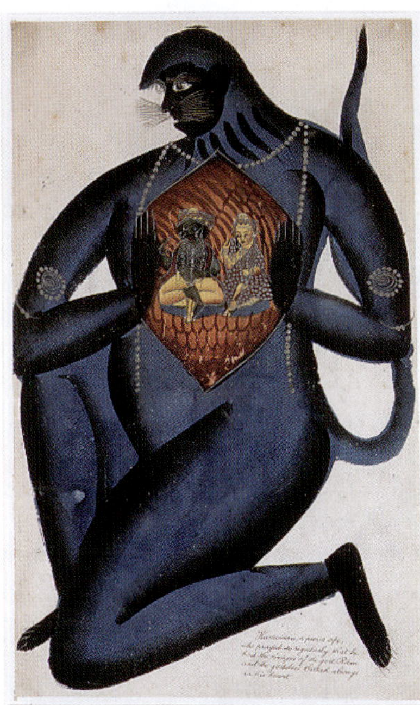

Figure 7. Hanuman Baring His Breast to Show Rama and Sita on His Heart. *National Museums and Galleries of Wales, 11666. Ultramarine blue has been used for the body and red lead on the heart.*

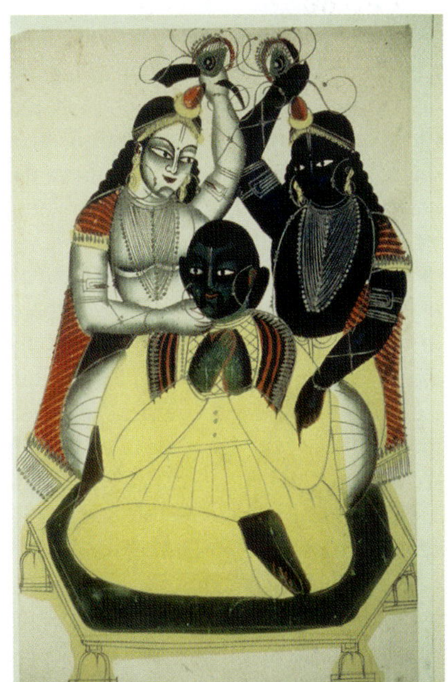

Figure 8. Krishna and Balaram Attack Kangsha. *National Museums and Galleries of Wales, 11674. The yellow taken from the foot of the chair shows a remarkable similarity to the "earth from the west" collected in Naya in 2002.*

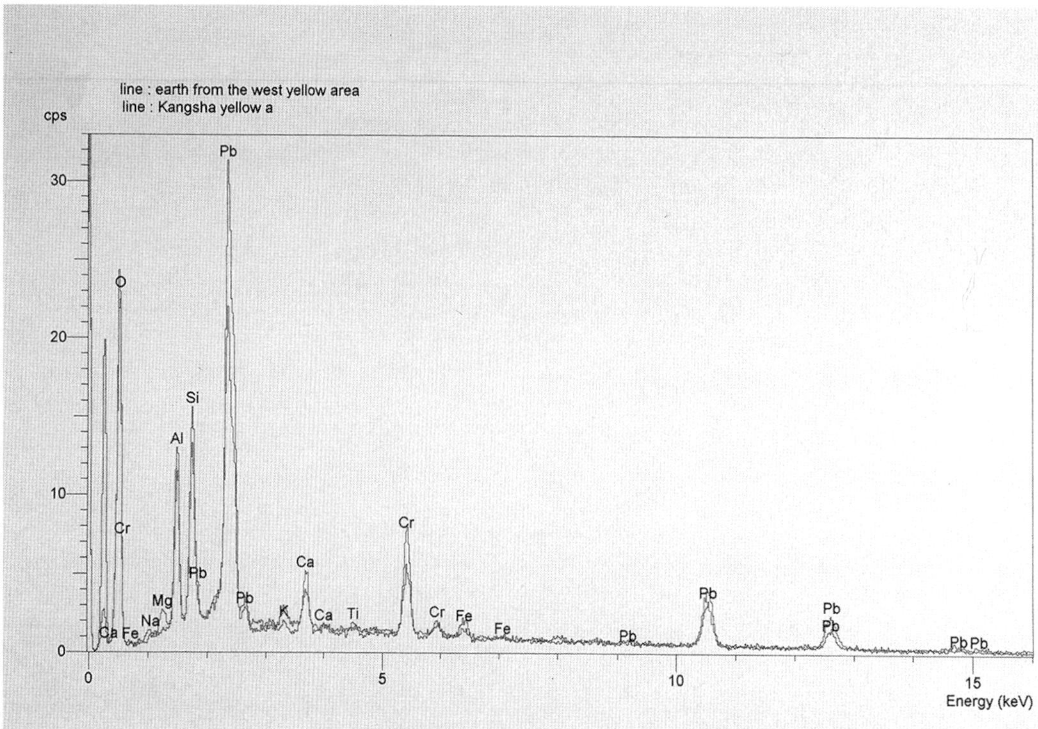

Figure 9. *A comparison of the two EDS profiles. The unbroken line is the sample from figure 8, while the dotted line is "earth from the west."*

being added for the sculptural definition of features. Examination using SEM and EDS found mainly barium sulfate. Barium sulfate, under the names of Permanent or Constant White, stayed a popular pigment even after the introduction of zinc white, but tests using Fourier transform infrared spectrometry (FTIR) to discover whether an organic color such as turmeric had been added were inconclusive.

In the other yellow, a lemon shade often used for chairs and cushions, the SEM identified a large lead content with chromium and some small amounts of silicon and aluminum also detected. During the same session, it was decided to identify the only inorganic sample brought back from the village of Naya. Termed "earth from the west" in the translation from Bengali ("west" implying to the *patua*s anywhere from Bihar onwards), the sample contained small clods of brown earth that resembled clay in consistency, but in the center of some of the lumps there were little clusters of saffron yellow particles. Overall, a clod was found to have elements typical of clay (hydrated aluminum silicates plus oxides of elements such as iron, magnesium, potassium, and titanium) but with a high lead content. When the SEM was focused on the yellow particles, the profile of minerals was discovered to be remarkably similar to the lemon yellow sample from the Kalighat picture (fig. 9). A sample of the yellow particles from the clay was subsequently analyzed by x-ray diffraction to determine its crystallography, and this analysis confirmed the presence of the mineral crocoite, lead chromate, $PbCrO_4$.[11]

Chemists made the first lead chromate pigment in 1816, and the discovery of chrome deposits in the United States soon afterward led to large-scale manufacture by 1820, but as George Field, the man who did so much to improve the range and permanence of artists colors, wrote in his *Treatise on Colour*: "This substance was known as a native pigment long before it was distinguished as a chemical element" (Field 1840, 143). Naturally occurring deposits were first found in Siberia and named after the saffron-stamened crocus flower.

While there is no geological record of sizable deposits of crocoite ever being found closer to West Bengal than Afghanistan, "earth from the west" or *paschimer mati* (clay), like other types of *mati*, is dug up from just below the surface, to be sold quite cheaply in rural grocery stores for assorted uses including textile dyeing and embellishing the body for rituals. The sample from Naya is very similar to the skin tone pigment on Kalighat *pat*s, so had the SEM profile matched this sample, it might not have been such a surprise.

This Kalighat pigment had previously roused interest. Some months ago, a master craftsman image-maker[12] visiting Wales from West Bengal had been looking through the collection of *pat*s at the National Museum and Gallery, Cardiff. Suddenly his attention was caught by the lemon yellow; he said he could remember his grandfather using it sixty years earlier. He assumed it was an earth pigment as he had seen it being ground down, but despite following his grandfather's profession, never once had he seen this particular pigment during the intervening years. This incident poses the question as to whether the lemon color was easily available a century ago but the particular source is exhausted.

Conclusions

The pleasure the Kalighat *patua*s found in painting with the new medium is very apparent in the vivacious free brushwork and the boldness of the designs. Like their forebears, they remained anonymous.[13] The tradition of having various members of the family helping with the production continued, though organized differently. Unlike the scrolls, the Kalighat *pat*s were painted for sale, and vast numbers had to be rapidly produced to meet the endless demand; this situation resulted in some images becoming pitiful pastiches of the original design.[14] Their generous application of pigment is impressive as the *patua*s had to do everything possible to keep costs down; not only were they living on the margins of poverty but so were their main customers, the pilgrims. Investigations of their materials suggest that the *patua*s exercised far more ingenuity in finding sources for them than has yet been credited by art historians. Initially their output would have been quite low, but as demand grew over the decades, so would the difficulties in finding enough paper and paints at an appropriate price. The cost of machine-made paper would have remained comparatively high to cover shipment until the Bally Mill was in full production, by which date colored lithographs were taking over the market. While the *patua*s could easily have used prepared pigments, as some of their favorites, like French Ultramarine and Chrome Yellow, were among the cheapest of manufactured cakes, it seems that they continued to exercise resourcefulness. It is possible that they bought dry powder to prepare their own, as there are many examples of works in which a color glistens in raking light due to excess binder, while on other pictures the same color appears flaky and abraded, suggesting that too little binder was used. The *patua*s may also have extended prepared pigments with appropriate binders. Without a doubt, they continued to use the knowledge gathered by countless generations of picture showmen.

Acknowledgments

The authors would like to thank Sandrine Decoux, fellow paper conservator in the Art Department; Mary Davis and Penny Hill, conservators in the Departments of Archaeology and Numismatics; and members of the Geology Department of the National Museums and Galleries of Wales for their help in identifying pigments. In Kolkata, they would like to acknowledge the invaluable assistance of the Laha family and Ruby Palchoudhury of the Crafts Council of West Bengal, and, above all, the *patua*s of Naya village, Medinipur.

Notes

1. This observation is not accurate; the image makers model clay rather than carve it.
2. A portrait of Siva bearing all the hallmarks of a Kalighat painting hangs on the wall of *An Interior of a Native Hut*, plate 14 of *Twenty-Four Plates Illustrative of Hindoo and European Manners in Bengal*, drawn by Mrs S. C. Belnos. Although Belnos had her volume printed in London in 1832, her husband ran a studio-gallery in the Chitpore Road close to another temple dedicated to Kali.
3. There is no record of how sixty-nine Kalighat paintings entered the collections of the National Museums and Galleries of Wales. The earliest letter on file from the keeper of art to the director, dated February 1954, states that a "bundle" was handed to the Art Department by the Archaeology Department, adding "they may have come from the Trinity Street Museum." This museum predates the founding of the National Museum, but no mention of these paintings is to be found in the records of the transfer of collections in 1922. According Mukharji's standards, they are all of the "old style," being "done with minute care and considerable taste" (Archer 1971, 9).
4. On a visit to Naya in 2002, the authors were given samples of many of the pigments. They have been analyzed and the spectra entered on the FTIR (PerkinElmer Spectrum One system using a universal attenuated total reflection [ATR]) database of National Museums and Galleries of Wales.
5. Dard Hunter noted that in Bengal a combination of paper cuttings and jute fibers was used for handmade paper (Hunter 1939). All the examples in the collection at National Museums and Galleries of Wales were backed many years ago with heat-set tissue, making the examination of sheet formation to determine whether the paper is machine made virtually impossible.
6. The original inspiration for the *patua*s' use of tin is not clear. They may have copied the idea from Rajasthani miniatures, in which tin is used to represent water. Powdered tin was easy to obtain, and they probably needed a stronger binder than gum arabic.
7. The Victoria Memorial Hall, Kolkata, holds two volumes (R 2609 and R 2610) commissioned by John Fleming who returned to England in 1811. Apart from being a record of these menageries, they are a fascinating source for the technical development of the Indian natural history artist.
8. H. J. Noltie (2002) observes that this characteristic seems common to Indian botanical artists throughout the subcontinent.
9. Generally, the use of chiaroscuro in *pat*s is rare. However, there is a striking similarity of the method of shadowing seen on the lithographic plates drawn by the first generation of students of the Government Art College to illustrate Fayrer's *Thanatophidia of India: Venomous Snakes of the Indian Peninsula* (Fayrer 1872) and on *Snake Swallowing a Fish*, National Museums and Galleries of Wales, 11611. Another example of the same subject is in the Monier-William Collection, Bodleian Library, Oxford.
10. Elemental analysis was carried out using a CamScan MaXim 2040 analytical scanning electron microscope (SEM) with a low vacuum chamber, plus an Oxford Instruments Link Isis energy-dispersive x-ray spectrometer (EDS). The samples were analyzed for 100 live seconds using a working distance of 35 mm and an accelerating voltage of 20 kV. The totals were semiquantitative.
11. Debye Scherrer powder camera diffractometer using a Solus Schall generator with Copper Kα radiation operated at 40 kV and 30 mA.
12. Sri Nimai Chandra Pal, who works in Krishnanagar, north of Kolkata, started working with his grandfather and father at the age of eight before receiving further training in Kumartoli. He has worked abroad molding clay images for the Peabody Essex Museum, Salem, Massachusetts, the Royal Museum of Scotland, and the Wales Puja Committee.
13. Their anonymity is in contrast to the examples in the British Library with a lithographic outline that nearly all carry the name of Becaram Das Pande.
14. Jyotindra Jain suggests that four or five women working together could color two to three hundred *pat*s within an hour (Jain 1999). Although smudging of the watercolor pigments is not uncommon, such a number is debatable due to practical reasons of space and working conditions. It is very rare to find an example with the tin smudged.

References

Archer, W. G. 1971. *Kalighat paintings*. London: Her Majesty's Stationery Office.

Balston, T. 1957. *James Whatman: Father and son*. London: Methuen and Co.

Basham, A. L. 1951. *History and doctrines of the Ajivikasa vanished Indian religion*. London: Luzac.

Carey, E. 1836. *Memoir of William Carey, D. D.* London: Jackson and Walford.

Coomaraswamy, A. K. 1929. Picture showmen. *Indian Historical Quarterly* 5:182–84.

Fayrer, J. 1872. *The Thanatophidia of India: Venomous snakes of the Indian peninsula*. London: J. & A. Churchill.

Field, G. 1840. *Chromatography: A treatise on colours and pigments*. London: Winsor & Newton.

G. C. Laha. ca. 1950. *The catalogues of G. C. Laha Ltd.* Calcutta: G. C. Laha.

Ghosh, S. 1990. Calcutta's industrial archaeology. In *Calcutta: The living city*, ed. S. Chaudhuri, 1:34, 246. Calcutta: Oxford University Press.

Goodwin, M. 1966. *Artist and colourman*. London: Michael Goodwin.

Hardie M., and M. Clayton. 1932. *Thomas Daniell, William Daniell*. London: Walker's Quarterly.

Hunter, D. 1939. *Papermaking by hand in India*. New York: Pynson Printers.

Jain, J. 1999. *Kalighat painting: Images from a changing world*. Ahmedabad: Mapin.

Mukharji, T. N. 1888. *Art manufactures of India*. Calcutta: [s.n.].

Noltie, H. J. 2002. *The Dapuri drawings: Alexander Gibson and the Bombay Botanic Gardens*. London: Antique Collectors' Club.

Palchoudhury, R. 1988. Natural dyes used by Patuas. In *Patua art: Development of the scroll paintings of Bengal, commemorating the bicentenary of the French Revolution*, 44. Calcutta: Alliance Française de Calcutta and Crafts Council of West Bengal.

Ray Choudhury, R. 1988. *Calcutta: A hundred years ago, 1880–1890*. Calcutta: Statesman.

Shaw, G. 1981. *Printing in Calcutta to 1800*. London: Bibliographical Society.

Sloman, S. 2002. *Gainsborough in Bath*. New Haven and London: Yale University Press.

Solvyns, F. B. 1796–99. *A collection of two hundred and fifty coloured etchings: Descriptive of the manners, customs and dresses of the Hindoos*. Calcutta: [s.n.].

Solvyns, F. B. 1808–12. *Les Hindous*. Paris: L'auteur.

Usher, G. 1974. *A dictionary of plants used by man*. London: Constable.

The Conservation of the Cultural Property of India: Board Games Depicted as Murals

Rangachar Vasantha

ABSTRACT Murals in the Sri Jayachamarajendra Art Gallery, Mysore, depict board games, the genealogy of the Mysore kings, and other palace matters. This paper details studies carried out on these nineteenth-century murals and presents a comprehensive picture of their materials and technique. This information is used as a basis for determining the causes of their deterioration. At the same time, art-historical studies relate the artistic techniques to the ancient text, *Manasollasa*, written by the twelfth-century king Someswara of the Chalukya dynasty, who ruled from Karnataka.

Introduction

Mysore is situated about 140 kilometers from Bangalore, the present capital of Karnataka, a state in southern India (fig. 1). The city extends from latitude 12°18′ in the north to longitude 76°42′ in the east. It is 776 meters above sea level. Mysore is well connected through road and rail networks with other parts of Karnataka and India (Rice 1887 [2001]). The spoken language is Kannada. A city of palaces, people, and fragrance, the princely city of Mysore is worth a visit whatever the month or season (Buchanan 1807). Mysore has always fascinated with its quaint charm, rich heritage, magnificent palaces, beautiful gardens, imposing buildings, broad shady avenues, and sacred temples. Even today, it retains much of the carefully built character of a maharaja's capital. The city's meticulous planning testifies to the benevolence of the Wodeyar kings who ruled from Mysore from the fourteenth century onward, for nearly six centuries (Hayavadana Rao 1946). The Wodeyar dynasty molded the city's identity as a "native" princely state. Beyond Mysore's architectural heritage, the city's long-held tradition of music and dance, art and literature, scholarship, and time-honored crafts was nurtured under their patronage. Mysore also stands next to Rajasthan in having the largest collection of murals based on *puaranic* (Indian mythological) themes. The murals of Mysore bear the stamp of uniqueness in aesthetic composition and techniques. Most were painted between the fifteenth and nineteenth centuries, although many date back to the tenth century.

Sri Jayachamarajendra Art Gallery

The Sri Jayachamarajendra Art Gallery (fig. 2), popularly known as Jaganmohan Palace, is one of the oldest buildings in Mysore. Built in 1861, it housed the royal family when the old wooden palace was gutted in a fire in 1897 (Hayavadana Rao 1946). The visual effect of the edifice is enhanced by a deep yard, and the rich and eye-catching facade is composed of the three large openings with graduated end-bays punctuated with decorative pilasters. Among the surviving original features of the interior of the hall are cusp-arched balconies and stained-glass window shutters and ventilators. The hall was also a royal auditorium for staging drama and other cultural activities exclusively for the members of the royal family.

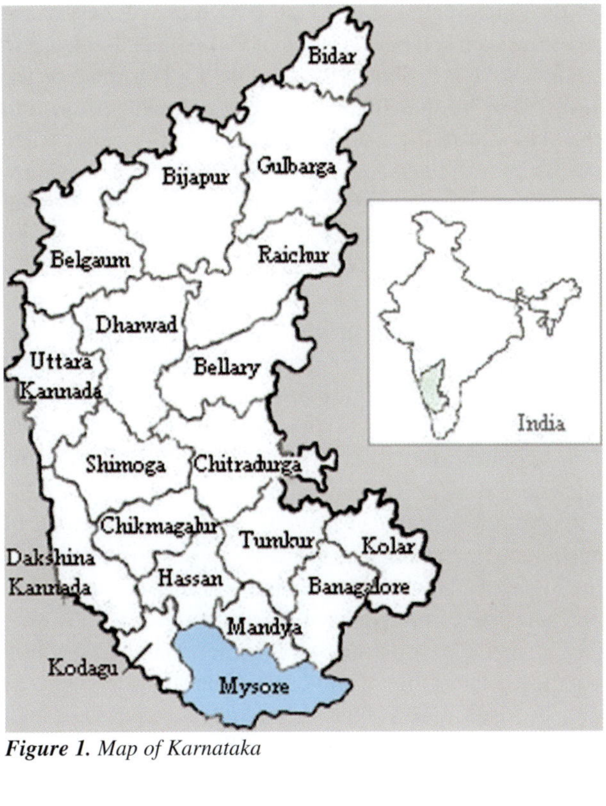

Figure 1. Map of Karnataka

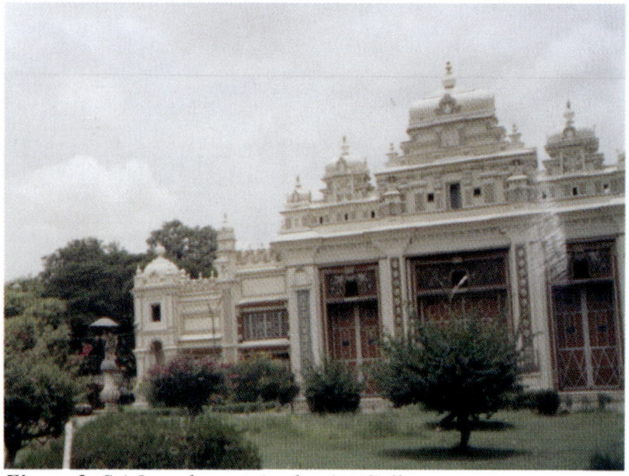

Figure 2. *Sri Jayachamarajendra Art Gallery*

Murals

People have always enjoyed devising puzzles, either for fun or as an intellectual challenge. As the different branches of mathematics have developed, so, too, has the range of puzzles (Kraitchik 1943). These brainteasers have helped to advance our understanding in many areas from logic to the study of shapes found in nature. Number games—and more—are depicted as murals in the Sri Jayachamarajendra Art Gallery, Mysore (Vasantha 2002) (fig. 3). The art gallery also has murals giving the genealogy of the Mysore kings and other palace matters. These inscriptions, as well as inscriptions on some of the portraits, deserve to be copied and preserved or printed. Some of the board games painted on the walls—for example, *devisayujyam* and *srikantasayujyam*, which are calculated to direct the thoughts of the players heavenward, are full of interest.

Many visitors to the palace in Mysore skip the last room on the topmost floor, but if they visit, they will be amazed to see a large number of multicolored boards divided into squares (of 6 × 6, 8 × 8, 10 × 10 or 12 × 12) painted on the walls. Many of these paintings present pleasing patterns to the eye or have the figure of some animal or geometrical shapes inscribed in them. A closer scrutiny reveals that numbers appear to be written randomly to cover all the cells with figures, while no number appears in the cells not covered by the figures (Vasantha 2002).

This paper details studies carried out on the nineteenth-century murals found in the Sri Jayachamarajendra Art Gallery and presents a comprehensive picture of their materials and technique. This information is used as a basis for determining the causes of their deterioration. At the same time, art-historical studies relate the artistic techniques to the ancient text, *Manasollasa.*[1] The *Manasollasa*, or *Abhilashitarthachintamani of Someswara Deva*, is a work of encyclopedic character. Its authorship has been attributed to the Western Chalukya king Somesvara (reigned ca. 1127–38) who ruled from Karnataka. Shrigondekar (1939) concludes, on the basis of internal evidence, that the work was composed under the supervision of the king by a prodigiously well-informed man thoroughly conversant with royal whims, household affairs, and the necessities of the princes. Apparently, the book was written particularly for the instruction of members of the royal family.

The section on painting in the *Manasollasa* discusses the characteristics of a painter, preparation of the ground for painting, preparation of crayons and brushes, substances from which pigments of pure colors are obtained, preparation of mixed colors, and the application of colors including gold in painting. It is apparent from the text that the author had a firsthand knowledge of the technique of painting, for he not only enlightens us on the method of painting but also notes the process of preparing various accessories and media for it. Moreover, he deals with such delicate topics as the method of shading, correction of primary drawing, and removal of color spots—topics that can be discussed only by a person with a firsthand practical experience in the art.

Materials and Technique

The quality of a mural depends on an effective technique. An excellent painting is one that is durable. After centuries of neglect and depredations, the fact that murals are still extant in India testifies to the soundness and efficacy of the methods and processes devised by the artists of old. The longevity of these paintings justifies an inquiry into the techniques their artists developed and followed. Scientific analysis of the murals, to understand the technical processes involved in their production, has just begun in India, owing to the difficulty of access to the originals. Recent scientific analysis of the pigments, plaster, and binding media has enabled us to estimate more precisely the materials used for the murals and, in turn, has enabled conservators to use the appropriate chemicals and materials for conservation treatments.

Carrier and Ground

The carrier of these paintings is brick masonry, which is sound and free from structural defects. Examination of the painted surface reveals that these paintings were executed on a carefully prepared ground, which was laid on the brick

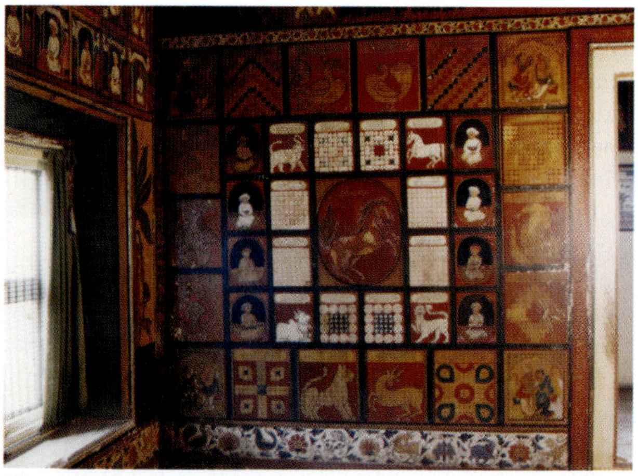

Figure 3. *Paintings on the palace wall*

masonry in three steps by the traditional method. Lime was used for plastering the walls. Typically, plaster was mixed with fillers such as sand, marble grit, jute, and casein (Gettens and Stout 1966). First, a coarse to medium clay was applied to the wall. Next, a thin layer of lime plaster was laid on it. Finally, the surface was rendered smooth by a careful application of a thin lime paste. Although the plaster is fairly hard, occasional disintegration is noticeable here and there where the plaster has cleaved between the ground layers. This imperfection may be attributed to application by junior painters or due to the percolation of moisture.

According to the *Manasollasa*, the wall should be smeared with a lime plaster. Then white earth mixed with glue should be applied thrice on the dry wall. Then powdered conch shell, mixed with sugar as adhesive, should be applied until the surface becomes smooth. The moon-bright white earth collected from the Nilgiri Mountains should be pulverized and mixed with glue and applied gradually and gently to make the surface crystal smooth.

The number of layers in the plasters found in other parts of Mysore varies, and sometimes it is two instead of three. In some cases the painters have used organic materials like paddy husk, straw, or coconut fibers with the clay plaster. It is fascinating to note that if lime had to be brought from a different place or a long distance, clay has been used. Probably the painter wanted to use the locally available materials.

One more interesting finding has been that the plasters of the early period in the Mysore region are very hard. They are much harder and less porous than even recent cement plaster, indicating that apart from lime and sand the painters must have used other additives for the preparation of the plaster and/or have given it a different treatment. Often the surface is very smooth and shiny. This type of lime plaster is very similar to that which is described in the *Manasollasa*.

Pigments

The pigments were prepared using locally available natural materials. The painters favored bright luminous colors, perhaps because the halls were poorly lighted. The colors and hues may have been chosen by the painters to match the theme, situation, and place.

The principal colors used are four: red, yellow, black, and white. These pigments were examined and identified. The red is red ocher; yellow is yellow ocher; black is lampblack (carbon soot of lamps); and white is kaolin and lime. The text *Manasollasa* refers to the above four colors as primary, whereas other texts include blue and state that there are five primary colors.

The intermediate colors used have been identified. The green is terra verte (or a mixture of yellow and blue); blue is ultramarine and indigo; deep red is lac; and orange-red is madder. A limited use of brown, yellowish green, greenish blue, and sky blue was also seen.

Gold was also used, in particular to decorate ornaments and dresses. Here it was used in the form of gold leaf or gold powder. Being a malleable metal, gold leaf was prepared by beating and rolling gold foils into ever thinner sheets. The spreading of gold leaf on the surface is a skill that comes only through long practice. The leaf was pasted down with a specially prepared gum known as *bat*.

Lemon juice or a solution of *thurisu* (copper sulfate) was applied on the surface to decrease the alkalinity of the lime before the surface was painted. Painting was followed by an overcoat composed of pine resin and linseed oil to provide luster and durability.

Pigments in the Manasollasa

White

Though white is mentioned as a principal color, none of the *silpa* (treatise for artisans) texts specified what kind of white should be used in painting. The *Manasollasa* said that the white color to be used in mural plasters could be obtained from white kaolin clay and lime.

Black

Black was also considered one of the principal colors. It is used for black outlines, hair, and eyes. The *silpa* texts discuss the preparation of a lampblack pigment. The artist is told to smear fine cow dung dust on the interior of an earthen pot. After extending the wick, the pot is placed over a lit oil lamp to collect the soot. The soot is scraped from the bowl and ground with pure water, then allowed to dry. This step is repeated several times. Finally, the juice of the *margosa* tree is added, the whole purified by levigation and then allowed to dry.

It is interesting to note that the use of different oils gives diverse black shades. Mysore painters use castor oil, whereas Tanjore painters use til oil to obtain black paints. Slight tonal differences can be seen even by the naked eye.

Red

Red is a favorite color of nature: there are red stones, red clays, and red minerals. According to the *Manasollasa*, red lead is used for crimson, the juice of lac for blood red, and red ocher for pure red. Because of its easy availability, red ocher seems to be the most popular red pigment used in paintings. It is obtained from natural earth, which contains 95% ferric oxide. The red color is due to this ferric oxide. The *Manasollasa* directs that it be prepared by first pulverizing it on a grinding stone for a day and then washing it in pure water.

Blue

The blue pigment can be obtained from mineral or vegetable compounds. Texts mention only two types of blue pigments: *nila* (indigo) and *rajavarta* (lapis lazuli). The former is obtained from a vegetable source, the latter from a mineral source.

Yellow

Even though a number of yellow pigments are known to us, only two types are mentioned in *silpa* texts: *haritala* (orpiment) and *pita datu* (yellow ocher). Orpiment is chemically arsenic sulfide (As_2S_3). Yellow ocher is chemically goethite ($Fe_2O_3{\cdot}H_2O$). The *Manasollasa* directs that yellow ocher be gathered from mountains and riversides, washed with pure water, and ground to a powder. It further directs that the pigment be purified through repeated levigation and the purified pigment smeared on an earthen pot to dry in the summer sun.

Green
Green color is generally prepared by mixing blue and yellow. The text records the green (*harita* and *vilomata*) as a primary color, perhaps because the Indian muralists knew the terra verte green as a separate color. But none of the *silpa* texts mentions its preparation.

Binding Media

Mixing of colors was done in wooden utensils, and the binding media were tender coconut water, sugar, and exudate from Neem trees. A local painter, Mr. Ramasubbaiah, aged 84, opined that in preparing ground layers and in fastening colors on the ground, sugar (*jaggary*) may have been extensively used as the medium in the Mysore murals. According to texts, vegetable gum is basically starch, prepared by boiling or by treating it with alkalis. It is generally supplied in the form of a granular powder or in flakes, which are stirred into cold water and allowed to swell considerably. Then it is mixed until a fine paste is obtained. Vegetable gums, especially sugar, always remain soluble and are therefore readily affected by moisture. On aging, they lose their binding power, and hence flaking occurs (fig. 4).

The paint may be softened by water, indicating the presence of a water-soluble medium. Treatment of the paint layer with 0.5% hydrochloric acid resulted in complete dissolution of the white priming and the formation of tiny flakes of pigment. These floated for a while on the surface of the solution and then gradually sank to the bottom of the vessel, producing a fluffy gelatinous mass. The solution was tested for gum and glue, but the results were inconclusive. The exact nature of the binding medium thus could not be determined. Microscopic examination of polished sections of specimens of painted stucco revealed a distinct interface between the priming and the pigment layer. The pigments were found to be laying superimposed on the white priming layer in the form of a thin skin, and there was no diffusion of the pigments into the body of the plaster. Moreover, the pigment layer could be removed intact from the priming layer with a needle. These results indicate that an *a secco* technique was used for the execution of these paintings and that a water-soluble adhesive was used as a medium.

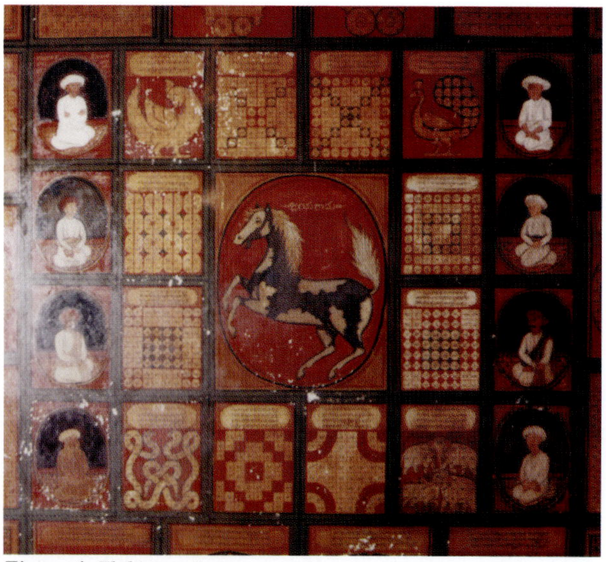

Figure 4. *Flaking paint*

It is noteworthy that the paintings of the early period in India are still intact, in spite of being exposed to heavy rain and sun, but the nineteenth-century paintings, only 150 years old, are slowly deteriorating due to moisture. The painting technique of the early period was obviously more secure and strong than that of the later period. Paintings executed in true fresco and *stucco lustro* are extremely durable and more resistant to the effects of weathering than paintings executed *a secco*.

Condition and Factors Responsible for the Deterioration of the Murals

These murals are slowly deteriorating due to pervasive external and internal agents of destruction. Thus, to keep them intact for future generations, they require special care. They need immediate, protective and preventive, and often restorative, attention. The carrier of brick masonry, although structurally sound and free from displacement, has been affected by moisture due to the seepage of water from the ceiling. This seepage has resulted in the bulging of the plaster at some places. The roof is in very bad condition, and the room is closed to visitors, for safety. Without ventilation, in all likelihood fungi will develop on the murals.

Although the traditional method or preparation of the ground is in three layers, mild cracks and cavities are seen on the painted surface. Due to the ingress of moisture, the ground is found here and there to have lost its adhesion to the masonry, and the gap so caused between the masonry and the ground is quite well defined. The plaster has become fragile and weak, particularly at higher levels just below the ceiling, presumably due to the attack of moisture. Microscopic examination of the painted surface showed that the paint layer is experiencing active decay, as the paint and the fine layer of priming have become loose and the bond between the fine layer of priming and the underlying coarse plaster has become weak. A network of minute cracks can be discerned at places on the paint layer, which also shows considerable cupping and curling.

Moisture is causing havoc in this instance and is the principal cause of degradation. The source of moisture is due to the infiltration of water owing to seepage through the ceiling (descending moisture) and crevices in the wall, rising dampness, and condensation.

Another form of damage that moisture is causing is salt efflorescence (fig. 5). The soluble salts present in the substrate are dissolved by moisture, and in the course of evaporation they are brought to the surface of the paintings and crystallize.

Chromatic alterations are also noticed in the murals. Very

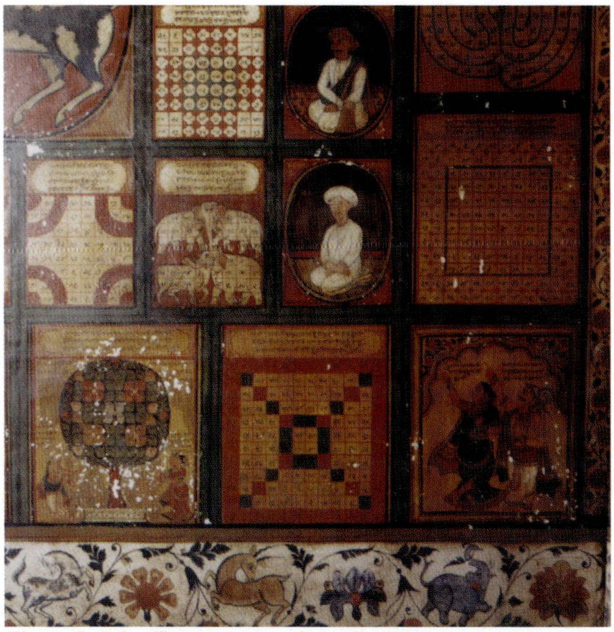

Figure 5. *Salt efflorescence on the wall painting*

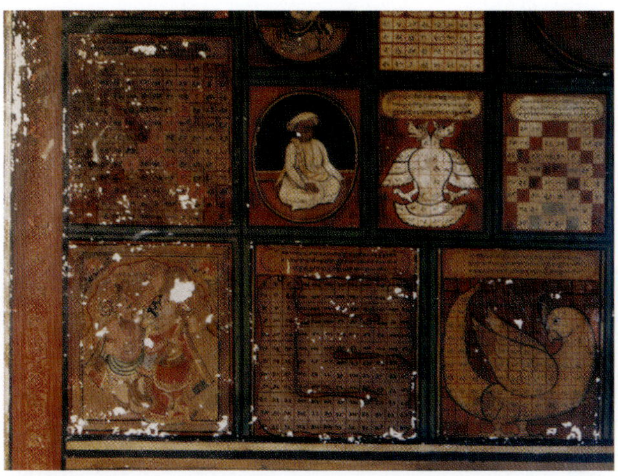

Figure 6. *Chromatic alteration. Note that the flesh tone of the faces has changed.*

commonly noticed are darkening or blackening of green, changes between red earth and yellow earth, and a darkening of flesh tones (fig. 6). In some spots white dresses have turned to gray and also to black. Detection of a chromatic alteration is a very complicated subject, and one must not prejudge the problem. A careful examination of a superficial paint layer as well as the underlying layer is required. A complete stratigraphic study in the laboratory, with preparation of cross-sections and thin-sections, and then a detailed analysis are necessary to ascertain a chromatic change. Also, the surface of the murals has darkened due to the deposition of extraneous material such as dust and dirt.

These paintings are slowly deteriorating on account of the flaking of paint, smudging of colors, and fading due to the action of moisture and extraneous accretions. In some places, details of the design cannot be made out. The paintings are fading, and the original colors are becoming dim. It is not known whether this condition is owing to ultraviolet radiation present in normal light in the hall, or to excessive flaking or alteration in the coating applied to the surface, or to other causes. Finally, human vandalism in the form of scratching or touching has caused serious, and at times, irreparable damage to the paintings.

Concluding Comments

These studies have related the artistic techniques used in the nineteenth-century Sri Jayachamarajendra Art Gallery murals to the twelfth-century text, *Manasollasa*. From the details given in this paper, it is apparent that the ancient technology was followed in the painting of these murals seven hundred years later.

The degradation processes discussed here show that we need an awareness campaign at once to stop the further decay of not only the murals but of everything in the art gallery. Each material is part of the ensemble of the museum, and all need protection. Otherwise the museum will soon disappear; already the structure of the palace has deteriorated significantly. This deterioration concerns health officials, economists, and all who care about South Asia's cultural heritage. The current population explosion translates into increasingly heavy demand for development of land, inevitably threatening important cultural properties. There are nearly a half million art galleries and museums in India that have gone unnoticed and unrecorded. Those residences associated with historical places are treasures. Our efforts to preserve this heritage continue to lack momentum due to scarce resources and limited public awareness.

Acknowledgments

I am highly thankful to Sri Kantadatta Narasimharaja Wodeyar, chairman of the trust, Sri Jaganmohana Art Gallery, Mysore, and Sri M. G. Narasimha, superintendent of the gallery, for permitting me to take photographs and study the content, material, and technique of the murals.

Note

1. In our present study of the section on painting, we have mainly depended on volume 2, edited by G. K. Shrigondekar and published in the Gaekwad's Oriental Series (no. 84) in 1939. Shrigondekar prepared this volume using four manuscripts. Of these, Mss A, C, and D are complete and belong to the Oriental Institute of Baroda, Biknir Durbar, and Bhandarkar Research Institute of Poona, respectively. Mss B, preserved in the library of the Bhandarkar Research Institute, is incomplete. We have, however, also taken into account an earlier edition of the text entitled as *Abhilashitarthachintamani of Someswara Deva* (Manasollasa) edited by R. Shama Sastry and published from Mysore in 1926.

References

Buchanan, H. F. 1807. *A journey from Madras through countries of Mysore, Canara and Malabar.* London: Directors of the East India Company.

Gettens, R. J., and G. L. Stout. 1966. *Painting materials.* New York: Dover Publications.

Hayavadana Rao, C. 1946. *History of Mysore.* 3 vols. Bangalore: Mysore Government Press.

Kraitchik, M. 1943. *Mathematical recreations.* London: Allen and Unwin.

Rice, B. L. 1887 [2001]. *Mysore: A gazetteer compiled for government.* Rev. ed. 2 vols. Westminster: Archibald Constable & Co. Reprint, New Delhi: Asian Educational Services, 2001.

Shama Sastry, R., ed. 1926. *Abhilashitarthachintamani of Someswara Deva.* Oriental Library Publications, Sanskrit Series 69. Mysore: Government Branch Press.

Shrigondekar, G. K., ed. 1939. *Manasollasa of King Someswara.* Vol. 2 (in Sanskrit). Gaekwad's Oriental Series 84.

Vasantha, R. 2002. Step by step. In *Proceedings of the Fourth Colloquium Board Games in Academia.* Fribourg: Editions Universitaires Fribourg Suisse.

Artistic Practices of the Bohol School of Painting: An Analytical and Archival Study of Nineteenth-Century Panel Paintings in the Philippines

Nicole Tse

ABSTRACT In the center of the Philippines, on the island of Bohol, a unique panel painting practice evolved linking Western artistic methods introduced by the Spanish with Filipino knowledge of materials and techniques. The scientific analysis of five nineteenth-century panel paintings belonging to the Immaculate Conception Parish, Baclayon, was undertaken and combined with an investigation of the records in the parish archives to develop a better understanding of their provenance. The results show that the construction methods based on Western pictorial techniques used in the panel paintings along with an oil medium as well as the utilization of local materials such as Pagsahingin (trade name Kedondong) wood for the panel support, cotton and bast fiber paper for a gap filler between the wood panels, and beeswax for the ground layer. Some of the pigments identified correlate with the geological deposits from the region, and others correspond with the archival church records. Other identified pigments were not referenced in the archives or found locally. The latter indicate the importation of high-quality pigments, not of Filipino origin. Further, the good condition of the panels highlights their sound preparation and an environment suited for these works.

Introduction

In the Philippines there are numerous nineteenth-century panel paintings by "unknown Bohol masters" (Pilar 2000, 1). Bohol is an island in the Visayas region located in the center of the Philippines, where contact with the Spanish colonists began in 1596 (Jose 2001). The paintings by "Bohol masters" have sound provenance supported by parish archives and local history; five of them are located in the Immaculate Conception Parish of Baclayon in Bohol. According to the parish's Libros de Cargo y Data, 1856–1909 (Book of Income and Accounts, 1856–1909), the paintings were commissioned in 1859 by Father Antonio Ubeda, the parish priest from 1856 to 1859 (Immaculate Conception Parish n.d.).[1] There is little information, however, about the origin of the panel paintings. Recently, Regalado Trota Jose, a well-known historian of colonial church art, proposed that the paintings were either commissioned and transported to Bohol or were painted in Bohol by local artists (personal communication 1998). This paper explores the possible origins of the five paintings using both archival and analytical studies.

Contained in the parish archives are detailed references to art materials and the payment of wages to painters and carpenters, as the Spanish colonists were thorough record keepers and ensured their accounts were well maintained (Jose 1992). These data provided an area of inquiry against which technical examination could be compared in order to determine whether the paintings were painted *in situ*. The five panel paintings in this study are the *Ascension of the Lord* (fig. 1), *Pentecost*, *San Gregorio Magno*, *San Ambrosio*, and *San Geronimo Maximo*. The first two works are unsigned, and the artist is unknown. The latter three

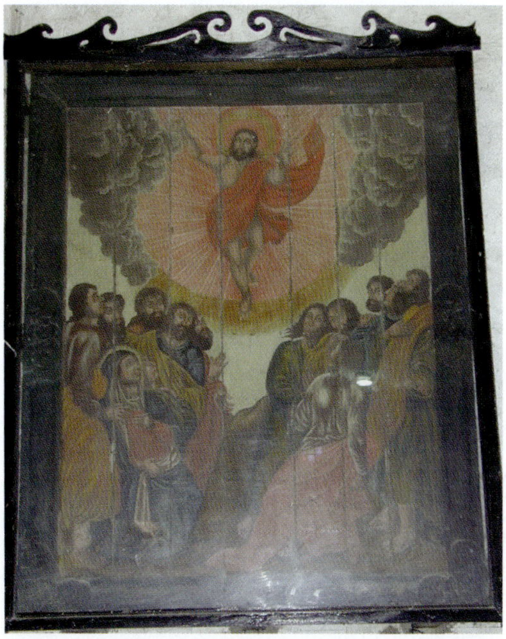

Figure 1. *Artist unknown,* Ascension of the Lord, *ca. 1859. Immaculate Conception Parish, Baclayon, Bohol (This image has been printed with permission from the Roman Catholic Bishop of Tagbilaran.)*

panels are part of a series depicting the four *Augustine Recollect Doctors of the Church* (one of the panels is missing). During recent treatment, the author uncovered an inscription by the artist reading "Liberato Gachalian lo pinto pr. Manda del MRP Antonio Ubeda Año 1859" (Liberato Gachalian painted this as commissioned by Very Reverend Father Antonio Ubeda year 1859).

General History of the Use of Western Painting Materials in the Philippines and Bohol

The use of Western art materials in the Philippines is well documented. With Spanish colonization in 1521, Western art forms were introduced to help convert Filipinos to the Catholic faith (Santiago 1992). Conversion to Catholicism began in 1596 in Bohol, as Cebu, the main island close to Bohol, was the site of the first Spanish settlement and the capital of the colony from 1565 to 1571 (Jose 2001).

During the early colonial period, artists undertook their training locally in the Philippines (Pilar 1992). For example, as early as 1609, students of the Jesuit school in Loboc in Bohol requested examples of paintings from Santa Maria Maggiore in Rome to copy (Jose 1992). Pilar (1994, 66) states further that Jesuit Father Antonio Sedeno "coached Chinese painters in the early 1580s," and from then on there was a tendency to train *mestiza* artists of mixed Chinese and *indio* or Spanish parentage. The Chinese were assigned the Parian district in Manila, where they set up workshops, and by 1741 they founded their own guild, known as Gremio de Mestizos de Binondo (Guild of Mestizos from Binondo).

Liberato Gachalian, the artist of three of the panels in this study, is thought to be of Chinese origin according to Pilar, and may even be a descendant of the "Guanyin Master" from Bohol (Pilar 2000, 4). Jose also reports that Gachalian is not a local Visayan name but a Tagalog or Kapampangan name, which is indigenous to the Luzon region where Manila is located (Jose 2001).

Research Methods

The three areas of inquiry that informed this study included the visual examination of the panel paintings, research in the parish archives, and technical analysis of the panel paintings. The Diocese of Tagbilaran and the Catholic Bishops' Conference Philippines Episcopal Committee for the Cultural Heritage of the Church initiated conservation treatment of the panel paintings. They asked the National Museum of the Philippines, Manila, to undertake treatment, and the author joined the project as a conservator and researcher. A research project was developed jointly by the University of Melbourne Conservation Service, Regalado Trota Jose, and Father Milan Ted Torralba, parochial vicar of the Immaculate Conception Parish of Baclayon.

Visual Examination

The author conducted two conservation treatment field trips in October and November 1998 in collaboration with the Conservation Department of the National Museum in the Philippines. Preliminary treatment was undertaken when the paintings were still attached to the wall, and hence examination of the reverse was not possible.

Archival Research

Jose collected data from the Libro de Cargo y Data, 1807–56 and 1856–1909, in the parish archives, Baclayon, and they provided a basis of inquiry for the analysis. Relevant to this study were references to work undertaken by artisans and the purchase and cost of art materials prior to 1859 when the panels were painted (Immaculate Conception Parish n.d.). During the 1850s in Bohol, the monthly wage was two pesos, so the cost of materials sheds some light on their relative value and purpose (Jose, personal communication 1998). This study compared the pigments, fillers, papers, and wood recorded in the parish archives with the materials analytically identified.

Technical Analysis

During field visits in 1998, samples were removed from the *Ascension of the Lord*, *Pentecost*, *San Gregorio Magno*, and *San Geronimo Maximo* for analysis at the University of Melbourne. Technical analysis was not undertaken for *San Ambrosio*; however, similar conclusions were made for this work as it has the same artist, date, provenance, and style as the other two signed works.

Wood Identification

The Commonwealth Science and Industry Research Organization (CSIRO), Forestry and Wood Division, undertook the wood identification. Thin whittlings from larger splinters were prepared and examined at 50× low-power magnification and 400× high-power magnification.

Pigment and Inert Material Identification

The pigments were analyzed using a combination of techniques. First, pigment samples were prepared in Meltmount for analysis with polarizing light microscopy (PLM). The slides were viewed under 63× to 400× magnifications. Under plane polarized light and between crossed polars, characteristics as outlined by McCrone (1982) were observed. Second, samples were prepared as cross-sections mounted in polyester resin and examined under reflected and ultraviolet (UV) light for interpretation of the layered structure. Further confirmation of pigments was then undertaken with microchemical tests according to methods outlined by Plesters (1956) and by Hassell and Sheldon (1994).

When identification of pigments was uncertain, Raman spectroscopy and scanning electron microscopy–energy-dispersive spectroscopy (SEM-EDS) were utilized. Raman spectroscopy provided spectra identifiable to the pigment type, and SEM-EDS provided elemental analysis and images of the surface topography of the paint layers.

Binder Identification

First, staining techniques were used to identify the binding media, using a saturated solution of Sudan Black for oil and Amido Black AB3 for proteins (Gay 1976, 79; Martin 1977, 64). An oil binder was consistently identified in all samples; to confirm the results, gas chromatography–mass spectroscopy (GC-MS) was undertaken on selected pigmented samples (Schilling and Khanjian 1996). Jose also mentioned the use of coconut oil in paint films (Jose 1992), and GC-MS analysis was used to determine whether it was present. The filling material was identified with Fourier transform infrared spectroscopy (FTIR), GC-MS, and melting point encasements.

Results and Discussion

Provenance of the Panel Paintings

Between 7 March and 13 April 1859 the Libro de Cargo y Data records the payment of twenty-four pesos for the paintings of the *Doctors of the Church* series and twenty pesos for the two paintings of the *Ascension of the Lord* and *Saint John the Baptist*.[2] The *Saint John the Baptist* is no longer extant. However, the *Pentecost*, which is in the church, has the same dimensions as the *Ascension of the Lord*. The two have identical carved wooden frames with design motifs of pineapples and papayas, and they are companion pieces in terms of their iconography. It can be surmised that these are the paintings to which the parish archives refer. Jose has suggested that *Saint John the Baptist* was painted over with the image of the *Pentecost*, as the archives record the payment of wages to painters (personal communication 1998). Jose has also suggested that Father Ubeda commissioned the five paintings prior to his engagement at the Immaculate Conception Parish, Baclayon, in 1856 and then transported them, since the parish archives detail the crating of items from Cebu in 1858 (Immaculate Conception Parish n.d.; Jose 2003).[3]

Materials and Techniques

Wood Panel Support

According to Jose (1992), the wood panels were constructed from the hardwood Molave (*Vitex geniculata*) or Narra (*Pterocarpus santalinus/P. pallidus*). Records in the parish archives detail the purchase of Molave in December 1857, July to August 1857, May to June 1858, and July to August 1858, and also payments to a teacher of carpentry in December 1857 (Immaculate Conception Parish n.d.). The corresponding expenditure for each record is significant, indicating its relative importance and suggesting the wood may have been used for the wood panel painting support.

Small wood samples from the two panels, *Ascension of the Lord* and *San Geronimo Maximo*, were identified. Thin whittlings were prepared and examined under 50× low-power magnification and 400× high-power magnification. The specimens from both panels were consistent with the wood traded under the name Kedondong (*Canarium asperum* from the Burseraceae family) or Pagsahingin in Tagalog and did not correspond with the woods reported in parish archives as described above (Illic 2003). Interestingly, the Burseraceae family is also the source for Manila elemi (*Canarium luzonicum*), a well-known resin from the Philippines. Its use highlights the technical understanding of wood and its products (Mills and White 1987). Kedondong is a common, semihardwood in the Philippines with poor durability, readily attacked by termites and powder post beetle. Visual examination of the wood panels, however, revealed their stability in spite of the extreme humidity and temperatures on Bohol without any insect damage (fig. 2). The sound condition of the wood panels, despite the known characteristics of the wood, suggests they have been well cared for and housed in a stable environment.[4]

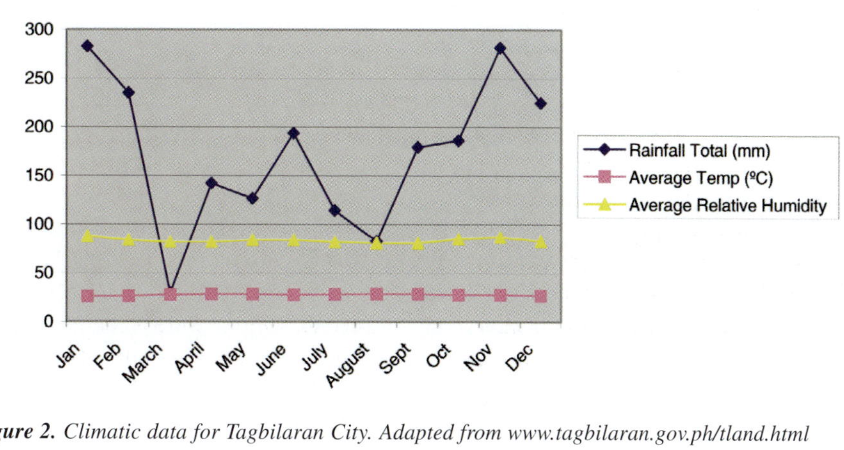

Figure 2. *Climatic data for Tagbilaran City. Adapted from www.tagbilaran.gov.ph/tland.html*

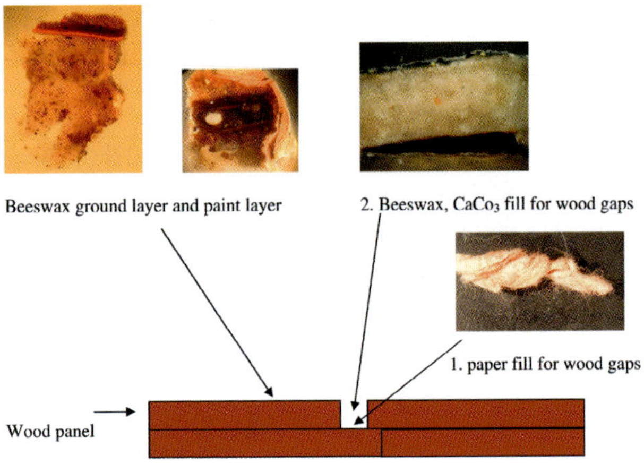

Figure 3. *Diagram of the structure of the five panel paintings examined belonging to the Immaculate Conception Parish, Baclayon, Bohol*

Wood Panel Construction

The construction method of the wood panel is based on traditional Western practice; however, the fills between the panels are unique (Veliz 1995). Each panel is 175–320 mm wide, joined with a plain lap join and 30–50 mm wooden pegs across the join. Gaps between the wood are filled with a combination of rolled paper, an unknown brown material, and calcium carbonate (fig. 3). Preliminary examination suggested an old repair. According to Jun Concepcion, paintings conservator, National Museum in the Philippines, however, many panel paintings have been constructed in this way, and it is thought to be a standard technique.

Wood Gap Filler

Analysis of the paper in-fill material from the *Ascension of the Lord* and *San Geronimo Maximo* suggested a composite cotton and bast (possibly *kōzo* of Japanese practice) and cotton fiber paper, as identified by PLM and Herzberg tests. Examination under magnification and a negative reaction to phloroglucinol suggested the bast or *kōzo* fiber; however, this classification was not conclusive. The fibers themselves are relatively intact, with few interweaving fibrils meshing the paper. The overall construction suggests a handmade paper with a heavy use of size and granular filling material (fig. 4). A positive reaction with the spot test of Amido Black AB3 suggested a gelatin size; however, these tests are indicative and not always reliable.

The surface of the paper is covered with a red chalk or pigment, implying that the paper had been used for another purpose (see fig. 4). PLM identified the pigment as vermilion (HgS), with its high red birefringence under crossed polars.

There are a number of possibilities regarding the source of the vermilion-coated paper, identified as a cotton and bast (possibly *kōzo*) fiber paper sized with gelatin. In 1858 the parish archives refer to *cola y papel de Japon* (glue and paper of Japan) (Immaculate Conception Parish n.d.). Today, *papel de Japon* is manufactured locally; however, it is not known when these papers were first manufactured in the Philippines (Jose, personal communication 2003). One possibility is that an artistic community of bookbinders and jewelers was producing it under Father Mariano Gutierrez, as recorded in the parish archives in Jagna in 1830, another parish in Bohol. Father Gutierrez taught parishioners how to obtain dyes from local plants and to make paper and parchment (Jose 2001). Later in the Baclayon Parish Archives, the purchase of four dozen buttons of vermilion for four pesos was documented in 1858, and this record may relate to the vermilion on the paper (Immaculate Conception Parish n.d.).[5] Alternatively, Jose states that *papel de arroz* from China was mainly used in the Philippines and was composed of bamboo or cotton (Jose 1992). The 1842 commerce records of the Philippines also record the highest number of paper imports from China (Mallat 1994). These Chinese papers may also relate to the composite paper identified. Overall, no positive conclusions can be made for the identification and provenance of the paper.

The brown transparent material used in the gap filler was identified as beeswax and will be discussed in the next section. The white filling material from all the panels was identified as calcium carbonate ($CaCO_3$). Limestone is

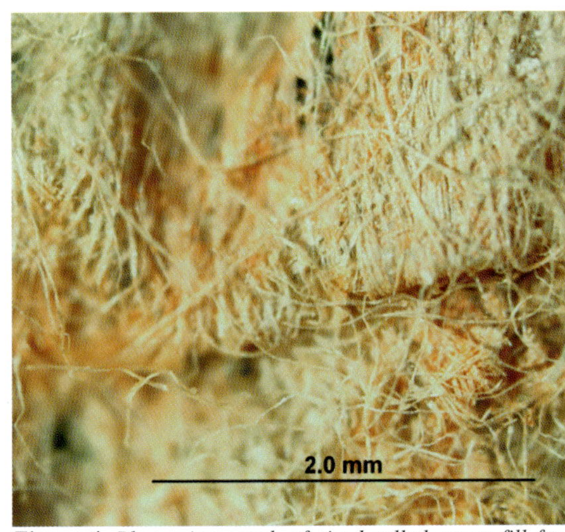

Figure 4. *Photomicrograph of sized rolled paper fill from* Ascension of the Lord *(fig. 1)*

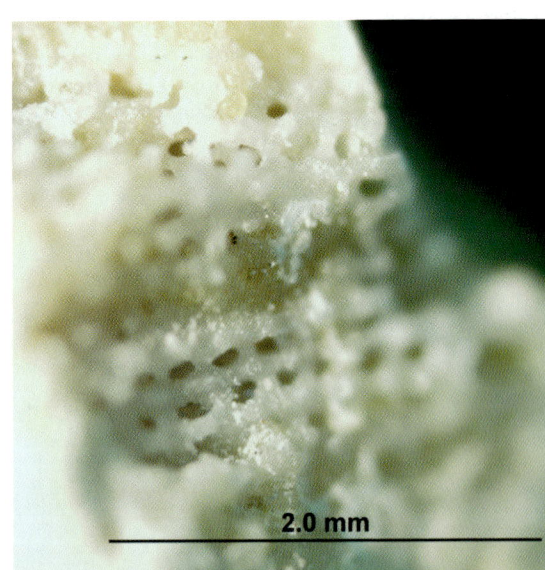

Figure 5. *Photomicrograph of calcium carbonate with a regular structure of biological origin from* Ascension of the Lord *(fig. 1)*

found naturally throughout the island of Bohol, and calcium carbonate was produced by cooking crushed shells from oysters (Jose 1986). Further, there are references in the archives to the purchase of lime ($CaCO_3$) in February 1857 at the minimal cost of three pesos, implying it was locally obtained (Immaculate Conception Parish n.d.).[6] For the *Ascension of the Lord*, particle sizes varied, and large uncrushed particles revealed a regular structure of biological origin, indicating the calcium carbonate was manually crushed (fig. 5). Its source is likely to be coral stone once known as *piedro de Visayas*, which is formed from a fossil coral reef and accretions of shells (Jose 2003). Another sample from *San Geronimo Maximo* also suggested crushed oyster shells under PLM, as strong cleavage and high birefringence were visible in crossed polars.

Preparatory Layers

Cross-section examination of the paint and wood support shows an unusual paint structure (fig. 6). There is no traditional ground layer directly on the surface of the wood panel even though calcium carbonate was purchased by the parish and used as a wood gap filling material in the panels (Immaculate Conception Paris n.d.).[7] Nor was there confirmation of a proteinaceous size layer with Amido Black AB3 stain for the panels, even though the archives record the purchase of skin and glue (Immaculate Conception Parish n.d.).[8] The cross-sections of the four analyzed paintings reveal a dark brown translucent layer in direct contact with the wood panel followed by the pigmented paint structure; the traditional whiting-based ground layer is lacking (fig. 7). The brown material was soft and flexible, did not fluoresce under UV light, and included a nonmelting black solid (Pellegrino 2003).

Identification of the brown substance was undertaken with FTIR and GC-MS for the *Ascension of the Lord* and *San Geronimo Maximo*. The FTIR spectrum produced the doublet at 2917 cm^{-1} and 2848 cm^{-1}, and peaks at 1735.5 cm^{-1} and 1175.92 cm^{-1}—results corresponding with the University of Canberra's database spectrum for beeswax. GC-MS analysis produced alternating fatty acid ester peaks with hydrocarbon compound peaks—results also characteristic of beeswax (Mills and White 1987). It appears, then, that beeswax was used as a preparatory layer on the wood panels and as a bulking material between the wood gaps. This method of preparing a panel for painting has not been reported in translations of treatises on Spanish artistic techniques by F. Numes (*Arts of Poetry, and of Painting and Symmetry, with Principles*, 1615), or by V. Carducho (*The Art of Painting*, 1649) (Veliz 1986). A more typical preparation technique involved the initial application of a layer of size followed by two gesso layers based on plaster of paris. Nor are there references to the purchase of wax in the archives even though "wax, and honey are produced there in great abundance" according to *Conquest of Luzon* (1569–76) (cited in Blair and Robertson 1909, 3:169). It appears that these panel paintings may represent an unusual practice.

Paint Layer Stratigraphy

Examination of the cross-sections for the *Ascension of the Lord* and the *Pentecost* shows a beeswax, calcium carbonate, and pigmented paint layer for the wood gaps (fig. 8), and then a beeswax and pigmented layer for the main image (see fig. 6). Generally the paint layers are thin, no greater than 600 µm, and are well bound by the media. The pigments are finely ground and composed of one to two pigment types.

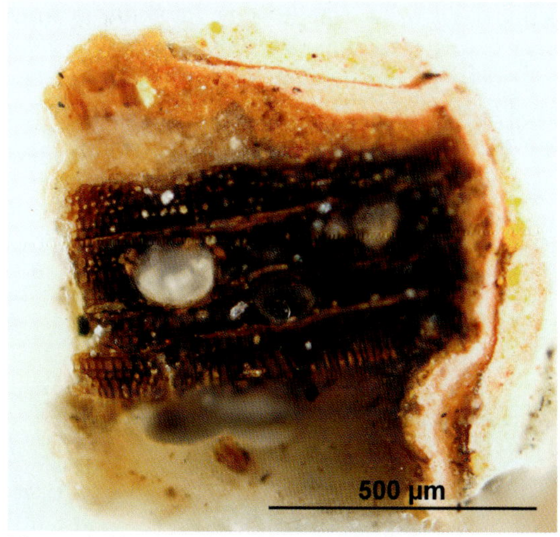

Figure 6. *Photomicrograph of cross-section of Kedondong wood, preparatory layer of beeswax, and paint layer from* Ascension of the Lord *(fig. 1)*

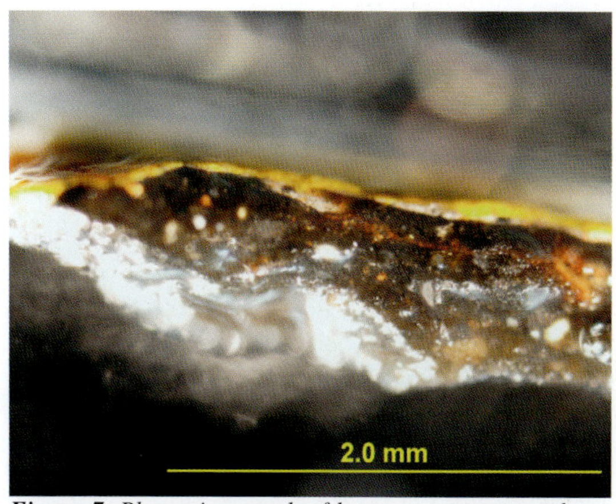

Figure 7. *Photomicrograph of beeswax preparatory layer and orpiment from* Ascension of the Lord *(fig. 1)*

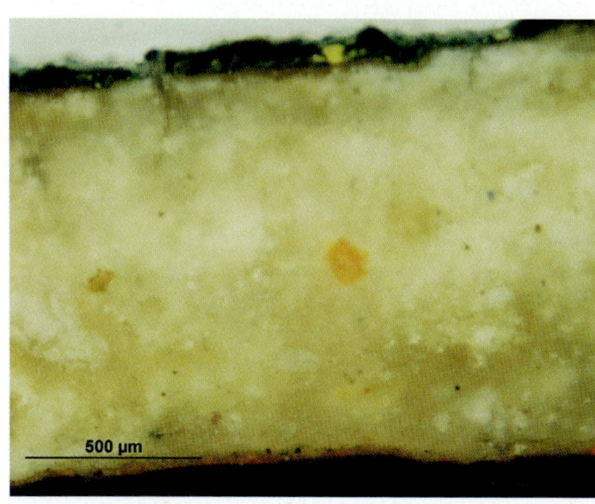

Figure 8. *Photomicrograph of wood gap fill composed of beeswax, calcium carbonate, and a paint layer from Artist unknown,* Pentecost, *ca. 1859*

Examination of the paint samples in the *Pentecost* indicate there is only one paint layer, and it is unlikely that *Saint John the Baptist* was painted over, as proposed by Jose. *The Doctors of the Church—San Gregorio Magno* and *San Geronimo Maximo*—have more complex paint structures and contain beeswax throughout the layers (figs. 9 and 10). Like the other two panels, their preparatory layer is composed of beeswax and not calcium carbonate, as would traditionally have been expected. SEM also identified a well-bound paint layer, and cracks resembled slow drying damages rather than hard-edged mechanical cracks (fig. 11). This finding implies that the paint is cohesive and has remained well adhered to both the beeswax preparatory layer and wood panel.

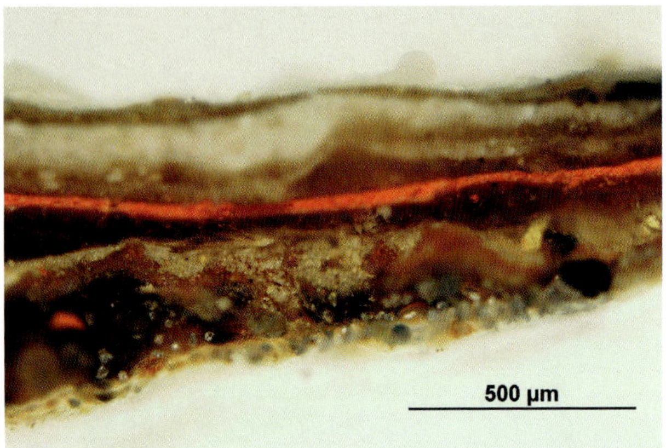

Figure 9. *Photomicrograph of cross-section from Liberato Gachalian,* San Geronimo Maximo, *1859*

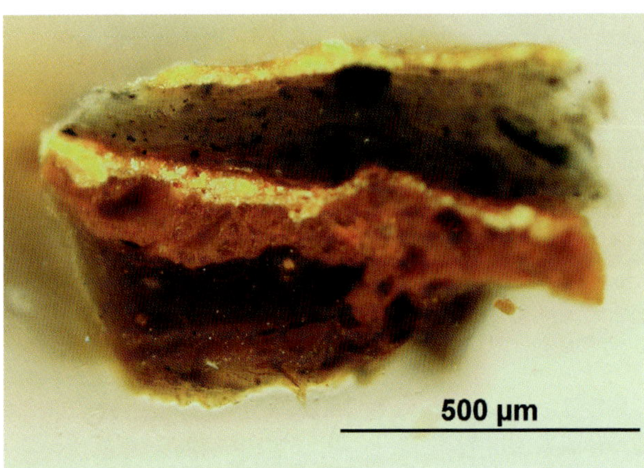

Figure 10. *Photomicrograph of cross-section from Liberato Gachalian,* San Gregorio Magno, *1859*

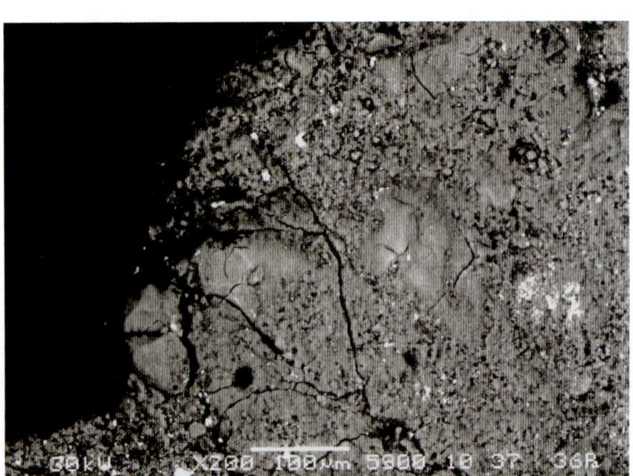

Figure 11. *SEM image of paint surface of* Ascension of the Lord *(fig. 1)*

For all works, UV fluorescence showed no natural resinous inclusions or surface layer. With GC-MS, there was no sign of methyl abietate or methyl dehydroabietate ions, and it is unlikely that resins are present in the paint or applied as a varnish layer. Interestingly, the use of *malapago* varnish for paintings has been recorded by Jose in the Cavite Parish Archives in the Philippines in 1876 (Jose 2003). *Malapago* is a resin from the Balao tree (*Dipterocarpus gracilis*) from the *Dipterocarpus* spp., where the Gujun balsams originate (Mills and White 1987). They are also part of the Dipterocarpaceae family, the same source for damar resin, which is widely used in traditional Western painting practice.

Pigments

Results for the analysis are recorded in table 1 together with their archival reference. A positive correlation between the archival and analytical results informs the discussion on the possible provenance of the paintings and whether they may have been painted *in situ*.

Red and White Pigments The identification of vermilion, red lead, lead white, and calcium carbonate is consistent with the parish archives and coincides with the proposal that the paintings were painted *in situ*. Further, the pigments are finely ground, suggesting they were imported, a suggestion that correlates with their high cost of seventy-five pesos. It was a common European practice to combine vermilion and red lead to lower the cost of producing vermilion pigments (FitzHugh 1985).

Blue Pigments The identification of ultramarine (from lapis lazuli) is significant, as it was historically the most expensive and celebrated pigment, reserved for important areas in a painting (Harley 1982). Further, the Spanish record keepers would have undoubtedly recorded it in the parish archives, and this assumption could lead to the conclusion that the *Ascension of the Lord* and *Pentecost* were executed elsewhere and not *in situ*. Jose (2003) has also suggested that the parish may have donated this pigment for use by the local painters, as no blue pigments are recorded in the parish archives. Spanish art treatises (Veliz 1986) refer to smalt and azurite, but test results did not identify these pigments. Indigo is found in the Philippines, but no results confirmed its presence. Finally, for the *San Geronimo Maximo* and *San Gregorio Magno* the blue particles were less consistently identified, and the results are not conclusive.

Yellow Pigments The identification of both realgar and orpiment correlates with a Spanish traditional palette for oil painting (Veliz 1986), though there are no references to these pigments in the parish archives. There are, however, records detailing the supply of realgar from Spain from 1569 (cited in Blair and Robertson 1909, 3:245). Gamboge is also not mentioned in the archives, and it is a pigment found

***Table 1.** Results of pigment analysis*

Painting	Color	Archival reference	Method of detection	Pigment identified	Archival to analytical
Unknown artist, *Ascension of the Lord*, ca. 1859	Red	(" 1858 … 10. Oil paint – paint verde, vermilion, red lead, lead white, resin from the Balao tree *(Dipterocarpus gracilis)*, lime. Food for the painters and carpenters all listed in the minutes that were kept.") ("1858 … Four dozens buttons of vermilion")	PLM: red-orange, circular florets, no pleochroism or cleavage, high relief, RI> 1.66, anisotropic, oblique extinction SEM EDS: HgMa line, SKa line detected @ 800×	Vermilion (HgS)	Correlates
			SEM EDS: Pb line, OKa line detected @ 800× Raman: very strong peak 545cm^{-1}, weak peaks 387 cm^{-1} and 307 cm^{-1} PLM: crossed polars showed green birefringence	Red lead (Pb_3O_4)	Correlates
	Blue	No record	PLM: isotropic, intense blue pigment, low relief, RI>1.66, no pleochroism or cleavage Microchemical tests: no Cu or $CaCO_3$, excludes azurite SEM-EDS: all elemental lines identified for $Na_8(Al_6Si_6O_{24})S_n$ Raman: very strong peak 548cm^{-1}	Ultramarine blue ($Na_8(Al_6Si_5O_{24})S_n$)	No correlation
			PLM: possibly cobalt blue, but characteristics not consistent with Becke line behavior		No correlation
	Yellow	No record	SEM-EDS: AsLa line, SKa line Raman: very strong peak at 351 cm^{-1}, strong peak at 306 cm^{-1}, medium peak at 287cm^{-1}, and weak peak at 378cm^{-1}	Orpiment (As_2S_3)	No correlation
			SEM-EDS: AsLa line, SKa line	Realgar (As_4S_4)	No correlation
			Raman: peaks at 1595 cm^{-1}, 1615 cm^{-1}, 1463 cm^{-1}, 1336 cm^{-1}, 1204 cm^{-1}, 1189 cm^{-1}, and 1163 cm^{-1} PLM: regular pebblelike clusters, isotropic, low relief, RI <1.66, no cleavage or pleochroism	Gamboge	No correlation
	White	See first record	Microchemical tests confirmed Pb, $CaCO_3$ PLM: parallel extinction: $CaCO_3$	Lead white ($2PbCO_3{\cdot}Pb(OH)_2$)	Correlates
			PLM and chemical tests	Calcium carbonate ($CaCO_3$)	Correlates
Unknown artist, *Pentecost*, ca. 1859	Red	See first record	PLM: red-orange, circular florets, no pleochroism or cleavage, high relief, RI> 1.66, anisotropic, oblique extinction SEM-EDS: HgMa line, SKa line detected @ 800× Madder?	Vermilion (HgS)	Correlates
			Raman: very strong peak 545 cm^{-1}, weak peaks 387 cm^{-1} and 307 cm^{-1}	Red lead (Pb_3O_4)	Correlates
	Blue	No record	PLM: isotropic, intense blue pigment, low relief, RI>1.66, no pleochroism or cleavage SEM-EDS: all elemental lines identified for $Na_8(Al_6Si_6O_{24})S_n$ Raman: very strong peak 548 cm^{-1} Chemical tests: no Cu or $CaCO_3$, excludes azurite	Ultramarine blue ($Na_8(Al_6Si_6O_{24})S_n$)	No correlation
	Yellow	No record	SEM-EDS: AsLa line, SKa line Raman: very strong peak 351 cm^{-1}, strong peak 306 cm^{-1}, medium peak 287 cm^{-1}, and weak peak 378 cm^{-1}	Orpiment (As_2S_3)	No correlation
			SEM-EDS: AsLa line, SKa line PLM: high relief, opaque, RI>1.6, more indicative of realgar with red-orange particles than orpiment with lemon yellow particles	Realgar (As_4S_4)	No correlation
			Raman: peaks at 1595 cm^{-1}, 1615 cm^{-1}, 1463 cm^{-1}, 1336 cm^{-1}, 1204 cm^{-1}, 1189 cm^{-1}, and 1163 cm^{-1} PLM: regular pebblelike clusters, isotropic, low relief, RI <1.66, no cleavage or pleochroism	Gamboge	No correlation
	White	See first record	PLM: small rounded particles, RI>1.6 Chemical tests: Pb^{2+} with needlelike, yellow particles when exposed to HNO_3 and KI	Lead white ($2PbCO_3{\cdot}Pb(OH)_2$)	Correlates
			PLM: translucent, low relief, symmetrical extinction, highly birefrigent Chemical tests: effervescence with dilute HCl	Calcium carbonate ($CaCO_3$)	Correlates

Table 1. *continued*

Painting	Color	Archival reference	Method of detection	Pigment identified	Archival to analytical
Liberato Gachalian, *San Geronimo Maximo,* 1859*	Red	See first record	PLM: red-orange, circular florets, no pleochroism or cleavage, high relief, RI> 1.66, anisotropic, oblique extinction SEM-EDS: HgMa line, SKa line detected @ 800×	Vermilion (HgS)	Correlates
			PLM: showed green birefringence and was deep orange in plane polarized light	Red lead (Pb_3O_4)	Correlates
	Blue	No record	PLM: smalt but RI does not match, indigo but shape is not needlelike, azurite but pleochroism not visible	No match	No correlation
	White	("February 1857 … 37 *savanes* of lime") Limestone is found naturally in Bohol and produced by cooking crushed oyster shells.	PLM: anisotropic, oblique extinction, high birefringence, high cleavage, and elongated shape	Gofun ($CaCO_3$)	Correlates
			PLM: round florets, parallel extinction Chemical test: Pb^{2+} with needlelike, yellow particles when exposed to HNO_3 and KI	Lead white ($2PbCO_3{\cdot}Pb(OH)_2$)	Correlates
Liberato Gachalian, *San Gregorio Magno*, 1859*	Red	See first record	PLM: red-orange, circular florets, no pleochroism or cleavage, high relief, RI> 1.66, anisotropic, oblique extinction	Vermilion (HgS)	Correlates
		See first record	PLM: showed green birefringence and was deep orange in plane polarized light	Red lead (Pb_3O_4)	Correlates
	Blue	No record	PLM: smalt but RI does not match, indigo but shape is not needlelike, azurite but pleochroism not visible	No match	No correlation
	White	See first record	PLM: translucent, low relief, highly birefrigent Chemical microscopy: effervescence with dilute HCl	Calcium carbonate ($CaCO_3$)	Correlates
			PLM: round florets, parallel extinction Chemical test: Pb^{2+} with needlelike, yellow particles when exposed to HNO_3 and KI	Lead white ($2PbCO_3{\cdot}Pb(OH)_2$)	Correlates

* No yellow pigments detected and yellow was not visible in the palette

mainly in Thailand (Harley 1982). Winter (1987) reports its use in East Asian paintings since the eighth century, and Harley (1982) reports its use as a watercolor paint in Europe from the seventeenth century. In this study, gamboge has been mixed with an oil binder not typical of its usage in Europe. It is not known whether the trees of the genus *Garcinia*, of which gamboge is derived, are found in Bohol, or whether Father Mariano Gutierrez from Jagna Parish was producing this colorant on the island. Its identification, therefore, does not contribute to the panel paintings' history.

Binders

Throughout the archives there are a number of entries for *aceite de pintar* (oil of the paintings) (Immaculate Conception Parish n.d.). Jose's publications further refer to *aceite de linaza* (linseed oil), *aceite de China* (tung or China oil), and the use of coconut oil in the paint films (Jose 2003, 1992). First, a Sudan Black stain gave a positive result for oil in all cross-sections; however, later GC-MS analysis did not identify any of the above oils according to the palmitic to stearic acid ratios (P/S) (Mills and White 1987). The peaks indicative of the saturated acids in drying oils did give a P/S ratio of 2.457, which corresponds either to pure walnut oil or to a combination of poppy and linseed oil, neither of which is recorded in the archives. Current literature states that it is not possible to differentiate between pure walnut oil and mixtures of linseed and poppy seed oils (Schilling and Khanjian 1996). These oils have traditionally been used in Western pictorial practices, and more specifically walnut oil was the preferred medium of southern Europe (Mills and White 1987). Further, the lack of yellowing in the analyzed paint films indicates walnut oil paint films.

Conclusions

A comparison of the lists of materials found in the Immaculate Conception Parish Archives, Baclayon, with the findings of the recent visual and technical analysis suggests these materials were not used for the panel paintings. There is some correlation of materials with regards to the identification of vermilion, red lead, lead white, and calcium carbonate, but there are many unrecorded materials identified in the panel paintings. These include natural ultramarine in the *Ascension of the Lord* and *Pentecost*, an expensive and valued pigment; realgar, a pigment used in a typical Spanish palette around this time; gamboge, an organic colorant found in Asia; and Pagsahingin or Kedondong wood instead of Molave. Further, the identified pigments were generally finely ground, indicating they were imported and not obtained locally. Comparative costs for pigments were also high, again indicating that they were mostly imported. This finding is also supported by the archival sources, which document the purchase of the five panel paintings in 1859 and their possible transportation. Overall, the current study cannot contribute more information on the actual origin of the five panel paintings except to state that they were not painted *in situ*.

The study does, however, raise some interesting issues about the materials and techniques of panel paintings during the mid-nineteenth century in Bohol. There appears to be an incorporation of some imported, high-quality pigments with locally ground pigments and the unusual use of beeswax as a primary ground layer in preference to a traditional whiting layer. This combination of a soft beeswax preparatory layer and wood panel may explain the structural stability and sound condition of the panel paintings. The fills between the wood are composed of layers of handmade paper with vermilion, beeswax, and calcium carbonate, and this construction is very different from that typical of Spanish panel paintings. The paint media was composed of oil, likely to be walnut, as reported in traditional Spanish practices. The presence of coconut oil was not found, even though Jose mentioned it in the literature. Technical examination also confirmed that there are no changes to the *Pentecost* and it has not been overpainted, even though the archives record a painting with a different title, *Saint John the Baptist*.

Experimental

Materials and Preparation

Meltmount Material (refractive index of 1.66) (Cargille Labs cat. 24160)

Saturated Sudan Black in 3:2 ethanol: water (Sigma cat. no. S0395)

Amido Black AB3 (for gelatin identification), glycerin, deionized water (Amidoschwartz 10B Merck)

Polyester embedding Resin 480 and MEKP activator (Boatsheath Resin and Service Polyesters)

Raman spectroscopy

A Renishaw 2000 Raman Spectroscopy Microscope with a 780nm Diode laser was utilized with a CCD detector for direct 2-D Raman imaging and Raman spectroscopy. Resulting Raman peaks were compared against a known Raman spectroscopic library (Bell et al. 1998). Although excitation wavelengths for this library are 632.8 nm and not 780 nm as for the instrumentation, characteristic peaks were within range and successfully compared.

SEM-EDS

A JEOL JSM-5900 scanning electron microscope with energy-dispersive spectrometer (SEM-EDS unit) was employed, using a backscattered electron (BSE) detector, an x-ray energy-dispersive spectrometer (EDS), and a cathodoluminescence (CL) detector. Samples were simply mounted on a glass slide and imaged at 100× to 800× magnifications.

GC-MS

A Hewlett Packard 5890 gas chromatograph fitted with a Hewlett Packard 5970 mass selective detector was utilized. A

Hewlett Packard Ultra 2 column (5% phenyl/95% methyl silicone) was used, and data were processed on a Hewlett Packard G1034C using Chemstation software. Samples analyzed were prepared using Meth-Prep II to produce volatile methyl esters that could be separated during the chromatographic process.

FTIR

A Perkin Elmer FTIR Spectrometer Spectrum 2000 used attenuated total reflection (ATR) with a germanium crystal for the characterization of the beeswax. Examination using this technique requires that a sample be crushed and rolled onto a potassium bromide plate.

Acknowledgments

Special thanks go to Mr. Regalado Trota Jose, researcher of Colonial Church Heritage and Permanent Committee for the Cultural Heritage of the Church, of the Catholic Bishops' Conference of the Philippines (CBCP), and Father Milan Ted Torralba, chair, Diocese of Tagbilaran Commission for the Cultural Heritage of the Church, and executive secretary of the Permanent Committee for the Cultural Heritage of the Church, of the Catholic Bishops' Conference of the Philippines (CBCP); the Immaculate Conception Parish, Baclayon, Bohol, under the pastoral jurisdiction of the Diocese of Tagbilaran; staff from the Centre for Cultural Materials Conservation, University of Melbourne; and the Biomedical Sciences Laboratory at the University of Canberra, which undertook GC-MS and FTIR.

Notes

Endnotes quote entries from the Libro de Cargo y Data, 1856–1909 held in the parish archives of the Immaculate Conception, Baclayon, Bohol. Listed alongside each record is the expense according to pesos (p)-reales (r)-granos (g).

1. Entry for 13 May 1859:
   18. Por los 4 Doctores *24-0-0*
   19. Por los cuadros de la Ascension del Sr y S Juan en el Bautisterio *20-0-0*
   In English:
   18. For the four Doctors. *24-0-0*
   19. For two panel paintings of the Ascension and Saint John the Baptist *20-0-0*

2. See note 1, above.

3. Entry for May and June 1857:
   8. Por cajones, cargadores, banca y flete a Cebu y de Cebu aqui *2-3-0*
   In English:
   8. For the crate, carrier, stand and charter to Cebu and Cebu return. *2-3-0*

4. Throughout the parish archives are records relating to the repair and care of artworks, including the following:
   Entry for 13 May 1859:
   21. para dos cuadros de Ascension u S. Juan, mil nipas, coco negros, para continas de las images ... de passion
   In English:
   21. for the two cuadros of the Ascension and S. Juan, 1000 nipas, coco Negro for the screen of the images of the passion

5. Entry for July and August 1858:
   4. 4 docenas de sonboltonios de vermilion *4-0-0*
   In English:
   4. Four dozens buttons of vermilion *4-0-0*

6. Entry for February 1857:
   4. 37 savanes de cal *3-4-6*
   In English
   4. Thirty-seven savanes of lime *3-4-6*

7. Entry for July and August 1858:
   15. 21 1/2 ficos de yeso de Dauis y 600 pilonsitas del de China *27-6-6*
   In English:
   15. Twenty-one and a half ficos *of gesso from Dauis and 600 jars from China* *27-6-6*

8. Entry for July and August 1858:
   16. 6 1/2 ficos de cueros pa cola *6-6-9*
   In English:
   16. Six and a half ficos *of skin and glue* *6-6-9*

References

Bell, I. M., R. J. H. Clark, and P. J. Gibbs. 1998. Raman spectroscopic library of natural and synthetic pigments (pre-1850 AD). Christopher Ingold Laboratories, University College, London. Available from www.chem.ucl.ac.uk/resources/raman/speclib.html.

Blair, E. H., and J. A. Robertson. 1909. *The Philippines Islands, 1493–1898*. Cleveland, Ohio: Arthur H. Clark Company.

FitzHugh, E. 1986. Red lead and minium. In *Artists' pigments: A handbook of their history and characteristics*, vol. 1, ed. R. Feller, 109–39. Cambridge: Cambridge University Press.

Gay, M. C. 1976. Application of the staining method to cross-sections in the study of media of various Italian paintings of the fourteenth and fifteenth centuries. In *Conservation and restoration of pictorial art*, ed. N. Brommelle and P. Smith, 78–83. London: Butterworth.

Harley, R. 1982. *Artists' pigments c. 1600–1835*. London: Butterworth Scientific.

Hassell, C., and L. Sheldon. 1994. Pigment notebook from workshop conducted at the University of Melbourne, November 1994. UCL Painting Analysis Ltd, London.

Illic, J. 2003. Wood identification. Forestry and Wood Division, Commonwealth Science and Industry Research Organisation (CSIRO), Melbourne.

Immaculate Conception Parish. N.d. Libro de Cargo y Data, 1856–1909 [Book of Income and Accounts, 1856–1909]. Archives of the Immaculate Conception Parish, Baclayon, Bohol, Philippines.

Jose, R. T. 1986. Church expenses in 19th century Cavite: A cursory look at account books. *Philippine Studies* 34(3):360–73.

Jose, R. T. 1992. *Simbahan: Church art in colonial Philippines, 1565–1898*. Manila: Ayala Museum.

Jose, R. T. 2001. *Bohol visita iglesia: A guide to historic churches.* Manila: National Commission for Culture and the Arts.

Jose, R. T. 2003. Palitada: Skin of the church. In *Zero in 2003*, 6–53. Manila: Ayala Foundation, Eugenio Lopez Foundation, and Museo Pambata Foundation.

Mallat, J. 1994. *The Philippines: History, geography, customs, agriculture, industry and commerce.* Trans. L. Castrence. Manila: National Historical Institute. Original ed., *Les Philippines Historie, Geographie, Moeurs*, 1846.

Martin, E. 1977. Some improvements in techniques of analysis of paint media. *Studies in Conservation* 22:63–67.

McCrone, W. 1982. The microscopical identification of artists' pigments. *Journal IIC-Covodeder Group* 7 (1–2):11–34.

Mills, J., and R. White. 1987. *The organic chemistry of museum objects.* London: Butterworth-Heinemann.

Pellegrino, F. 2003. GCMS and FTIR analysis. Biomedical Sciences Laboratory, University of Canberra, Canberra.

Pilar, S. A. 1992. *Pamana: The Jorge B. Vargas Art Collection.* Quezon City: Committee on Arts and Culture.

Pilar, S. A. 1994. Philippine painting: The early Chinese heritage. *Arts of Asia* 24(6):62–72.

Pilar, S. A. 2000. The untold story of the Bohol masters. *Philippines Star.* 13 June.

Plesters, J. 1956. Cross sections and chemical analysis of paint samples. *Studies in Conservation* 2:110–57.

Santiago, L. 1992. The loom of colonial art. In *Art Philippine: A history, 1521–present*, ed. L. Roa, 11–64. Manila: Crucible Workshop.

Schilling, M., and H. Khanjian. 1996. Gas chromatographic determination of the fatty acid and glycerol content of lipids, pt. 1, The effects of pigments and aging on the composition of oil paints. In *Preprints: ICOM Committee for Conservation*, ed. S. Bridgland, 220–27. 11th Triennial Meeting, Edinburgh. London: James and James.

Veliz, Z. 1995. Wooden panels and their preparation for painting from the Middle Ages to the seventeenth century in Spain. Paper presented at the conference on The Structural Conservation of Panel Paintings, J. Paul Getty Museum, Los Angeles.

Veliz, Z., ed. 1986. *Artists' techniques in Golden Age Spain : Six treatises in translation.* Cambridge and New York: Cambridge University Press.

Winter, J. 1987. Organic colours in East Asian paintings: The identification of gamboge. In *Preprints: ICOM Committee for Conservation*, ed. K. Grimstad, 111–17. 8th Triennial Meeting, Sydney. Marina del Rey, Calif.: Getty Conservation Institute.

Wall Decoration Systems in the Temples of Angkor

Esther von Plehwe-Leisen and Hans Leisen

ABSTRACT In the region that now makes up Cambodia and Thailand, the wealthy and powerful empire of the god-kings of Angkor reigned from the ninth to the fourteenth century. Today we can admire the many different temples that date from this period of the Khmer. The construction materials were brick, laterite, and sandstone. Excellent carvings and a highly impressive architecture illustrate the high skill of the Khmer specialists. Nearly all parts of the temples still display the original color of the building materials. But in the time of the Angkor empire the temple walls would have been covered with paint systems, stucco, and plasters. Very little of these decoration materials has survived. Long periods of neglect and restoration measures like overpainting, cleaning, and dismantling, have destroyed most of the historic wall finishing systems. Hardly anything is known about the techniques and materials. During the last three years a survey of the different decoration systems on the interior and exterior temple walls has been started under the auspices of the German Apsara Conservation Project, Cologne. Stucco and plaster techniques have been assessed, and different paint campaigns as well as preliminary drawings have been identified. Small samples of paint, mortar, plaster, and stucco have been taken, documented, and analyzed. This paper is intended to focus attention on the value of this part of the Khmer heritage, so far neglected, and to give support to restorers in their conservation work.

Introduction

The Angkor Heritage Site in the center of Cambodia is one of the most exciting monuments on the UNESCO World Heritage List. With the remains of more than a hundred temples, it reminds us impressively of the mighty past of the Khmer people (fig. 1). From the ninth to the fourteenth century the Khmer reigned over large parts of the Southeast Asian region.

A highlight of engineering art, a very efficient irrigation system combined with enormous ponds for water storage enabled the Khmer to establish a highly developed rice-producing system in the very fertile region (Jacques 1990). The Khmer rulers' wealth and power relied on a thriving rice exportation trade.

Thus the Khmer kings were able to enlarge their territory constantly and to plan and construct enormous temples in brick and stone architecture. They had the status of god-kings. The temples were consecrated to individual gods and served as mausoleums for the kings and their families. Here the gods were worshiped and served. Temples alone were constructed in stone; all other buildings were built in wood. Today only the brick and stone structures survive as witnesses to the Khmer world empire.

The Khmer were not only good architects but also excellent stonemasons. They created stone sculptures as well as

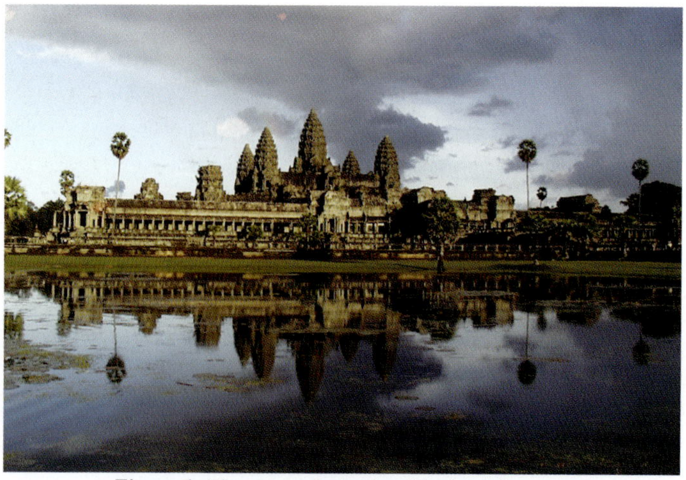

Figure 1. *Khmer temple Angkor Wat from the west*

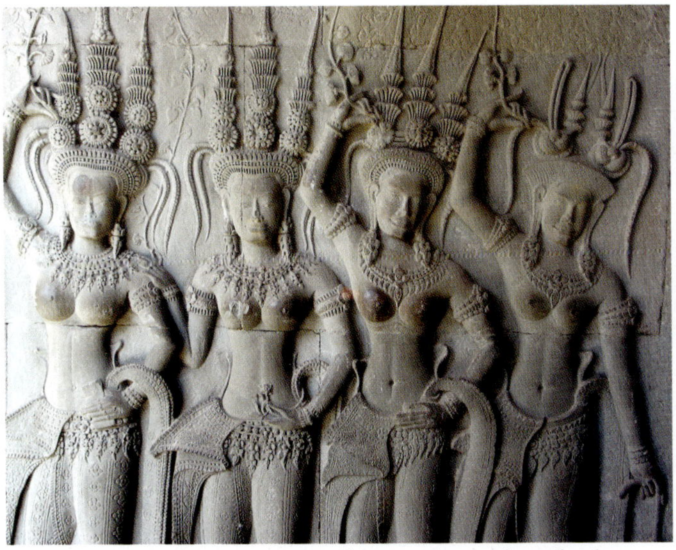

Figure 2. *Reliefs of dancing* apsaras *decorate the temple walls of Angkor Wat.*

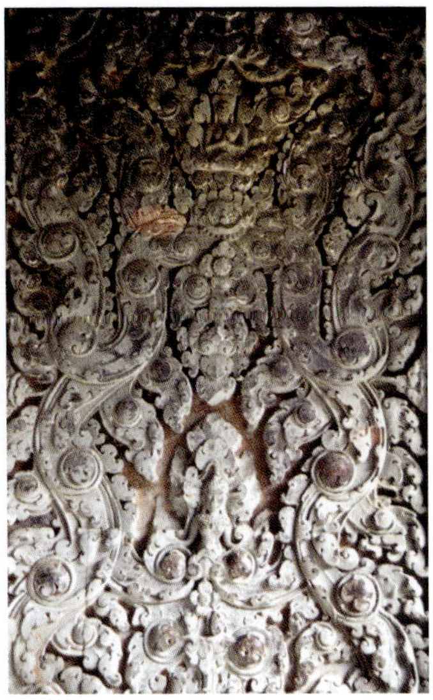

Figure 3. *Decorative carving at Angkor Wat Temple*

bas-reliefs and decoration elements of outstanding quality (figs. 2 and 3). Construction and sculptural techniques were highly influenced by wood construction techniques that were simply transferred to the new materials: brick and stone.

German Apsara Conservation Project

The temples and their precious sculpted reliefs are threatened by natural weathering as well as by intrusive and unsuitable interventions. Not only static disintegration but also material degradation processes endanger the cultural heritage. Parts exposed to weathering show frightening deterioration. Degradation processes often lead to a total loss of the carved stone surfaces.

Due to the extent of deterioration, the German Apsara Conservation Project (GACP) was established in Cologne in 1995 to prepare and to perform conservation measures as well as to carry out research work and to teach conservation to Cambodian workers and students. In the course of stone preservation work, the authors became aware of the remains of wall finishing systems and the bad preservation state of these rare relics. Therefore studies of wall decoration systems including stucco, plaster, and polychromy were initiated. This paper gives an overview on the actual status of the investigation done as part of the GACP (Kroner 2002; Kiesewetter et al. 2001; Dürre 2001; Plehwe-Leisen and Leisen in press).

Building Materials and Techniques

The Khmer building technique differs considerably from most building techniques known. First a "monolithic rock" was created by piling up the prepared stone blocks or bricks. To achieve this aim, single stone blocks were roughly smoothed and then stone ashlars as well as bricks were

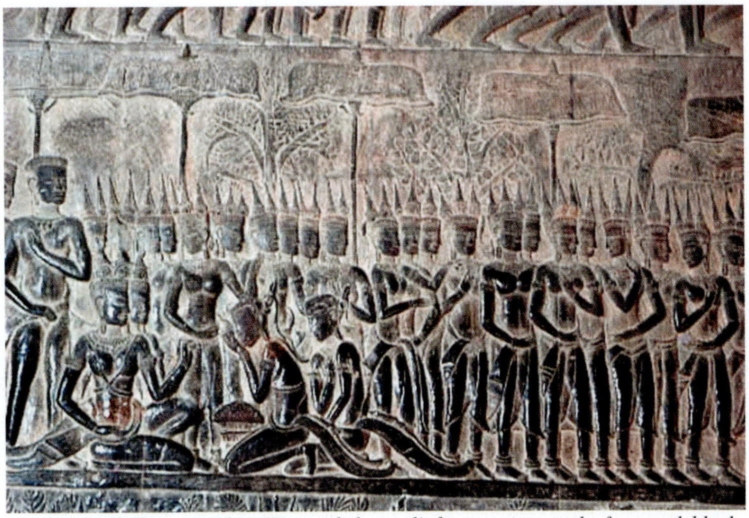

Figure 4. *Walls at Angkor Wat with bas-reliefs are composed of several blocks perfectly fitted together*

rubbed against each other to produce flat joints. The Khmer did not use joint mortars. By this technique they produced a homogeneous temple surface that could subsequently be carved like a single block (fig. 4). Thus the Khmer temples are better characterized as sculptures than as buildings.

In the early times of the Khmer empire, brick was the common construction material for the temples. Architectural details like door frames, lintels, pediments, and columns as well as guardian (*dvarapala*) and divinity (*devata*) sculptures were often carved in sandstone. Later, sandstone replaced brick as the principal building material for the temples. Substructures and outer walls, and in rare instances prominent temple parts, were constructed in laterite, a stone that could be broken near the construction sites and was easy to cut while it was still wet. Toward the end of the Khmer empire materials were no longer selected carefully, and a "material mixture" was used in order to finish the temples.

Wall Decoration Systems

Many details point to the fact that all the surfaces in the finished construction works were once covered by decoration systems.

To start with the early brick constructions, the monolithic brick body was carved on site. By this process the sintered outer layer was damaged or even removed. The bricks lost their sheltering skin and were exposed to rain, sun, and microbiological colonization. Thus they needed a new protective layer.

Unprotected sandstone, in contrast, is able to resist weathering for some time. In this case, aesthetic reasons may account for the original coatings on walls and sculptures. The surface aspects of the sandstone blocks vary considerably. There are greenish, grayish, yellow, brown, and red types of stone, as well as blocks with a distinct layering. The mixture of very different blocks is also common in the extremely fine carved reliefs.

Holes were chiseled in the stone blocks to insert wooden sticks for carrying the blocks and grinding the joints. While carving the temple walls on site, the stonemasons often hit these holes. These marks had to be filled later by a mortar differing completely from the surrounding stone in color and texture. Sometimes the sandstone pieces were not properly smoothed and needed an equalizing layer. To present their masterpieces without distractions the Khmer artists had to mask the unintentional diversity of the support (fig. 5).

Laterite is a very coarse-grained type of stone with many pores and different components. It is not a surface for inside walls and even less for stone carvings. This uneven surface would have had to be smoothed by a coating.

All the facts cited, combined with the small remains of surface coatings found in the Cambodian temples, imply that most of the surfaces, if not all, were coated by different wall decoration systems. Today there is only very little evidence of the whole technology left. Unfortunately, most of the rare relics are already in a deteriorated and endangered state of preservation.

Stucco

Stucco remains are found mostly on brick walls and reliefs (fig. 6). Shell lime was the binder. Shells from the Tonle Sap (Great Lake) and the surrounding rivers were burned and then slaked. As filler, river sands were used, and brick particles and dust were added (fig. 7). Vegetable fibers reduced the shrinking and increased the tensile strength of the stucco. Organic additives like polysaccharides and proteins have also been identified (Lujan Lunsford 1994).

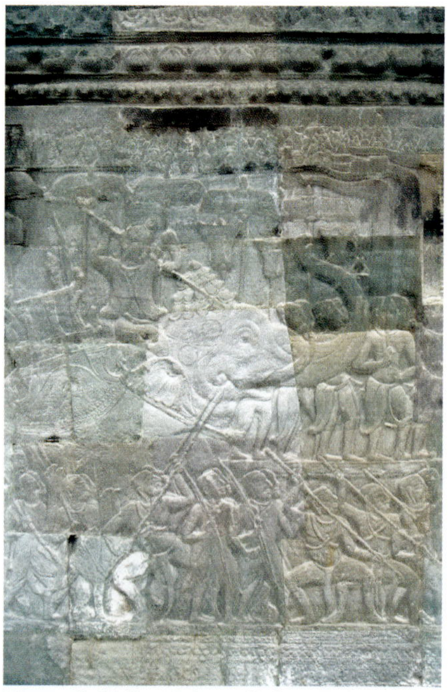

Figure 5. *Wall with bas-reliefs at Bayon is composed of differing sandstone varieties.*

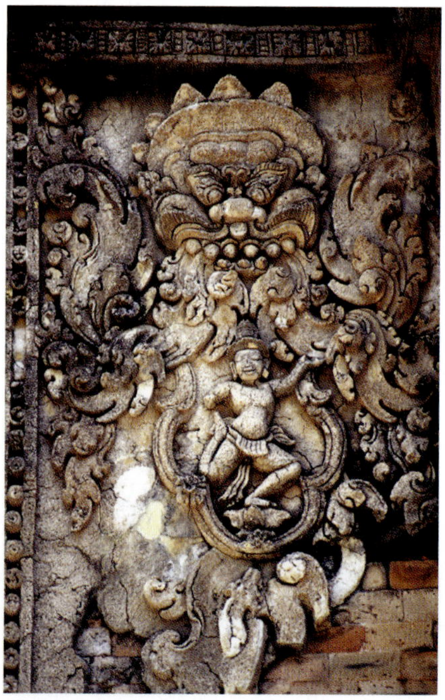

Figure 6. *Stucco detail on the brick wall of the Preah Ko Temple*

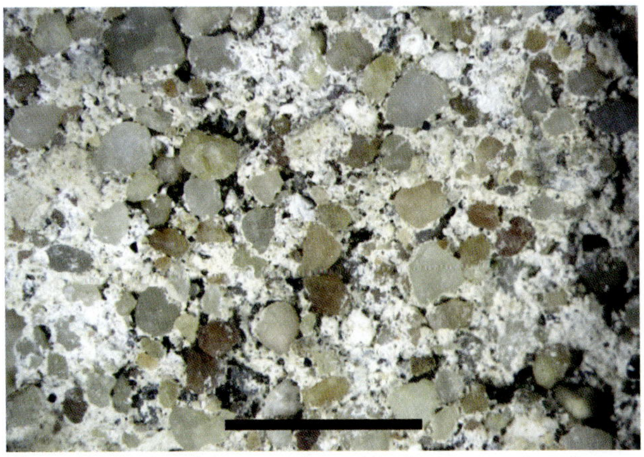

Figure 7. *Weathered stucco surface with a high proportion of quartz grains as aggregates in a lime matrix. Scale bar = 5 mm*

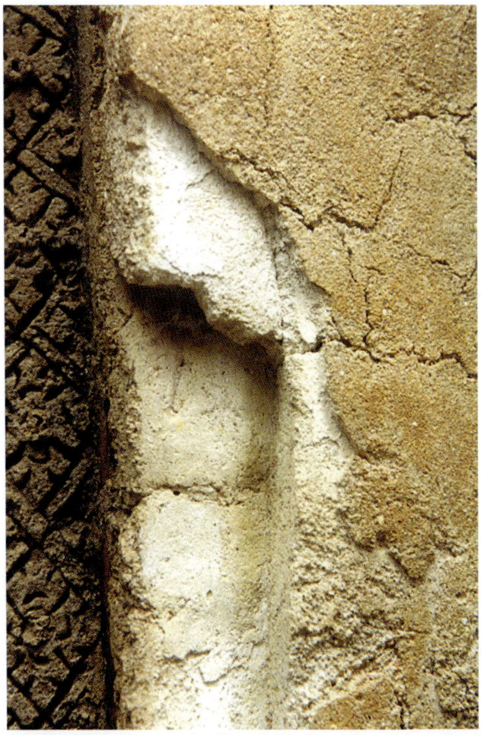

Figure 8. *Stucco at the Preah Ko Temple was applied in several layers, each one perfectly smoothed.*

The walls and the reliefs were often chipped or scratched before application of the stucco in order to increase adhesion. Sometimes there is a pattern of little holes that may have been created for the insertion of wooden supports for big and heavy areas of stucco.[1] The rough shape of the reliefs showing divinities or guards and other decoration elements was first modeled in the brick and then covered by stucco.

The stucco sequence was performed in several layers (fig. 8). First, there could be a very fine-grained, sometimes pink, layer rich in binder as a type of primer. Then two or three layers of lime stucco followed. Each layer was very carefully prepared, as if it were the uppermost one, and then chipped or scratched again while still fresh in order to allow the next layer to adhere. Floral applications were most probably prefabricated and attached later to the stucco plaster. The right positions were marked freehand. There are still a few strokes to be seen. The big reliefs seem to have been modeled directly at the wall. A lime wash was used to coat the stucco reliefs in order to smooth and protect them. As a final step, paint was applied: ocher, red, and black are still to be seen in traces, but blue and gold have also been reported (Groslier 1960). The pigments of the paint remains found on the stucco are mostly iron pigments like hematite and ocher (limonite and goethite).

It is not clear whether all stucco applications were part of the original construction plans of the temples. Sometimes the supporting walls had been carved very carefully and covered by a thin red paint before they were chipped and decorated by stucco. Historic restoration activities or changes in design led to different finished surfaces on top of each other.

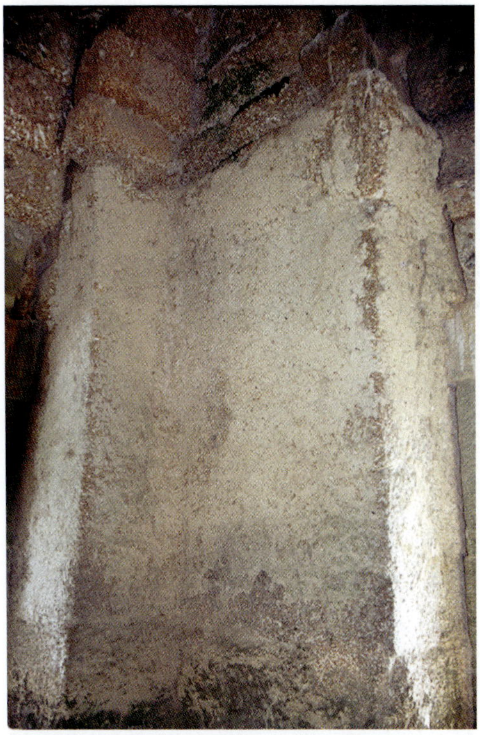

Figure 9. *Loam plaster at Prasat Suor Prat covers only the visible part of the laterite wall.*

Plaster

Inside and outside walls were protected by renderings in Khmer times. They were applied on all building materials, sandstone, laterite, and brick. Remains are to be found only in sheltered situations. Interior plasters were applied only at the visible parts of the walls up to the ceilings (figs. 9 and 10).

Different types of plasters and renderings can be distinguished with regard to their composition. Binders used were lime and loam (figs. 11 and 12). As filler, river sand was used. Further ingredients were brick and laterite particles, vegetable fibers, and charcoal. Where remains of the still complete sequence are found, it becomes clear that two or three layers of plaster normally were applied. The layers become finer from the first to the last one, finishing with a final layer that

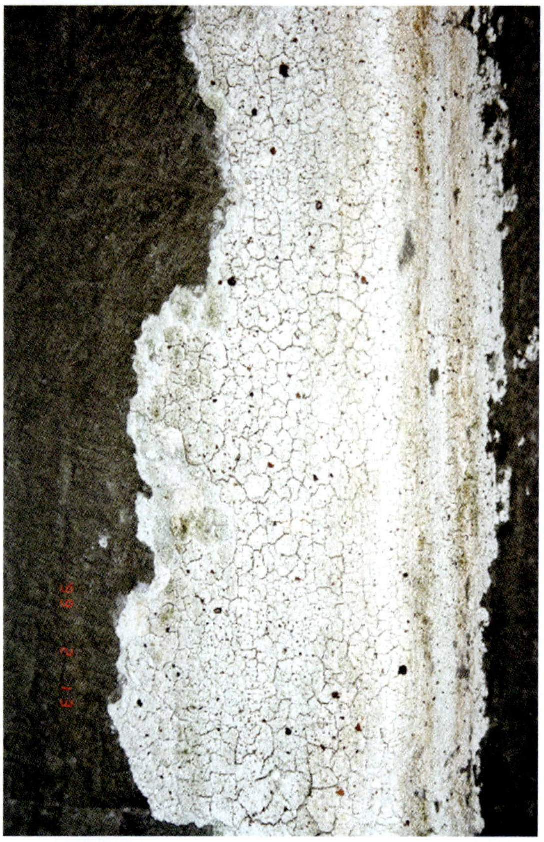

Figure 10. *Lime plaster at Ta Nei Temple shows a typical craquelure pattern on a sandstone wall*

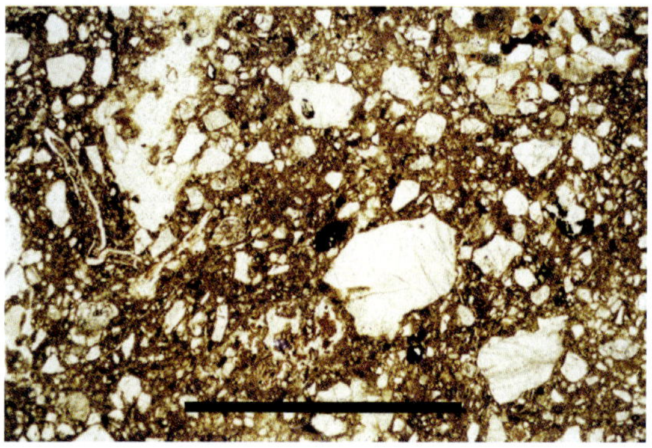

Figure 11. *Thin-section of loam plaster, Prasat Suor Prat, showing quartz grains and black charcoal relics. Plane polarized light. Scale bar = 2.5 mm*

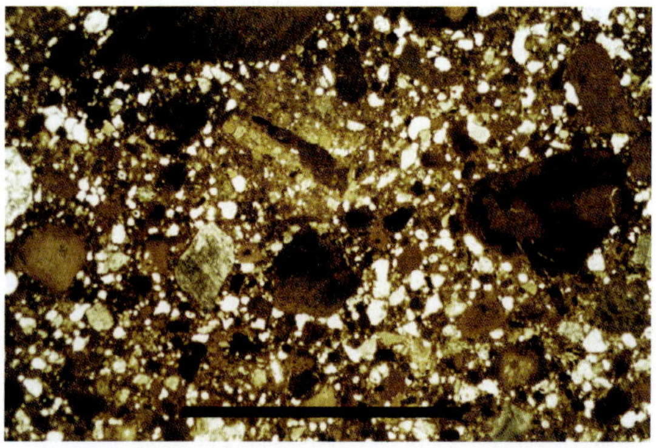

Figure 12. *Thin section of lime plaster, Preah Pithu, with calcite, laterite and quartz as fillers. Plane polarized light. Scale bar = 2.5 mm*

is very smooth and contains few fillers. This layer was painted and also most likely often polished.

Depending on the needs of the substratum, the ingredients of the plaster systems vary. Laterite with a very uneven surface requires a thick and equalizing plaster. These needs were met by applying a loam-based rendering. From the point of aesthetics, it is a quite simple type of plaster. Most probably it was covered by a lime wash that was painted.

Lime-based thick plasters were normally used on sandstone that was not very carefully carved or had defects. The holes for the sticks inserted in order to carry or rub the stone blocks were also filled with that type of mortar.[2] Due to their protected position, mortars in the holes are quite often preserved, and sometimes they also show polychromy.

Another type of plaster was found most often on brick walls. It is a thin layer, 1 to 2 mm thick, and very fine-grained with a binder consisting of lime. This very smooth surface was prepared to carry wall paintings. Here it is presumed that organic additives were admixed to the plaster (Lujan Lunsford 1994). The final layer, if not a geometrical or representational mural painting, was a simple paint layer.

Polychromy

Most if not all surfaces must have been polychromed, including both interior walls and facades. Today, with rare exceptions, the only remains of the polychromy are to be found in the temple interiors (fig. 13). The prevailing colors are reds in different tones, white, and ocher, but brown, black, and gold can be seen also. Colors like green and blue are restricted to more recent layers.

Paint was at times applied directly on the stone and brick surfaces. Walls, doors, and door frames or stone carvings were coated with color directly. Alternatively, stone and brick walls can also carry a wash or a thin plaster layer as a support for the paint layers. Entire walls were covered by one color paint only, or polychromy was used for accentuation of details. Even remains of mural paintings can still be detected (fig. 14).

Microscopic and analytical studies showed iron red and ocher, white, red lead, cinnabar, and gold to be the prevailing pigments used (fig. 15). As the binder, calcium carbonate, perhaps in combination with organic additives (proteins), was identified (Kiesewetter et al. 2001; Jägers 1996).

The walls in the Khmer temples are for the most part covered by encrustations formed by evaporation of water from solutions coming from the higher areas of the walls and by salt crusts. These make the detection of polychrome remains at the site and in samples very difficult.

Preliminary Sketches and Metal Plates

Careful investigation of the carvings at the temple Angkor Wat showed that they were planned by preliminary sketches in black and red (Dürre 2001). Today remains of these sketches are preserved only in the temple interior (see fig.

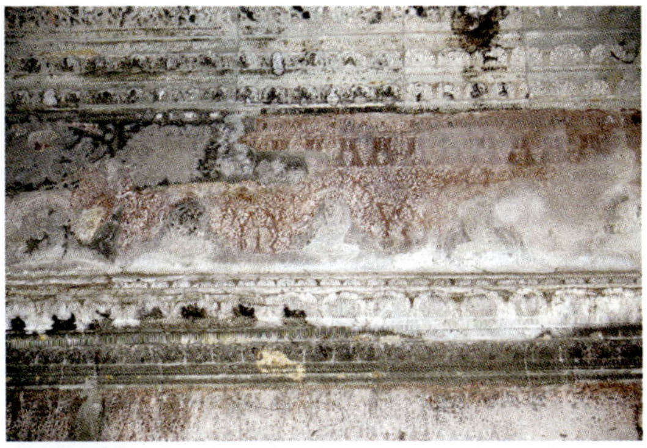

Figure 13. *Remains of polychromy in the galleries of Angkor Wat*

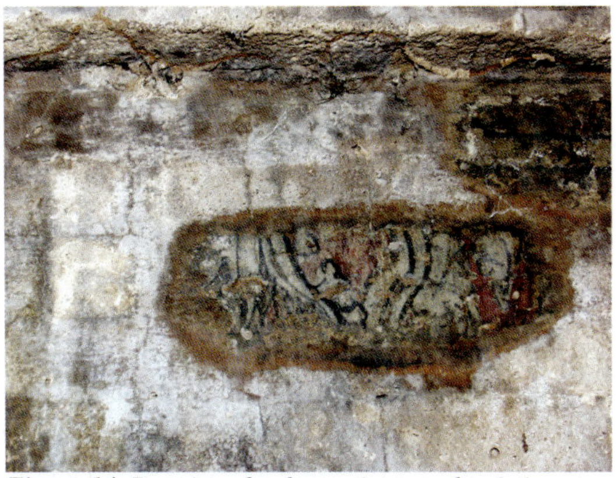

Figure 14. *Remains of a decorative mural painting on a brick wall at Preah Ko*

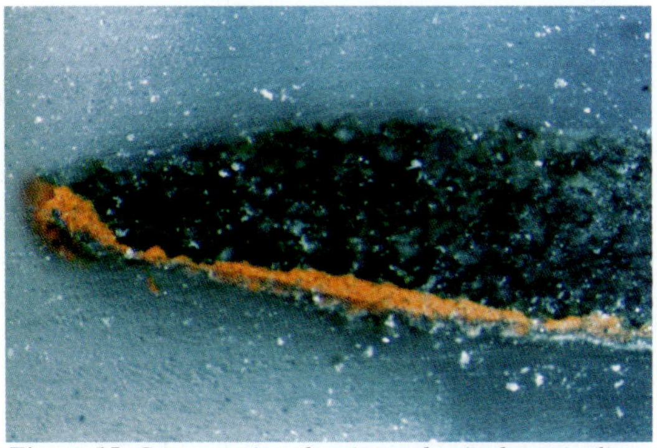

Figure 15. *Cross-section showing red paint layer in direct contact with sandstone at Angkor Wat*

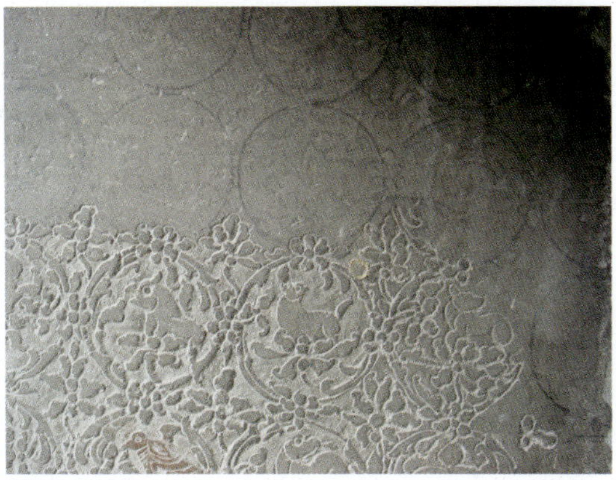

Figure 16. *Preliminary sketches showing shape and filling of a tapestry-type decoration, Angkor Wat*

15). For tapestry-type decoration, construction lines were drawn in order to divide the area into equal parts. Grids allowed the regular grouping of decorative ornaments. Round decorations were planned either by freehand drawing or by some type of compass (fig. 16). The shaping of the tapestry-type circles was also sketched before carving. Whether or not stencils were used is still uncertain. Larger reliefs like the *apsara* dancers or the famous bas-reliefs in the galleries of Angkor Wat also began with preliminary sketches. The composition of the up to 100 m long reliefs was impossible without orientation lines.

It is not known whether the Khmer also used metal plates for decoration, as is described for the site of Ayuthia in Thailand (Boisselier and Beurdeley 1974). Some temples show a regular hole pattern in the walls that could have been used for fixing such plates (fig. 17).

Conclusions

The investigation presented is the first systematic step toward a detailed inventory of the preservation and the technology of Khmer wall finishing systems. In the course of research, it became clear that all temples show to varying degrees signs of former wall finishing systems. All projects that deal with the preservation of Khmer temples have, therefore, to be planned by thorough investigations into the whole monument, of which wall-finishing systems are an essential part. All steps of preservation work, like cleaning procedures, application of coatings, replacement of material, and intrusive methods of dismantling, are a danger to the rare relics that are still left from this highly developed technology. Wall finishing systems have to be incorporated

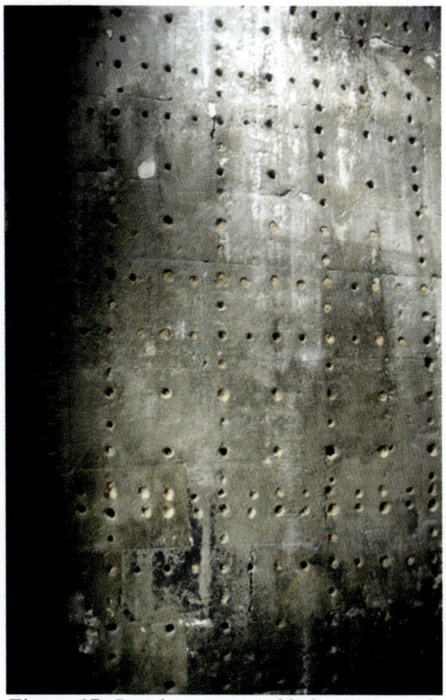

Figure 17. *Regular pattern of holes in a temple wall, Preah Khan*

into preservation plans of Khmer heritage. They must be protected during interventions, and adequate conservation techniques for these systems have to be developed and performed.

Acknowledgments

The German Apsara Conservation Project is funded by the German Ministry of Foreign Affairs and several private sponsors. Headquarters for GACP is at the University of Applied Sciences in Cologne, Germany. The Institute of Conservation Science is in charge of the project work, supported by the Institute of Imaging Sciences. GACP activities are coordinated with the Cambodian institutions, the APSARA Authority, and the Angkor Conservation Office (ACO), and are part of the community of international projects coordinated by UNESCO that make up possibly the world's biggest preservation effort for a single site.

Notes

1. We have not found remains of wood or similar substances, but the purpose of the holes is deduced from the general practice of stucco use and from a similar use in Thailand.
2. Plaster and rendering are also mortar systems. The material is similar, but the purpose is different. Plaster and rendering are mortars that are applied on a wall or a ceiling in order to coat the whole area. In this case the mortar was used for repair by filling deep cavities. Therefore the term "mortar" should be used.

References

Boisselier, J., and J.-M. Beurdeley. 1974. *Kunst in Thailand*. German ed. Stuttgart: Kohlhammer.

Dürre, St. 2001. Bildhauer am Angkor Wat [Stonecarving at Angkor Wat]. University of Applied Sciences, Cologne, Germany.

Groslier, B. P. 1960. *Hinterindien*. Zurich: Schweizer Druck- und Verlagshaus.

Jacques, C. 1990. *Angkor*. Paris: Bordas.

Jägers, E. 1996. Angkor: Identification of the painting materials in two samples. University of Applied Sciences, Cologne, Germany.

Kiesewetter, A., H. Leisen, and E. von Plehwe-Leisen. 2001. On the polychromy of Angkor Wat: Results of initial paint color investigations. *UDAYA Journal of Khmer Studies* 2:57–66.

Kroner, J. 2002. Dokumentation des Kalkstucks am Tempel Preah Ko in Kambodscha und Versuche zu seiner Konservierung [Documentation of the lime stucco at Preah Ko Temple in Cambodia and experiments on its preservation]. University of Applied Sciences, Cologne, Germany.

Lujan Lunsford, R. 1994. Conservation of external stuccos and brick. Technical report to the Royal Angkor Foundation, Phnom Penh.

Plehwe-Leisen, E. v., and H. Leisen. In press. Kalkstuck am Tempel Preah Ko: Kambodscha [The lime stucco at Preah Ko Temple, Cambodia]. Paper presented at IfB workshop Stuckkonservierung, Erhaltung von Ornamenten und Figuren. Trebsen, Germany, 27–28 June 2003.